AF841842

Nihed CHAÂBANE
and Frédéric SCHUSTER

AMETIS

Advanced Manufacturing for Energy and Transportation International School

Science Press | edp sciences

EDP Sciences – ISBN(print): 978-2-7598-2446-5 – ISBN(ebook): 978-2-7598-2588-2
DOI: 10.1051/978-2-7598-2446-5

Introduction

Advanced Manufacturing for Energy and Transportation International School

Nihed CHAÂBANE[1] and Frédéric SCHUSTER[2]

[1]*Université Paris Saclay, CEA, Institut National des Sciences et Techniques Nucléaires, 91191 Gif-sur-Yvette, France*
[2]*Université Paris Saclay, CEA, Cross-cutting Program on Materials Science and Engineering, 91191 Gif-sur-Yvette, France*

Abstract

The objective of Advanced Manufacturing for Energy and Transportation International School (AMETIS) is to present an integrated vision of advanced manufacturing dedicated to materials and components for low-carbon energy and transportation. The proposed approach essentially focuses on three main families of emerging processes, as well as their possible synergies to generate an often disruptive innovation framework. The option is also to present new methodologies that allow faster discovery or design of new materials of interest or faster optimization of complex emerging processes. These methodologies are often based on the significant progress made by digital technologies that have a considerable impact on the acceleration of Materials Science and Engineering. In particular, they make it possible to considerably shorten development time and minimize costs. Finally, this approach cannot be global without taking into account the drivers of sustainability in terms of resources, environment, safety and society. Therefore, AMETIS proposes a holistic approach.

1.1 Introduction

The impact of Materials Science and Engineering on innovation in the field of components for low-carbon energy and transportation is considerable. Issues related to the durability of solutions, such as economic sustainability, are often strongly linked to the choice of materials and their production process, which have a very important impact on performance during use all along the lifetime of the

DOI: 10.1051/978-2-7598-2446-5.c901

components. Therefore, the advanced manufacturing concept integrates the entire value chain, from eco-design to recycling, and even upcycling, taking into account a number of environmental, social and resource management factors including energy sobriety. This systemic presentation is the core of the AMETIS (figure 1).

Among the large families of emerging processes, AMETIS focuses on three of them due to the high potential of each one, and it is also their possible synergies that can initiate breakthrough innovations or incremental advances, as shown in figure 2.

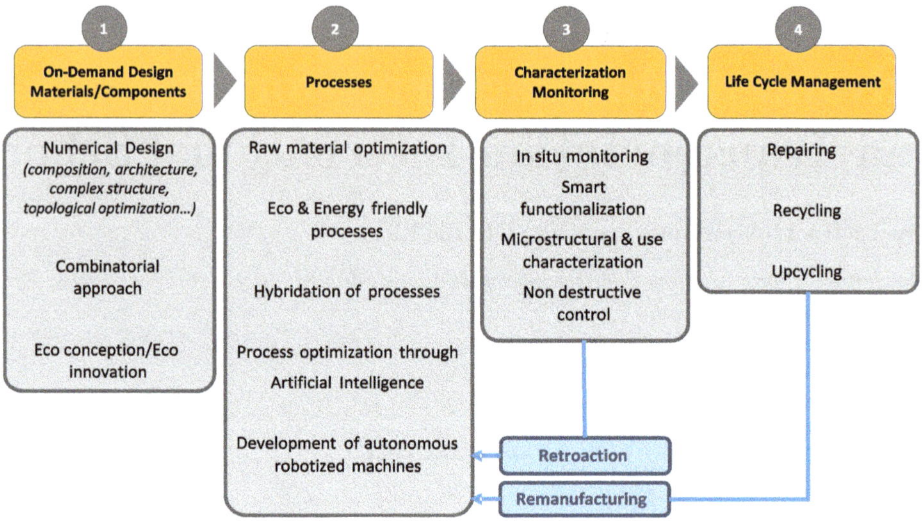

FIG. 1 – Systemic approach of advanced manufacturing.

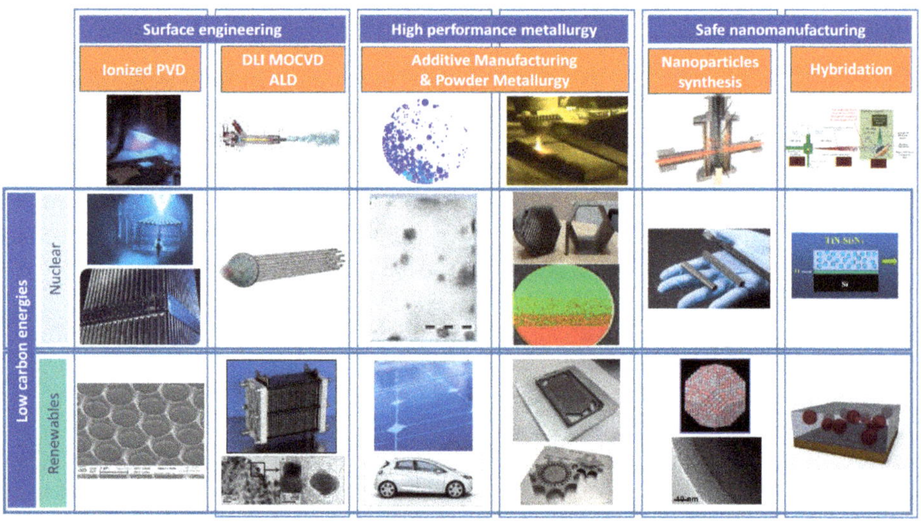

FIG. 2 – Examples of emerging processes for low carbon energies.

Therefore, after presenting an integrated vision of additive manufacturing, the latest advancements in surface engineering and nanomanufacturing technologies, AMETIS will focus on the convergence of these technologies as a source of innovation for advanced energy manufacturing and transport.

1.2 Holistic Overview of Additive Manufacturing

Regarding additive manufacturing, AMETIS will first give a presentation of an integrated vision of this family of technologies, presenting in particular the different processes, the entire value chain, the different types of materials processed, as well as examples in particular in the field of transport and low carbon energies. Then, AMETIS will focus on several trends that are generally not presented in technology-only instruction. AMETIS will first illustrate the relationships between process, microstructure and properties and also the importance of mastering raw materials from the start to the end result in terms of product performance. Next, AMETIS will present the design of on-demand architectures according to the required properties. This approach is based on digital design in a first step and the synthesis of these architectures through additive manufacturing in a second step, to compare reality with prediction, thus illustrating the link that can be established between numerical design and manufacturing of on-demand high-performance materials and components. Finally, AMETIS will focus on the possible smart integration of the electronic functions "Package Electronics Additive Manufacturing" that allows the technology to enter the era of the Internet of Things, with, for example, the possibility of monitoring the performance of the components remotely and their durability during the life time.

1.3 Integrated Approach of Nanomanufacturing: From Nanoobject Synthesis to Applications

The second "pillar process", also with a very generic character, consists of nanofabrication. In fact, the synthesis and safe integration of nanoobjects (nanopowders, nanowires, and nanotubes) into components for energy and transport is a fairly unique way to develop innovation in these sectors. AMETIS will present the main technologies for the development and implementation of nanoobjects, as well as the resulting applications. In addition to these purely scientific and technological aspects, the contribution of nanotechnologies to the development of sustainable solutions will also be presented. In fact, in some cases, nanotechnologies can lead to a reduction in the use of critical metals or strategic materials, and sometimes even allow their replacement. The last, but not least, consideration will be devoted to managing risks using nanomaterials as part of an integrated approach from production to consumer.

1.4 Emerging Surface Engineering Processes

Surface engineering has made remarkable progress in the past decade and new applications, particularly in the energy field, have become possible and are currently under industrial development in part due to this progress. Following a presentation based on the principles of thin films and thermal spray technologies, AMETIS will focus on a more detailed presentation of the progress made in recent years. In the Physical Vapor Deposition (PVD) sector, the development of highly ionized technologies and, in particular, HIPIMS (High Power Impulse Magnetron Sputtering) technology has enabled significant progress to be made, for example, in nuclear applications (EATF: Enhanced Accident Tolerant Fuel and also in the reprocessing sector). In terms of Chemical Vapor Deposition (CVD), the penetration of technologies such as DLI-MOCVD (Direct Liquid Injection MOCVD) and ALD (Atomic Layer Deposition) in the low carbon energy sector exploits all the richness and diversity of the chemistry of organometallic precursors, both for applications in extreme environments and for functional components for batteries, supercapacitors and high-temperature electrolyzers or fuel cell components. Regarding thermal spraying technologies, two very important advances are observed, firstly, the significant development of cold spraying technology, which can also be considered in some aspects as a repair or additive manufacturing technology. Another important innovation refers to the thermal plasma spraying of nanoparticle precursor solutions or suspensions that offers very interesting perspectives for the new generations of thermal and environmental barriers, in particular for the aeronautical and space sector.

1.5 Possible Synergies Between Technologies

After the presentation of the three main families of emerging processes, AMETIS will also present their possible synergies and the resulting innovations (figure 3), in particular:

- At the interfaces between 2D and 3D processes

This is usually the case when, for example, 3D manufacturing processes are combined with surface functionalization processes that give these architectures new functionalities (catalytic, anti-corrosion, electrical, etc.).

- At the border between 3D printing and nanofabrication

This is, for example, the case of the use of nanoparticles as raw material for 3D printing technologies, or for the *in situ* generation of nanocomposite materials in additive manufacturing machines (*in situ* generation of ODS (Oxide dispersion strengthened) steels, for example).

- With a combination of nanotechnologies and surface engineering

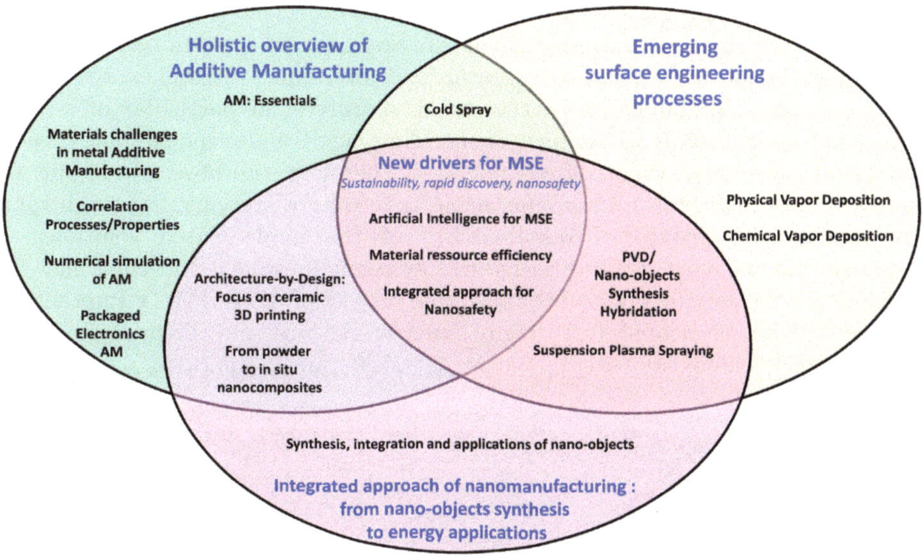

FIG. 3 – The three families of emerging processes, their possible synergies, the new generic methodologies for rapid discovery and optimization and the new drivers for MSE.

It is a completely new generation of innovative processes that combine thin film science and engineering with nanoparticle synthesis technologies. This new approach enables the production of innovative nanocomposite materials, in which the choice of the nature of the matrix is completely independent of the nature of the incorporated nanoobjects, which greatly opens the spectrum of applications for this type of new materials (solar thermal energy, photovoltaic energy, batteries...).

1.6 New Methodologies for Rapid Discovery of Materials and New Drivers for MSE

These three major families of processes also contribute to an accelerated discovery of new materials, components and solutions, either because they often intrinsically allow rapid screening of compositions, microstructures and architectures, or because they give access to new concepts of materials and components hitherto not accessible through more conventional production routes.

This accelerated discovery of solutions also benefits from considerable progress in digital technologies, both for data mining, numerical design of materials, components and architectures, numerical simulation of processes, as well as their optimization using Artificial Intelligence approaches. These new approaches are also intended to be introduced during this first AMETIS.

The final viewpoint to address in this systemic overview of advanced manufacturing is about the new drivers for the development of a sustainable Materials Science and Engineering. These drivers consist in environmental and energy sobriety of the processes, minimization of the use of resources, in particular of strategic metals and, in general, in the impact of emerging processes on the circular economy of materials and recyclability of components. Therefore, the objective of this first summer school on advanced manufacturing is to offer a systemic approach taking into account the drivers of sustainability of the implemented solutions and eco-innovation. In more general terms, this systemic approach and these new convergences are the core business of the scientific project of the IMPACT International Chair of INSTN (National Institute of Nuclear Science and Technology) at the University of Paris-Saclay (https://www.materials-impact-chair.org/).

Contents

Chapter 1

Additive Manufacturing: Development of Sustainable Industrial Processes for Circular Economy Improvement

Abstract

Additive manufacturing is undergoing strong development for various applications in different industrial fields. This global material efficiency addresses various manufacturing fields, taking into account new processes which make it possible to minimize the use of resources thanks to digital ecodesign. This fact allows these technologies to be framed within the circular economy, since it allows the efficient use of resources and introduces the concept of post-manufacturing recycling of raw materials, for instance, of non-melted powders in powder bed processes. This fact permits to reduce the impact of these technologies thus promoting the industrial ecology, since it contributes for example to the better management of critical metals. In this chapter, a holistic approach regarding the sustainable additive manufacturing in the frame of the circular economy is presented. The latter is based on the description of the existing whole value chain. On the one hand, the main technologies, the materials challenges, the innovative powder metallurgy production, the correlation of the process parameters/corrosion resistance, the process simulation and final microstructure prediction are described. On the other hand, the intelligent digital design of architectures associated with the desired final properties, as well as, the full monitoring of final parts by embedded electronic sensors integrated during processing are presented as the most promising developing routes.

DOI: 10.1051/978-2-7598-2446-5.c001

1.1 Additive Manufacturing: Essentials

Fernando LOMELLO

Université Paris-Saclay, CEA, Service d'Etudes Analytiques et de Réactivité des Surfaces (SEARS), 91191 Gif-sur-Yvette, France

Abstract

Additive manufacturing (AM) or commonly called 3D printing is a set of processes that allow the manufacture of a physical component from a virtual object. The progress made in recent years has taken them from rapid prototyping processes to real technological routes for the production of components with high added value. These technologies allow a significant reduction in the material and energy used, while minimizing the so-called subtractive manufacturing steps, notably related to machining. This is a fundamental point for the development of environmentally efficient manufacturing processes for the French industry of the future.

The reduction in the stock thanks to the possibility of manufacturing parts of great complexity, as well as, with the reduction in the supply chain, allows expecting a diffusion in several fields of high added value. This is the specific case of the aeronautical industry, which fostered the evolution of these rapid prototyping technologies towards true production routes.

Nowadays, technological maturity allows us to have a global vision ranging from the supply of primary materials to recycling aspects. Thanks to these 3D printing technologies, many existing primary materials manufacturing and recycling technologies were revisited. The main idea is to present an overview of the associated technologies and materials, as well as the evolution in the various fields of application in the following pages.

1.1.1 Overview of Additive Manufacturing: AM Common Terms

The term Additive Manufacturing as described by the "Union de Normalization de la Mécanique" (NF E 67-001) includes "all of the processes used to manufacture, layer by layer, by adding material, a physical object from 'a digital object' – see figure 1.1.1.

Computer-aided design (CAD) allows the creation of a 3D model that is an integral part of the additive manufacturing process. As shown in figure 1.1.1, the volume CAD modeler cuts the virtual object into layers of defined thickness according to a particular orientation. This digital processing requires the use of computer-aided manufacturing (CAM) software in order to save the details of each indexed layer.

The 3D model is stored using different computer formats such as IGES (Initial Graphics Exchange Standard), STEP (STandard for the Exchange of Product) and

DOI: 10.1051/978-2-7598-2446-5.c011

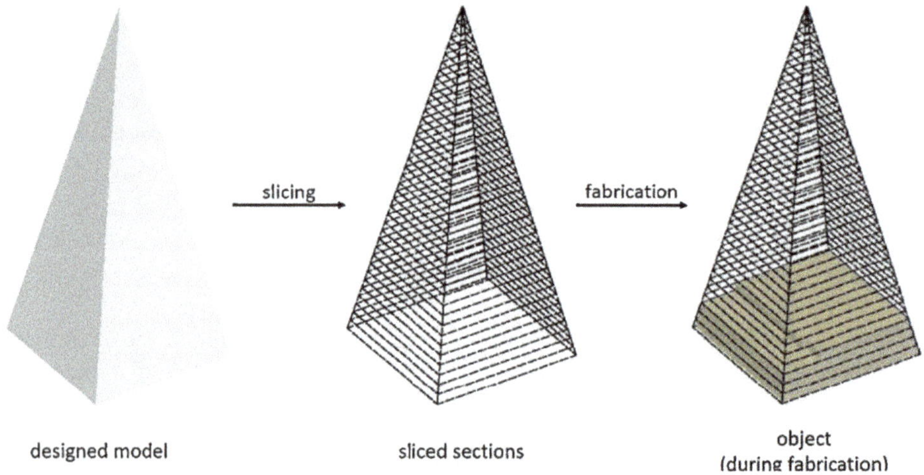

FIG. 1.1.1 – Production method: points common to all processes [1].

STL (Stereolithography Tessellation Language). The latter, the most used in the field of additive manufacturing was created by 3D Systems.

The progress made in recent years has shifted rapid prototyping processes to real technological processes for the production of components with high added value – see figure 1.1.2. Today, three types of materials are most commonly used: metals, polymers and ceramics (very upstream).

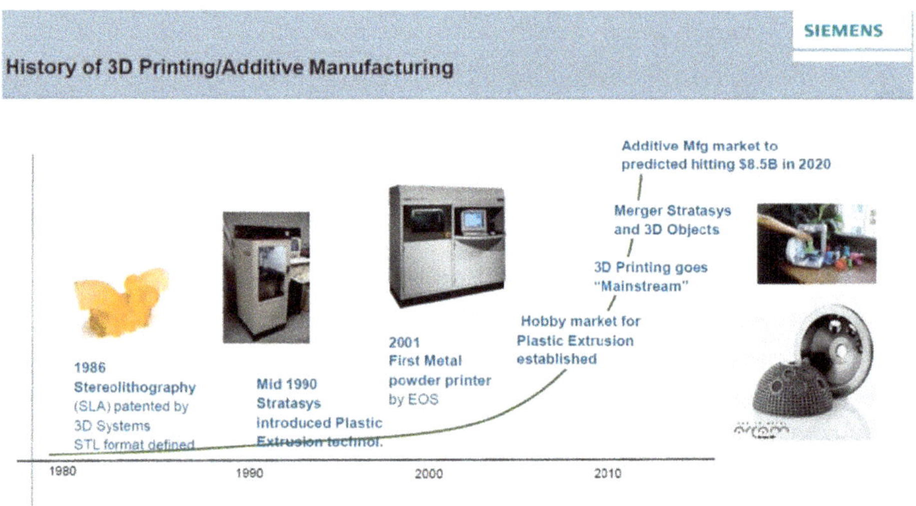

FIG. 1.1.2 – From rapid prototyping to additive manufacturing [2].

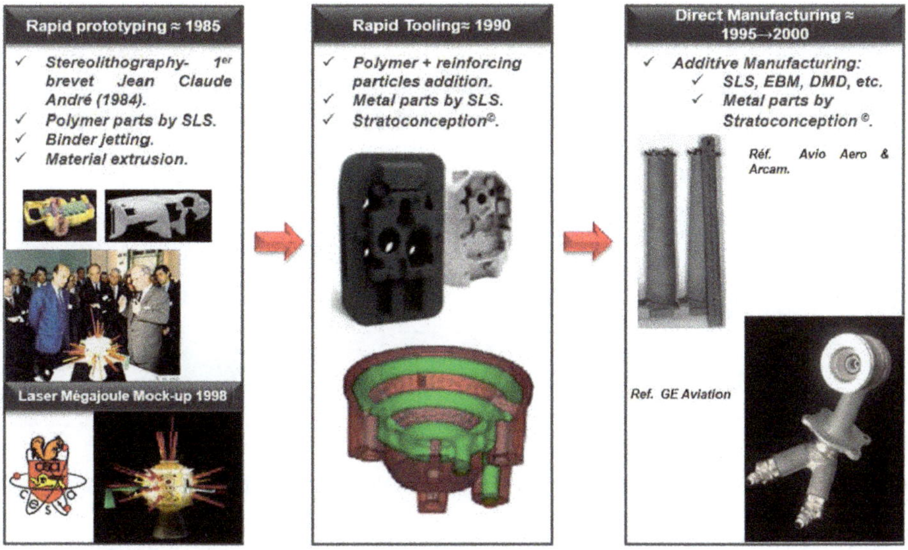

FIG. 1.1.3 – Additive manufacturing evolution.

The progress made by these new manufacturing technologies has enabled rapid prototyping processes to evolve into real new technological routes (figure 1.1.3) for the production of components with high added value. Different industrial sectors are considering new applications (nuclear, aerospace, medical, automotive, hydraulic, pneumatic, tooling, electronics, etc.). Since 1992, rapid prototyping (stereolithography) has been used by CEA CESTA for the production of mock-ups within the framework of the various programs – as shown in figure 1.1.3.

1.1.1.1 Description of Technology Capabilities in AM

In general, the concept of additive manufacturing according to standard ASTM F2792-12A developed by the commission F42 (2013) combines seven technologies (figure 1.1.4) [3]:

- *Stereolithography (SLA, Vat polymerization)*: it uses photopolymerizable resins which it is activated by a UV laser. This technique is used to process polymer-based composites reinforced with ceramics or metals. Stereolithography was born in the mid-1980s under the impetus of French and American researchers. The first patent was deposited by Jean-Claude André with CILAS (Compagnie Industrielle des Lasers) in 1984 followed by the American Charles W. Hull, the creator of the company, 3D Systems a month later. The first machine to be marketed was built in 1988 by 3D Systems. This technique also makes it possible to manufacture ceramic parts using resins loaded with 95% ceramic, followed by a sintering step – as is the case with the process developed by the laboratory "Science of Ceramic Processes and Surface Treatments"

FIG. 1.1.4 – Different additive manufacturing technologies depending on the energy involved.

(SPCTS UMR CNRS 7315) in Limoges. Today, the process is operated by the 3D Ceram. Examples of production: prototypes and dental prostheses. Manufacturer: 3D Systems.

- *Material Jetting*: Direct deposition of materials, layer by layer, which allows for example the manufacture of multi-material components. Two technologies are associated: inkjet printing and aerosol jet printing. This technique makes it possible to manufacture components based on polymer and/or reinforced with metals. Production examples: prototypes. Manufacturers: 3D Systems, Stratasys.
- *Binder Jetting (BJ)*: Deposit of liquid glue to associate grains of powder (metallic, polymer, ceramic) patented by MIT in 1989. The components manufactured by this technique have low mechanical resistance. This type of component requires thermal post-sintering or infiltration treatments in order to obtain an attractive final density. Production examples: prototypes. Manufacturers: Ex-one (digital part materialization), Z-Corp (3D Systems), Voxeljet.
- *Material Extrusion, Fused Deposition Modeling (FDM)*: Process developed in 1988 by Lisa and S. Scott Crump, one of the co-founders of the company Stratasys. In 1991, the technology was introduced to the market. The materials are extruded through a die which is heated. This technique makes it possible to manufacture parts which have low mechanical properties. Production examples: prototypes. Manufacturer: Stratasys.
- *Sheet Lamination – Stratoconception® – Laminated Object Manufacturing (LOM)*: this process was patented by Michael Feygin in 1988. In France, Prof. Claude Barlier in 1991 created the CIRTES R&D center which registered 19 patents on the stratoconception® process. The components are made from previously profiled sheets of preformed materials (polymers, metals, ceramics). The assembled sheets do not allow the production of structural components. Production examples: plastic injection mold. Manufacturers: CIRTES, Mcor Technologies.

- *Powder Bed Fusion (Electron Beam Melting: EBM/Selective Laser Melting: SLM/Selective Laser Sintering: SLS)*: Selective fusion of powders which are distributed as a form of a bed through a laser or electron beam. This technology allows the use of polymeric, metallic and ceramic powders. In particular, electron beam fusion (EBM) technology has several advantages in terms of oxygen content (under vacuum) and residual stresses (initial heating 700 °C). Examples of production: turbine blades, aeronautical engine injectors, stainless steel mesh and tools. Manufacturers: Phenix systems (France – today 3D Systems), 3D Systems, EOS, SLM Solution, MCP Realizer, Arcam.
- *Direct Energy Deposition (Construction Laser Additive Direct: CLAD®/direct metal deposition: DMD/laser metal deposition: LMD/direct energy deposition: DED/Laser Engineered Net Shape: LENS)*: materials as the form of powder or wire are deposited directly by focusing on a target. The geometric complexity is difficult. 5-axis systems need to be developed. Examples of production: composite dies, surface treatment, aeronautical parts and repair. Manufacturers: BeAM (France), POM, Optomec, Trumpf, InssTek, Efesto and Huffman.

As mentioned above, it is their implementation and their use for a layer-by-layer construction of objects which thus classify the 7 technologies. The base material can be in the form of liquid, powder, tape or wire. This material can be present at the start of the manufacturing process or deposited as you go. The material is shaped using a heat source: electron beam, laser, visible light, UV or IR. The manufacturing process is done *via* three different routes depending on the technique used: solidification, fusion and bonding of sheet materials (stratoconception®).

1.1.1.2 Categorization of Material Types Used in AM

The available printable materials for each AM technologies described in the previous lines is shown below in figure 1.1.5.

1.1.1.3 Additive Manufacturing Advantages for Decision Making: Discussion of the Impact of Both Process and Material Selection

As claimed in the Roland Berger report in 2017, the materials are extremely dependent on the decision making (figure 1.1.6).

The metal additive manufacturing processes could be classified into three main groups (figure 1.1.7): powder bed, powder projection and wire deposition.

First Group: Powder Bed Fusion

The first group, powder bed fusion technologies, allows the fabrication of a part by stacking successive layers. The scraper brings a layer of powder which is fused on the lower layer by a focused energy beam (laser, electron beam). These technologies have the advantage of making it possible to produce components having complex shapes with a good surface finish. On the other hand, they have technological limits in terms of working chamber size, volume of powder used and manufacturing speeds.

Materials	Example materials	Process categories						
		Vat photo-polymerization	Material jetting	Binder jetting	Powder bed fusion	Material extrusion	Directed energy deposition	Sheet lamination
Thermoset Polymers	Epoxies and acrylates	X	X					
Thermo-plastic polymers	Polyamide, ABS, PPSF		X	X	X	X		X
Wood	paper							X
Metals	Steel, Titanium alloys, Cobalt chromium			X	X		X	X
Industrial ceramic materials	Alumina, Zirconia, Silicone nitride	X		X	X			X
Structural ceramic materials	Cement, Foundry sand			X	X	X		
Note: Combinations of the above material classes, e.g. a composite, are possible								

FIG. 1.1.5 – Additive manufacturing process families & processes [4].

We can distinguish a first sub-group according to the ability to merge and/or sinter the materials:

- A first process, SLS (Selective Laser Sintering) was invented by R.F. Housholder who patented the concept in 1979, but without commercial exploitation – see figure 1.1.8. In addition, Dr. Carl Deckard at the University of Texas (Austin) invented and patented in 1987 a first SLS system for sintering PA 6.6 nylon powders. This patent contributed to the creation of the start-up DTM Corporation in Austin, Texas, which filed the main patent in 1992. In 1997, 3D Systems bought the patent filed which fell into the public domain in 2014. 3D Systems has signed an agreement with EOS in 1994 for the exploitation of the patent in exchange for the design of EOS stereolithography machines. This process is called by EOS "DMLS" (direct melting laser sintering). Historically, these systems allowed the selective sintering of powders with a certain porosity rate without achieving fusion due to the low powers of lasers (100 kW<). This is not the case today, the latest machines like ProX™ from 3D Systems which have two fiber lasers of 500 kW or one of 1 kW allowing the realization of dense parts. In France, the company Phenix Systems has developed its own machines since the year 2000. Today, the company belongs to the 3D Systems group.
- SLM (Selective Laser Melting) was developed and patented in 1995 at the Fraunhofer ILT Institute in Aachen, Germany by Dr. Dieter Schwarze and Dr. Matthias Fockele. Due to intellectual property issues, the patent will be used

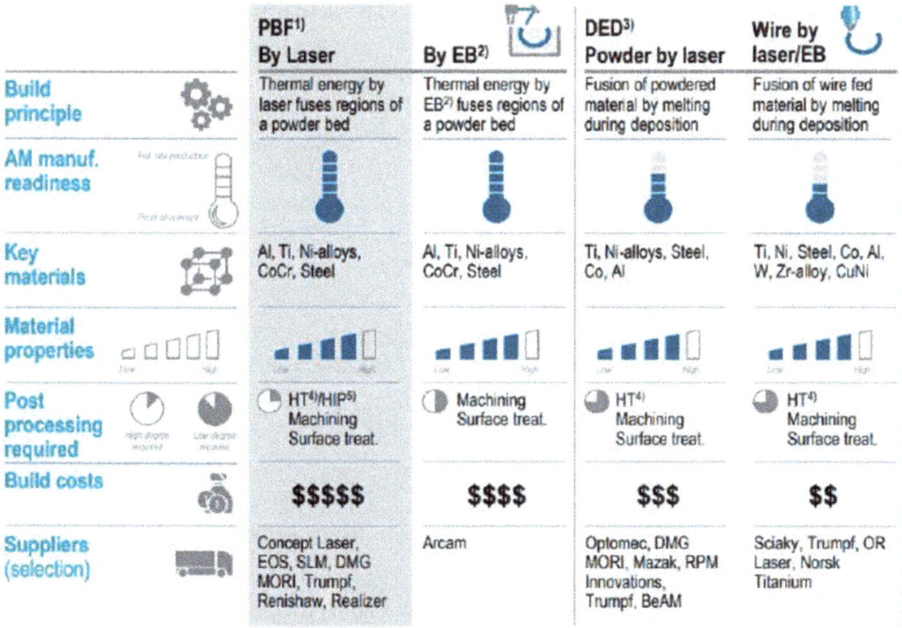

FIG. 1.1.6 – Additive manufacturing methods in the process decision [5].

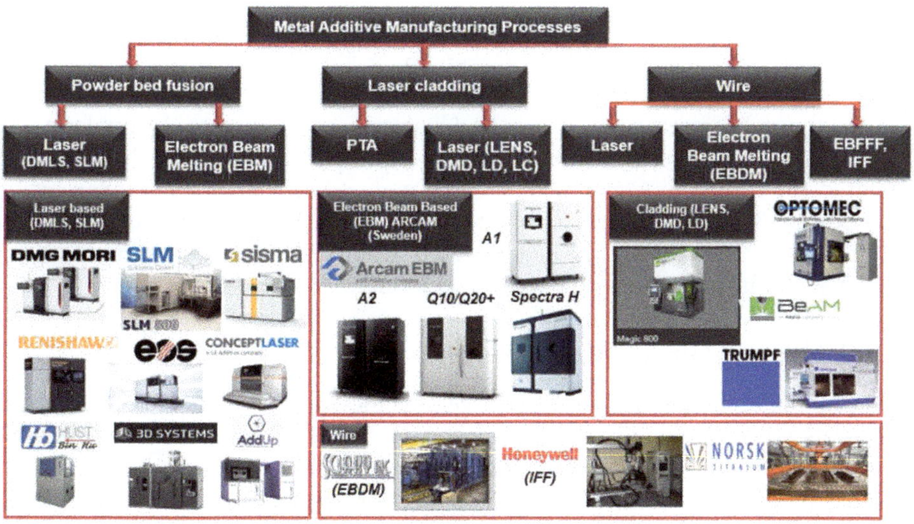

FIG. 1.1.7 – Different metal additive manufacturing processes.

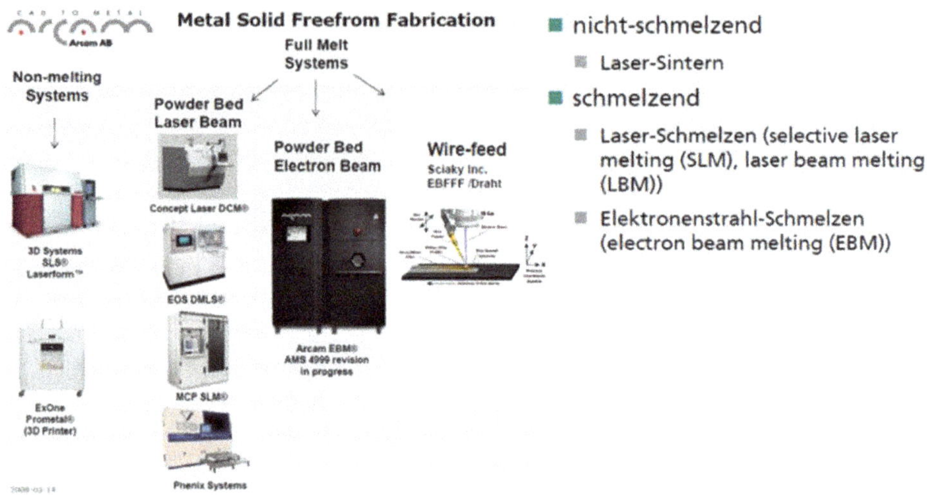

FIG. 1.1.8 – Powder bed fusion technologies [6].

by the Trumpf group and EOS. These allowed the patent to be used by MCP Realizer which, after its commercial separation, led to the birth of two new companies, MTT Technologies group (Germany) and MTT Technologies LTD (England). Today, the companies have been renamed SLM Solutions and Renishaw, respectively – see figure 1.1.9.

- LaserCUSING® is a technology very close to SLM patented in 2000 by Concept Laser GmbH. The first machine was sold in 2002. Today it is the manufacturer of the largest laser machine in the market, the X line 2000R.
- The electron beam melting (EBM) technology was patented in 1993 in Gothenburg by researchers from the Chalmers University of Technology – see figure 1.1.10. The same researchers with the help of the Swedish government created the company Arcam in 1997. The main advantage of an electron beam is the power (3.5 kW) used which melts different materials at high melting points (2700 ° C<), preheating the powder bed (reduction in stresses) and working under secondary vacuum (low oxygen content).

Second Group: Direct Metal Deposition

The second group, powder projection technologies are based on selective powder injection then laser fusion at the exit of a coaxial nozzle allowing the production of a part layer by layer or the reconstruction in service of worn areas of an existing room. One of the main obstacles to overcome is the development of new materials such as aluminum alloys, Inconel and materials with a functional gradient. Also, it is necessary to demonstrate technological maturity in terms of geometric resolution. Regarding this process, the first patent was filed in the United States by D.S. Gnanamuth in 1976 – see figure 1.1.11.

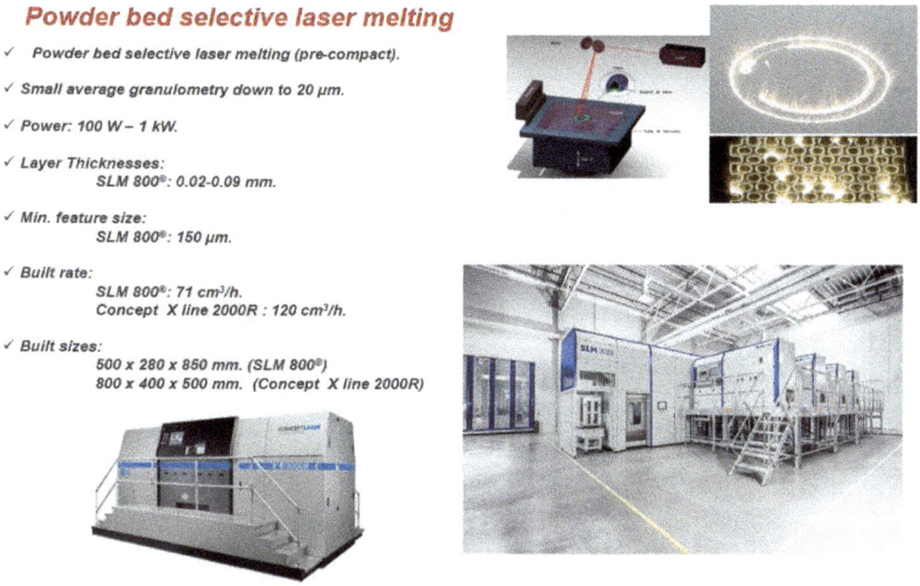

FIG. 1.1.9 – Powder bed fusion by laser: characteristics of the current machines.

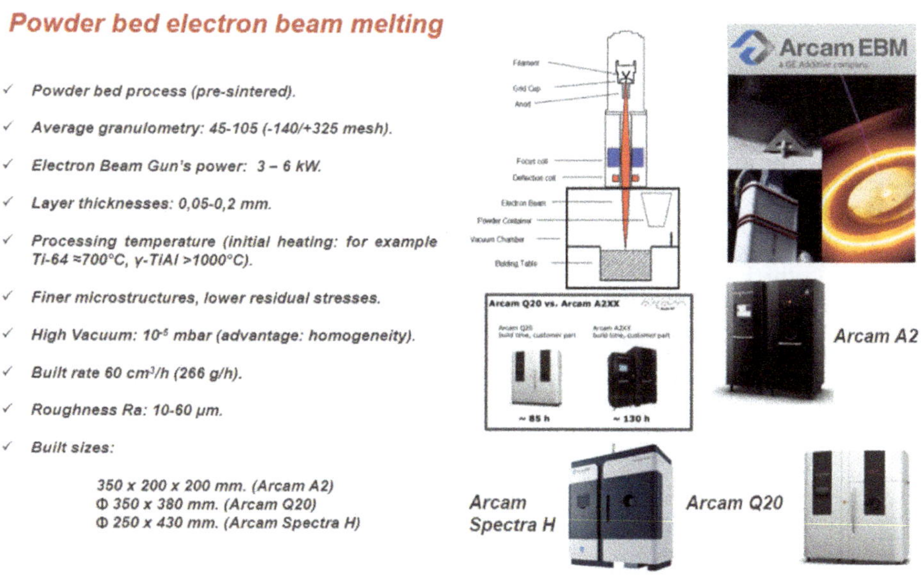

FIG. 1.1.10 – Powder bed fusion by electron beam: Arcam technology.

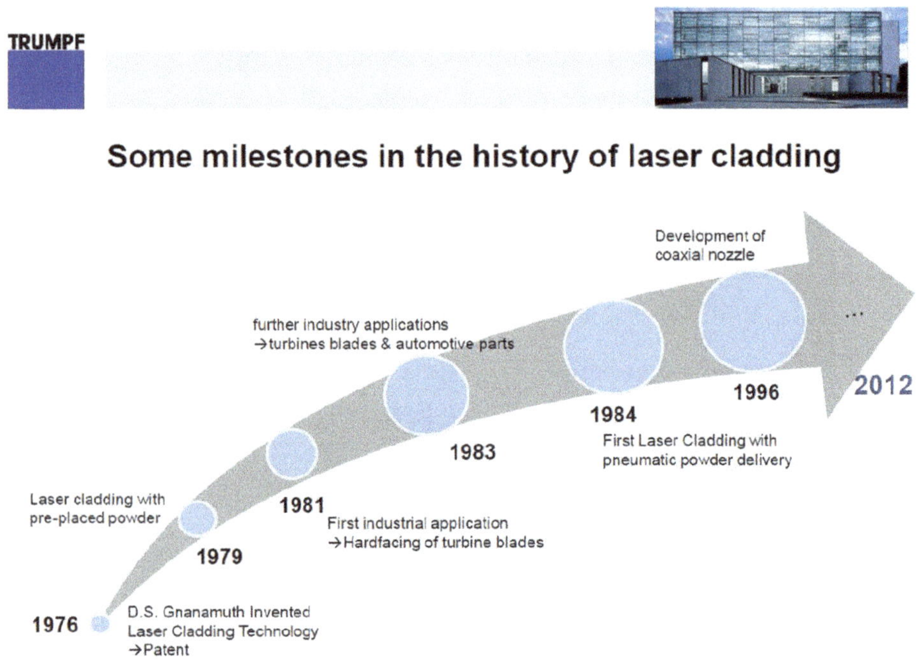

FIG. 1.1.11 – Direct metal deposition evolution [7].

Similar processes with certain technical variations were patented in the 1990s such as:

- Direct Metal Deposition (DMD – University of Illinois Urbana-Champaign, 1993): POM.
- Laser Engineered Net Shaping (LENS – Sandia National Laboratory, 1996): Optomec.
- Direct Light Fabrication (Los Alamos National Laboratory, 1996).
- Controlled Metal Build up (Fraunhofer ILT, 1996).

Third Group: Wire-Based Additive Manufacturing

The third group, wire fusion technologies uses an energy source to melt a metal wire on a metal support. The energy beams can be of different nature: electron beam, electric arc, etc. These technologies allow the local addition of material as well as the repair of components.

Technology Comparison

The table in figure 1.1.12 shows that the most suitable processes also for short-term industrialization are those based on manufacturing from metal powder – powder bed (figure 1.1.8).

++++ => best	Powder bed (laser)	Powder bed (EB)	Powder feed	Wire feed
Beam thickness	++++	+++	++	+
Layer thickness	++++	+++	++	+
Construction speed	+	++	+++	++++
Roughness	++++	+++	++	+
Internal stress level (less deformation)	+++	++++	++	++
Shape complexity	++++	+++	++	+
Repair capability	0	0	+++	+++
Multi material capability	0	0	++	0
Size of the part	++	+	+++	++++
Hollow and lattice structure capability	++++	++	0	0
Wall Thickness and precision	++++	+++	+	++

FIG. 1.1.12 – Metal additive manufacturing technologies comparison (Frost & Sullivan 2012).

Hybrid Technologies

We can add to this classification a fourth group, that of hybrid machines for additive manufacturing and by subtractive alternation based on powder bed fusion (Matsuura) and laser projection processes (Hamuel, DMG Mori, Mazak, WFL Millturn Technologies) – see figure 1.1.13. The Matsuura, Hamuel, DMG Mori and WFL M80 machines were introduced to the machine tool market the first two in 2013, the third in 2014 and the last in 2015, respectively. This group of processes is of interest for certain applications such as the production of tanks having improved interior surface roughness.

1.1.1.4 Additive Manufacturing Using Cases: List of Examples of Real-World Application on AM

1.1.1.4.1 Aeronautics

Since the installation of the Arcam A2 machine in the MATPERF joint laboratory (Mecachrome-CEA) in 2009, various prototypes for the aerospace industry have been produced. In particular, supports for the communication satellite Atlantic Bird 7 (figure 1.1.14) were manufactured in 2010 for the company Astrium – today Airbus Defense & Space. This satellite was transported to the geostationary transfer orbit by the Zenit-3SL rocket. This manufacturing route has allowed a significant reduction in weight (30%). Given the results, the European Space Agency (ESA) considers that additive manufacturing has a level of technological maturity equal to 9 (small-scale production).

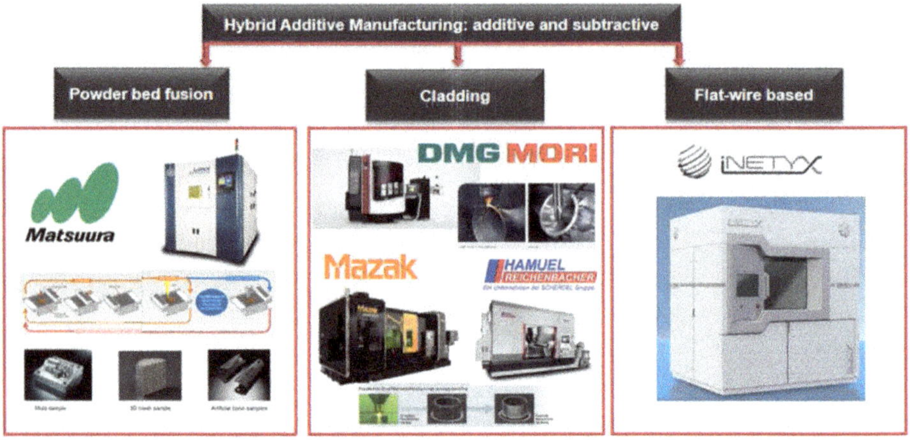

FIG. 1.1.13 – Hybrid manufacturing processes: additive and subtractive.

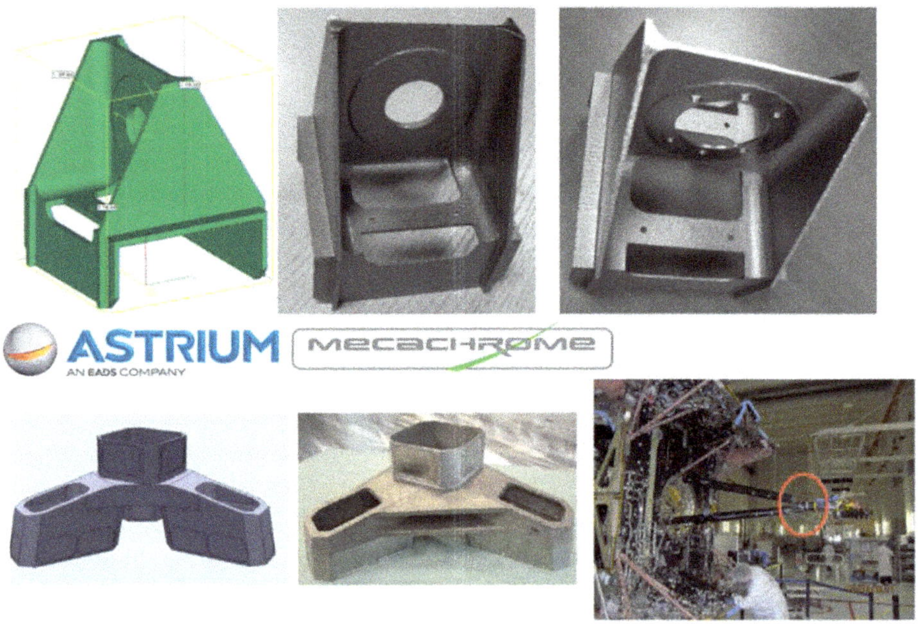

FIG. 1.1.14 – Atlantic Bird 7 satellite antenna supports [8, 9].

In the international context, a first study was carried out by the English subsidiary of EADS Innovation Works (today Airbus Group) in collaboration with the company EOS. The study was focused on the topological optimization of an A320 hinge (see figure 1.1.15) in TA6V in an M270 (DMLS) machine using Altair Hyperworks software. Topological optimization consists in defining the ideal

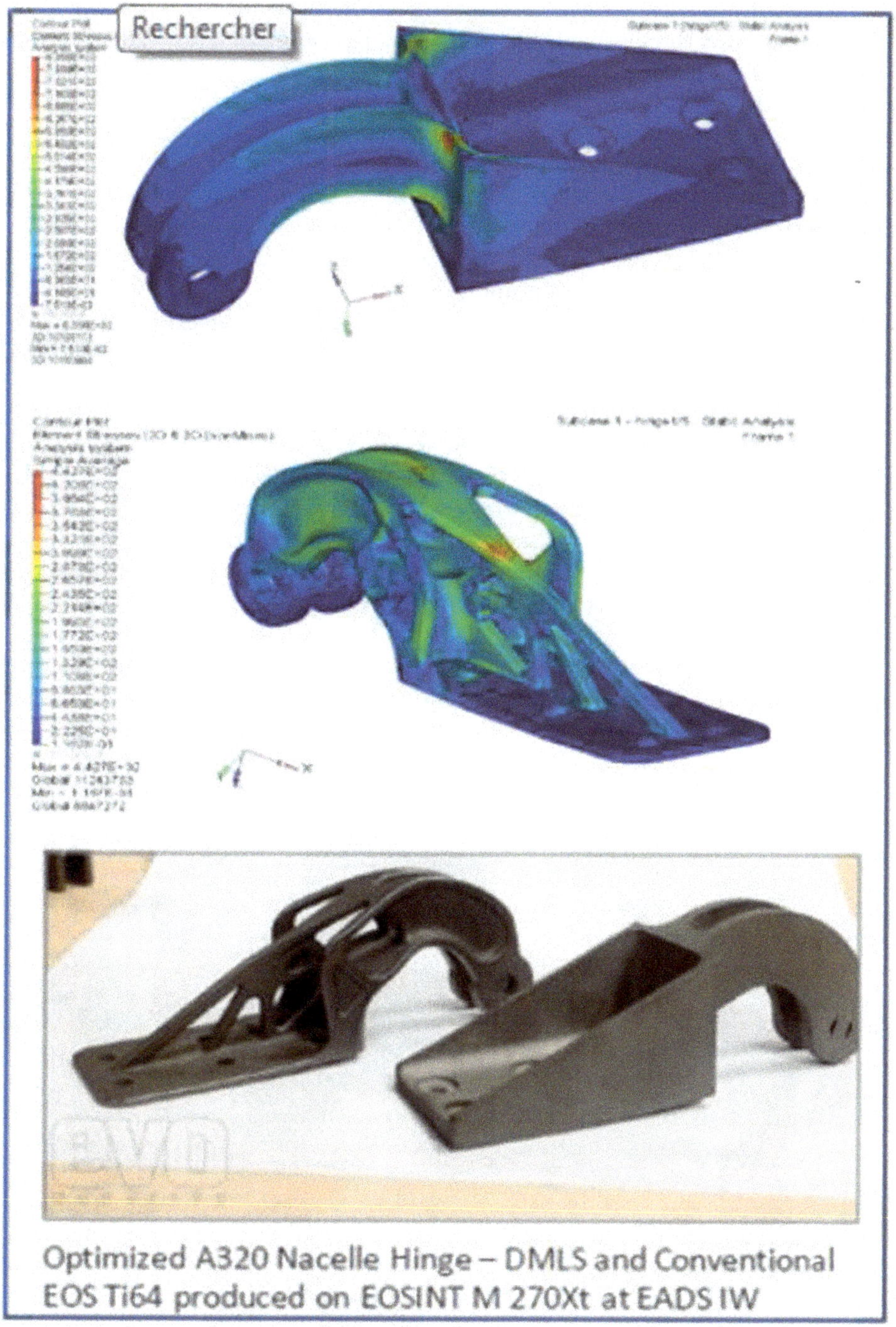

FIG. 1.1.15 – Topologically optimized hinge demonstrator for the A320 aircraft [10, 11].

material distribution in a room in the form of finite elements in a given volume taking into account the mechanical constraints. This approach allowed the creation of an optimized part 60% lighter.

A recent, a relevant example is that of the fuel injectors for the new LEAP-X engine from SAFRAN-GE (see figure 1.1.16) by laser fusion using a powder bed (EOS technology). This technology replaces the manufacturing and assembly of 20 parts by the production of a single part by additive manufacturing (19 injectors per engine). This induces a reduction of 25% and an increase in service life by a factor of 5). The GE (Morris Technologies) ALM research center is currently working on qualification, process optimization and non-destructive testing. GE announced an investment of 50 million euros in the Auburn and Alabama facilities to prepare the production lines (≈35–40 k injectors/year). Entry into service is scheduled for 2020.

Below is detailed the historical evolution of applications in the aeronautics/space industry according to the consulting company "Deloitte" – see figure 1.1.17.

In the same process, Avio Aero, an Italian subsidiary of the GE group in Cameri, has adapted electron beam technology (EBM) for the production of low pressure turbine blades made of titanium aluminide (γ-TiAl) for different LEAP-X, Genx and GE90 engines – in qualification stage. It has 11 Arcam machines, 2 EOS (DMLS) and an atomization tower (50 tonnes/year) to ensure the production of powders. Avio Aero forecasts growth in the number of machines installed at 60 – see figure 1.1.18.

The GE90 engine from the same group is the first engine to have composite materials in the blades (section of the Turbofan) and a part called T25 housing manufactured by ALM – see figure 1.1.19. The component is placed in the high

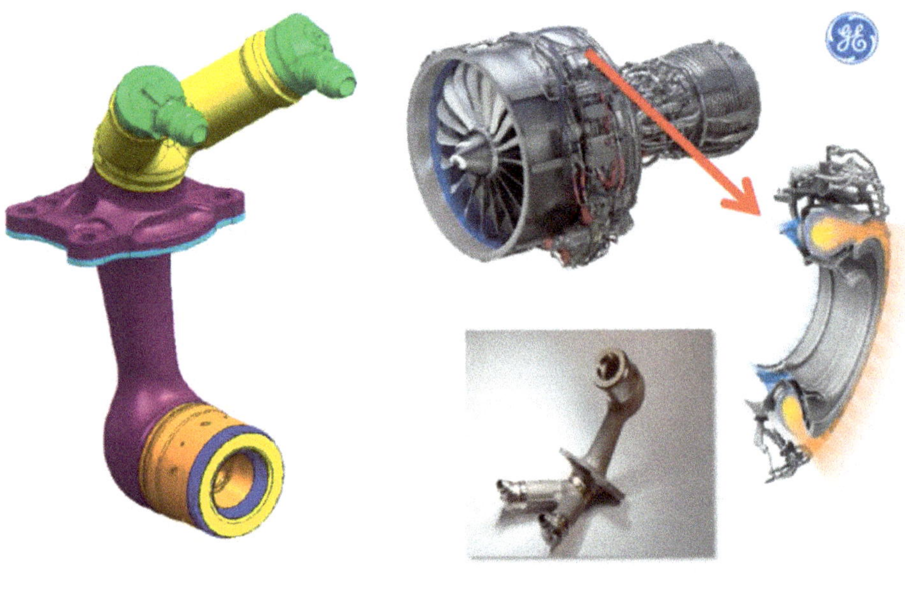

FIG. 1.1.16 – LEAP-X engine fuel injector produced by powder bed laser fusion (EOS) [12].

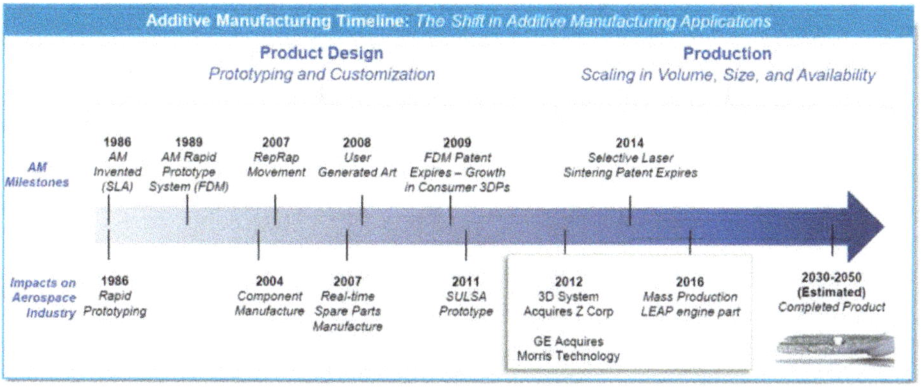

FIG. 1.1.17 – Chronology of the introduction of additive manufacturing technologies in the aeronautical field [13].

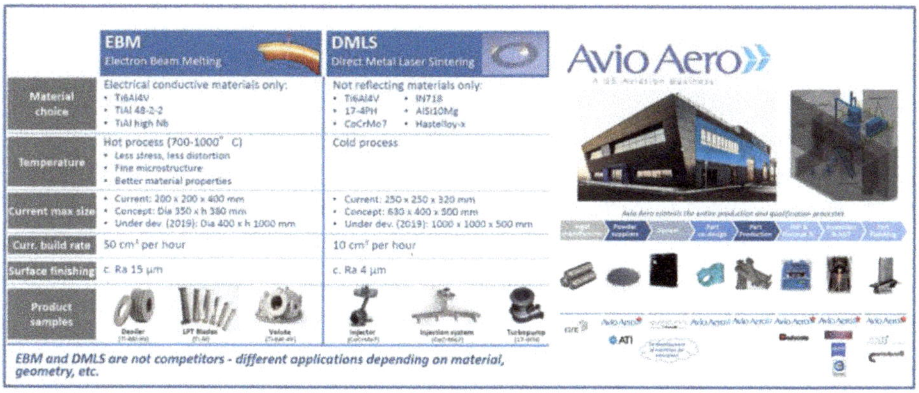

FIG. 1.1.18 – Avio Aero strategy [14].

pressure compressor allowing the protection of an electronic sensor – it was qualified in 2014.

Pratt & Whitney delivered Bombardier PurePower® PW15000G engines with components manufactured by ALM. Parts of the compressor stator and synchronization rings were produced by SLM.

Lockheed Martin explores the development of titanium-based components by powder-bed fusion by electron beam (EBM), wire deposition by free-form electron beam (EBF3 – the technology is shown in the following section) and by laser projection technologies.

Moreover, the British Air Force carried out the first flight test of various parts manufactured by ALM on the Tornado GR4 aircraft in 2014. The company BAE (British Aerospace) replaced several parts of the landing gear and cabin. This meant a reduction in manufacturing costs of £ 300 k.

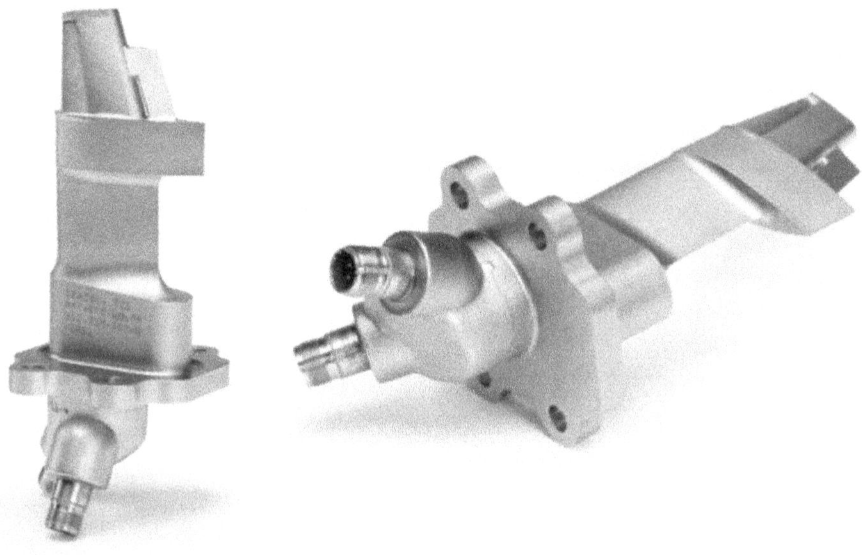

FIG. 1.1.19 – T25 housing component manufactured by SLM (GE Aviation).

In 2016, the SAFRAN group presented at the last "Paris International Air and Space Show" an auxiliary power unit of the Dassault Falcon 20 (APU SAPHIR 4.2) entirely manufactured by laser fusion in aluminum alloys and nickel. This project is piloted by the Monash Center for Additive Manufacturing (MCAM-Australia) in partnership with the Commonwealth Scientific and Industrial Research Oraganization (CSIRO), the University of Deakin and Microturbo – see figure 1.1.20.

Turbomeca announced in January 2015 that it will install fuel injectors manufactured by SLM on the Arrano and Ardiden 3 engines – see figure 1.1.21.

SAFRAN (Aircelle) has also planned for the application of Vinci cryogenic engines, the evaluation of ALM for the realization of collectors in Inconel 718. The SLM process is well adapted to the rates because it allows the reduction in the production cycle to 8 weeks between definition and the first part delivered. According to SAFRAN, the part has characteristics "close to those of the forged, much superior to those of the foundry". They are currently planning to build chambers based on copper alloys for the Vulcan and Vinci engines. They are in the development phase of different materials: Cu-Cr-Nb and refractory metals.

In 2014, NASA showed an example of the feasibility of producing injectors for Aerojet Rocketdyne RL-10 (liquid oxygen/hydrogen gas) engines. Indeed, different components have been manufactured by laser fusion (SLM) as well as by electron beam (EBM) – see figure 1.1.22.

1.1.1.4.2 Nuclear

In the nuclear field, there exist many potential applications for PWR (Pressurized Water Reactors) as well as for BWR (Boiled Water Reactors) as described in the

FIG. 1.1.20 – Auxiliary Power group SAPHIR 4.2.

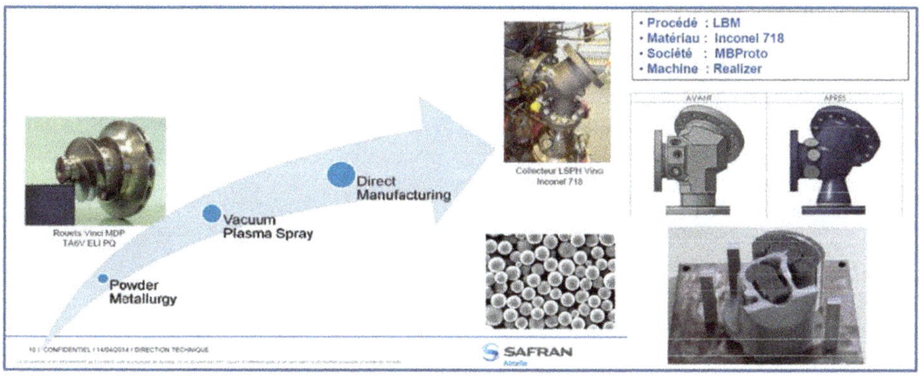

FIG. 1.1.21 – LSPH manifold for the VINCI engine [15].

figure 1.1.23. Majority of them are located in the core structure, which involves the stainless steels, nickel & zirconium based-materials.

In fact, there exist many challenges regarding the qualification that should be addressed in order to prove that additive manufacturing is suitable to the nuclear power operation plants as described in figure 1.1.24.

FIG. 1.1.22 – First test of an Aerojet Rocketdyne RL-10 engine injector manufactured by additive manufacturing (NASA Glenn Research Center) [16, 17].

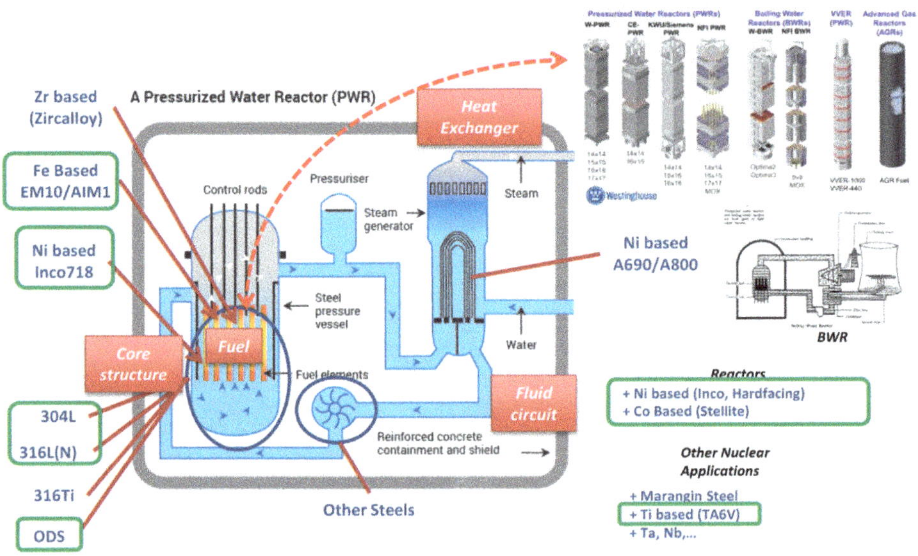

FIG. 1.1.23 – Potential application of additive manufacturing on the actual nuclear reactors.

With regard to the nuclear industry, in England Rolls & Royce and Sellafield launched a program on the application of additive manufacturing for the replacement of current components in the sanitation and dismantling sector. At the same time, the Nuclear National Laboratory in England was carrying out significant activity to assess new technologies such as powder bed fusion and spark plasma sintering for the manufacture of advanced nuclear fuels [18].

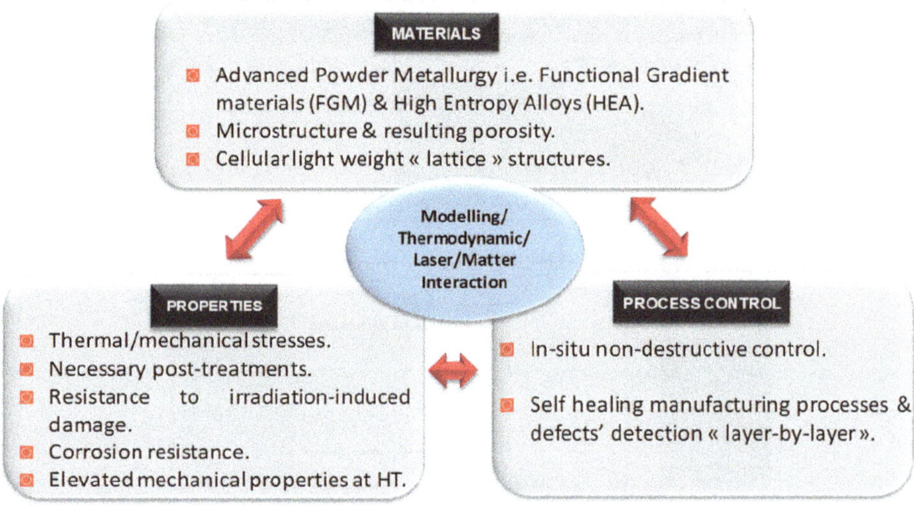

FIG. 1.1.24 – Principal challenges of additive manufacturing in the nuclear field.

Other more up-to-date metallurgical understanding studies are currently being supported by the United States Department of Energy (DOE) as part of a program called "Nuclear Energy Enabling Technologies (NEET): Advanced Manufacturing Methods (AMM)". Nuclear demonstrators, such as fuel assembly grids, have been manufactured by Lockheed Martin Space Systems Company using laser powder fusion technology (see figure 1.1.25). The final technical report of this study was published on December 31, 2015 [19].

DOE is also supporting additive manufacturing development of a new TRISO fuel concept for light water reactors (REL) through the company Free Form Fibers, LCC.

Another example in the nuclear field concerns the manufacture of components with complex shapes in FeCrAl steel reinforced by oxide dispersion (ODS) – see figure 1.1.26. This action is part of an approach led by the Oak Ridge National Laboratory, the Ames Laboratory and the University of Liverpool. Using the powder bed laser sintering (SLM) process, they demonstrated the technological feasibility of producing technological demonstrators.

The United States Department of Energy (DOE) also funded a study at the Lawrence Livermore National Laboratory (LLNL) for the evaluation of the SLM technique for the densification of ODS powders (Incoloy® MA956). The first results were published in 2015 in the "Journal of Nuclear Materials" 41. Weaker mechanical performance compared to forging due to the agglomeration of yttrium oxide particles was measured. A second publication by Chinese researchers on ODS steels produced by electron beam technique (EBM) was published in 2016 [24].

As mentioned in particular in the context of FP7, a second project called Oxigen (Oxide Dispersion Strengthened Materials for Additive manufacturing of High

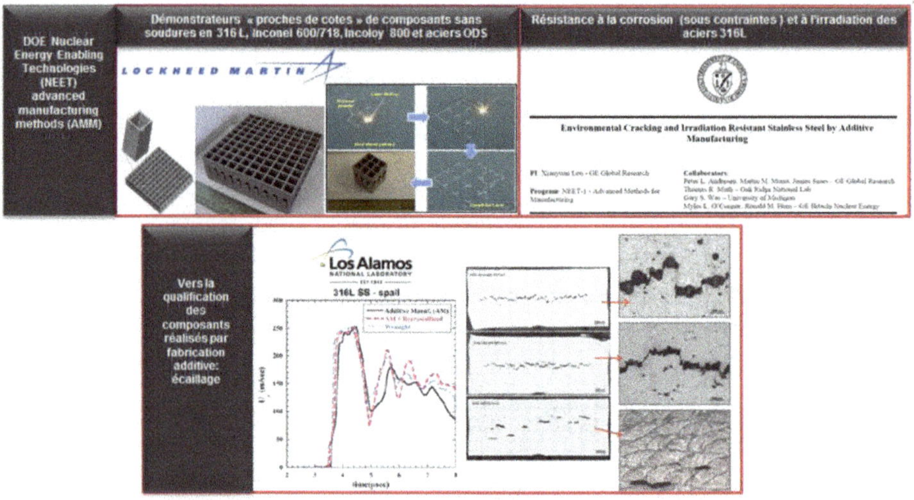

FIG. 1.1.25 – Examples of actions launched in the United States by the Department of Energy (DOE) as part of the Nuclear Energy Enabling Technologies (NEET): Advanced Manufacturing Methods (AMM) program [20, 21].

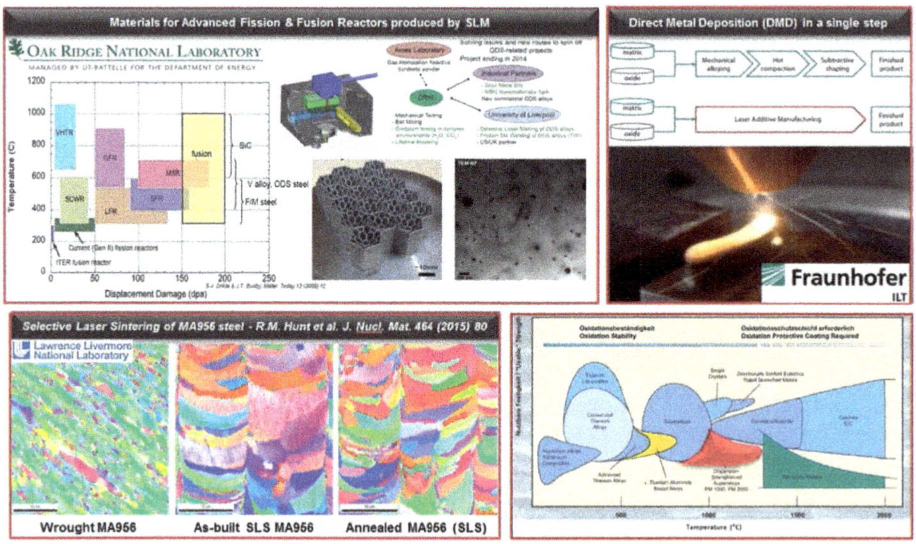

FIG. 1.1.26 – Manufacturing of technological mock-ups in ODS steels [22, 23].

Temperature in Power Generation), started in 2013, aimed to develop components in steels reinforced by dispersion of oxides based on Y_2O_3 (ODS) manufactured by additive laser fusion/projection manufacturing technologies – see figure 1.1.27.

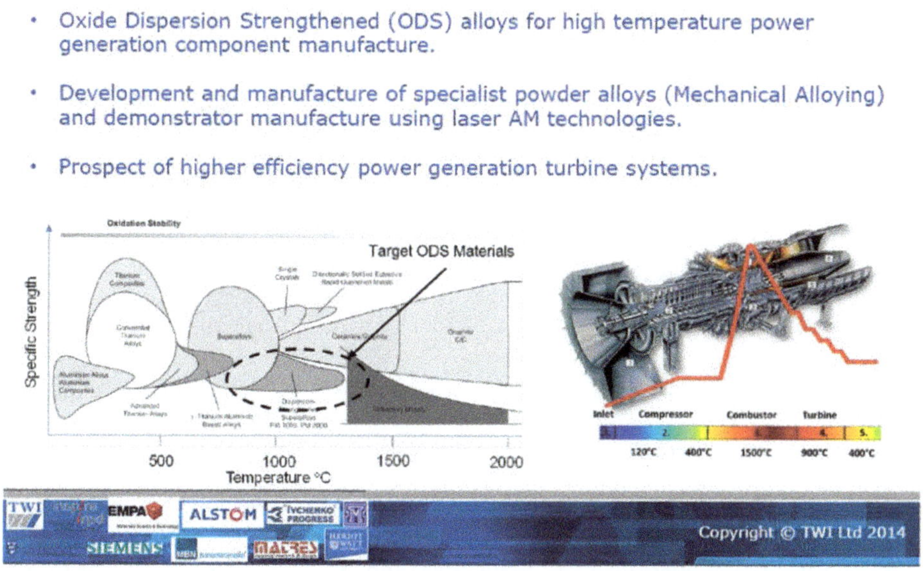

FIG. 1.1.27 – FP7 Project européen Oxigen (Oxide Dispersion Strengthened Materials for Additive Manufacturing of High Temperature in Power Generation) [25].

Research on direct manufacturing by laser projection (co-projection: metals and reinforcements) is also carried out by the Max-Planck and Fraunhofer ILT institutes (Aachen).

In India, "honeycomb" assembly grids for the Prototype Fast Breeder Reactor (Indian RNR-Na prototype) was manufactured by direct metal deposition – see figure 1.1.28a. Demonstrators of REP assembly grids for the CAP1400 reactor (Chinese evolution of the AP1000 "Westinghouse") produced by the laser fusion technique were also presented in January 2016 by the Chinese National Nuclear Company (CNNC) – see figure 1.1.28b.

In France, since 2016 CEA at the SAMANTA platform (Saclay's Advanced MANufacturing & Technological Applications) produced nuclear components in the frame of Sodium-cooled Fast Reactors (SFR) program. In the World Nuclear Event (WNE) 2016, CEA presented the first French additively manufacturing parts – see figure 1.1.29.

Thanks to additive manufacturing, it was possible to improve the manufacturing in one single step, thus avoiding milling and welding of complex nuclear grids.

In fact, the first prototype processed employing traditional techniques does not respect the dimensional tolerances (rails & chamfers manufacturing → difficult tool access). It is also possible to evolve into an internal conical cone.

CEA developed advanced nuclear materials such as ODS steels additively manufactured, such as Fe-14 Cr, traditionally processed by hot isostatic pressing and/or extrusion. As shown in figure 1.1.30, samples processed by selective laser melting exhibited a homogeneous distribution of only Y-Ti-O oxides ranging between few to 200 nm.

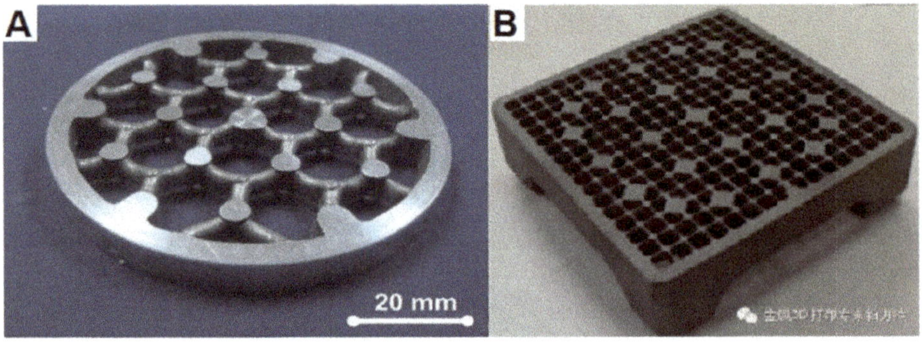

FIG. 1.1.28 – Nuclear component demonstrators produced by different additive manufacturing techniques: (a) "Honeycomb" assembly grid for the Indian concept of the future RNR-Na prototype and (b) Assembly grid belonging to the CAP1400 reactor [26, 27].

1.1.1.4.3 Alternative Energies

In the alternatives energy field, CEA has shown that is possible to produce *near-shape* components as developed at the POUDR'INNOV 2.0 platform, which proves that additive manufacturing is completely suitable with alternative energies (figure 1.1.31).

1.1.1.4.4 Defense

In the defense field, the National Nuclear Security Administration (NNSA) is currently conducting multidisciplinary research in the United States, piloted by the Lawrence Livermore, Sandia and Los Alamos laboratories for the "nuclear warhead" application – see figure 1.1.32. This study is centered on the understanding of metallurgy and on the development of new cellular materials (lattice structures). In particular, component manufacturing was transferred by Westinghouse to the Kansas City Plant (KCP) and Y12 sites. Thanks to the use of this new technology, they estimate a cost reduction of 85%, accompanied by a considerable improvement in mechanical properties. Similar studies are currently being carried out by the Naval Undersea Warfare Center (NUWC) belonging to the United States Navy.

Since the 90s, the American army (US Army) has used the direct metal deposition (laser engineered net shaping: LENS) developed by the Sandia National Laboratory, today marketed by the company Optomec, for repair (reloading) turbines of the Abraham M1 tank – see figure 1.1.33.

The United States Department of Energy (DOE) supported a collaborative study between the Oak Ridge National Laboratory (ORNL) and the company Lockheed Martin Aeronautics in order to evaluate the EBM technology for the realization of a support for the Lockheed Martin F-35 Lightning II aircraft: a reduction in the 33: 1 "buy-to-fly" ratio has been obtained, which means a reduction of 50% in terms of final cost (mainly related to machining) – see figure 1.1.34.

DOE also funds the ORNL for the enhancement of its neutronography skills for the observation of ALM components in its large research instruments, the spallation

FIG. 1.1.29 – Fuel Assembly Grid and Hexagonal Tube for Sodium-cooled Fast Reactors: iso-design.

neutron source (SNS) and the reactor (High Flux Isotope Reactor: HFIS). A stress measurement study as a function of the height of the ALM turbine blades was carried out in partnership with the Nasa-Langley Research Center and Morris Technologies (now GE group) to predict operation under nominal conditions – see figure 1.1.35. This activity was carried out in the VULCAN (Versatile nEutron imagiNg instrUment at Sns) line of the SNS.

Still in the military field, in France, the SME Poly-Shape manufactured various parts for Dassault Aviation. Following the qualification stage, these parts

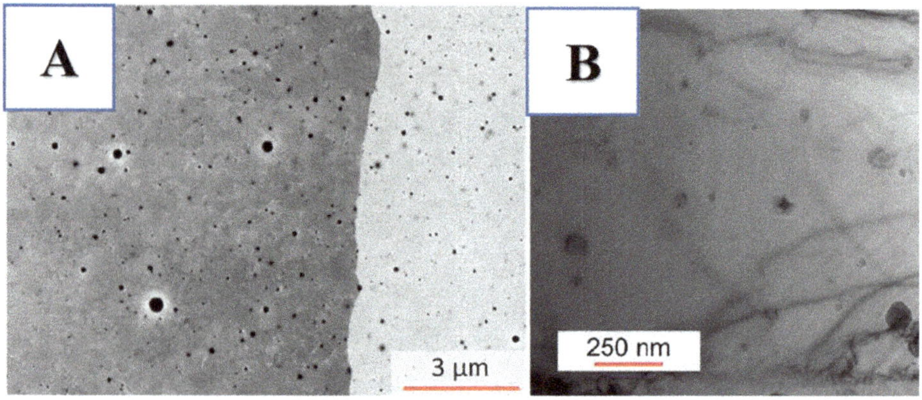

FIG. 1.1.30 – (A) SEM and (B) TEM images of the Y-Ti-O precipitation inside the SLM builds [28].

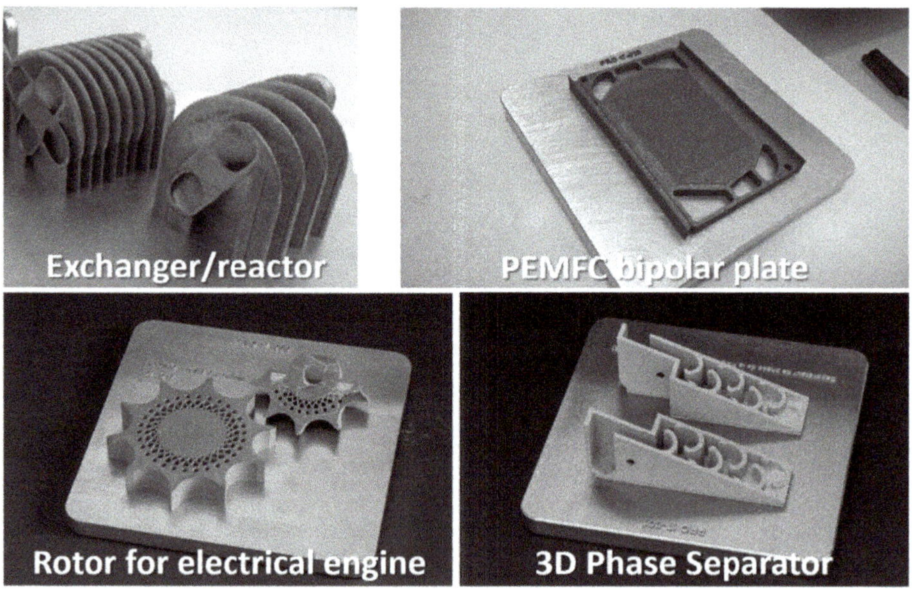

FIG. 1.1.31 – Examples of AM components for energy (CEA Liten POUDR'INNOV 2.0 platform).

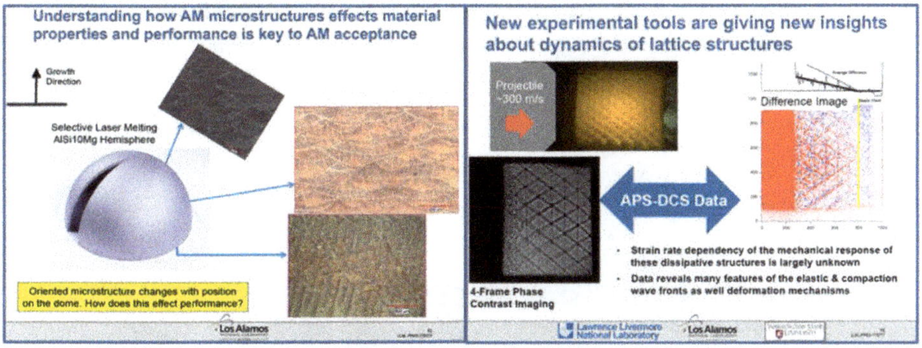

FIG. 1.1.32 – Research activity supported by the National Nuclear Security Administration (NNSA) [29].

FIG. 1.1.33 – Application of LENS technology for repairing components of Abraham M1 tanks [30].

FIG. 1.1.34 – Technological demonstrator of a Ti-6Al-4V support belonging to the Lockheed Martin F-35 Lightning II aircraft [31].

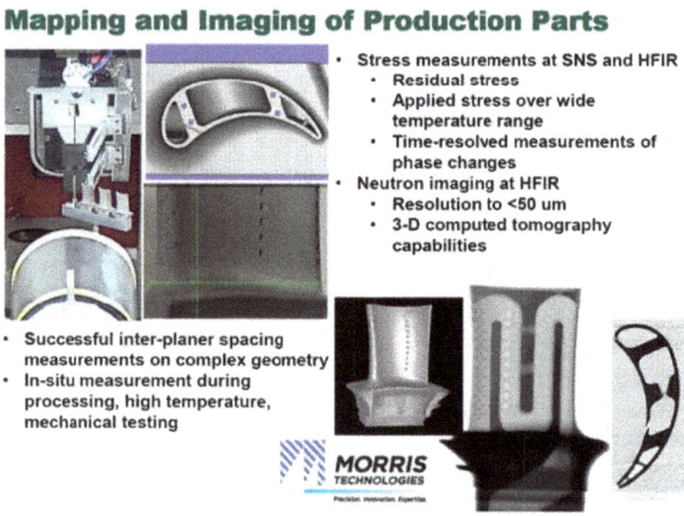

FIG. 1.1.35 – Oak Ridge Neutronography: turbine blade manufactured by ALM [32].

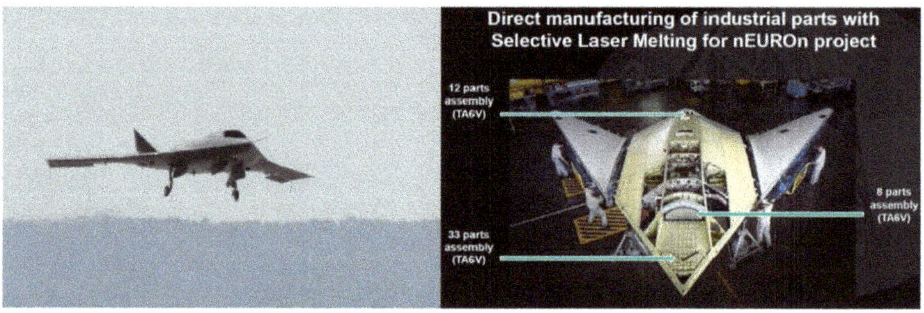

FIG. 1.1.36 – Different parts made for the nEUROn drone [33].

manufactured by AM are now fitted to the European stealth combat drone (UCAV) nEUROn – see figure 1.1.36.

1.1.1.5 The Current State of Additive Manufacturing: Industrial Up-Scalable Processes

Metal additive manufacturing is evolving with regard to possible applications mainly in the aerospace field. Unlike plastic additive manufacturing often used for rapid prototyping, metallic ALM finds numerous industrial applications – see figure 1.1.37.

Currently, additive manufacturing is well suited for the manufacture of components having complex shapes at low rates. On the other hand, the improvement of

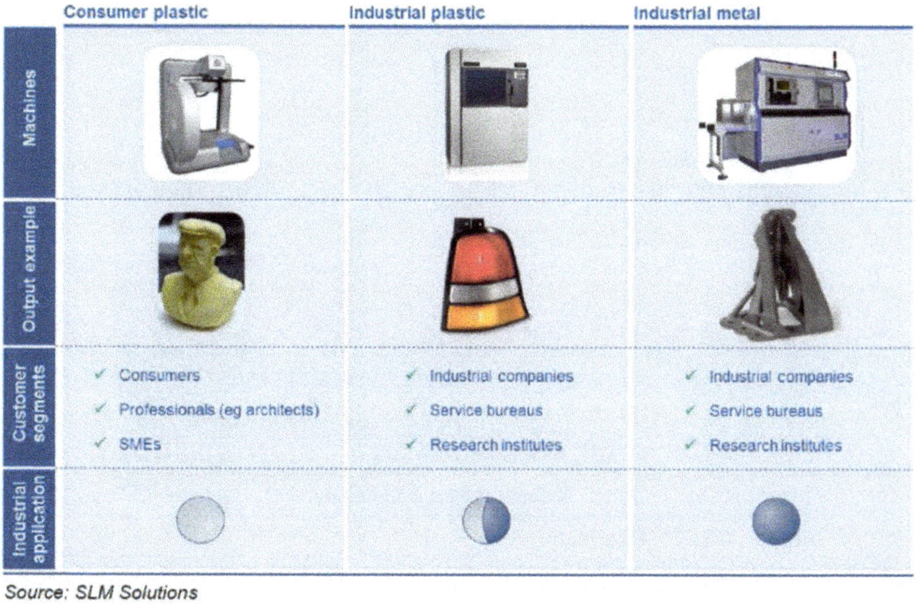

FIG. 1.1.37 – Applications of the various additive manufacturing processes [34].

current systems will allow the reduction in manufacturing prices compared to traditional methods such as forging and/or high speed machining – see figure 1.1.38.

Historically, additive manufacturing techniques have been adopted by research organizations (universities, R&D center, etc.) since the 1980s. Thanks to technological maturation, the main manufacturers have integrated ALM mainly for rapid prototyping. The industrialization of ALM can reduce the market for current suppliers, mainly manufacturers of machine tools ($93 billion), injection molds ($22 billion) and tool–tools ($5 billion) dollars). The major development is the disruption of the current supply chain with the introduction of new players (figure 1.1.39).

For this new industry, significant work on powder control is needed to make the powder supply more reliable and also to develop new shades as mentioned in the report on the future of French metallurgy and courses of action – see figure 1.1.40.

At the international level, numerous efforts are currently aimed at developing large-scale manufacturing systems for the aeronautical industry. In particular, South Africa is carrying out an activity to enhance the titanium production sector through the AEROSWIFT Project of CSIR South Africa in partnership with Airbus and Boeing – see figure 1.1.41. Large parts (support frames) of 2 m × 0.6 m × 0.6 m will be produced by laser technologies on a powder bed. An 8.3 times faster manufacturing speed (112.3 cm^3/min) is expected compared to a standard machine (EOS 270). The first scientific results were published in 2019 [37].

In the United States, NASA (Langley Research Center) has developed a double-wire electron beam projection system on a 6-axis platform (electron beam

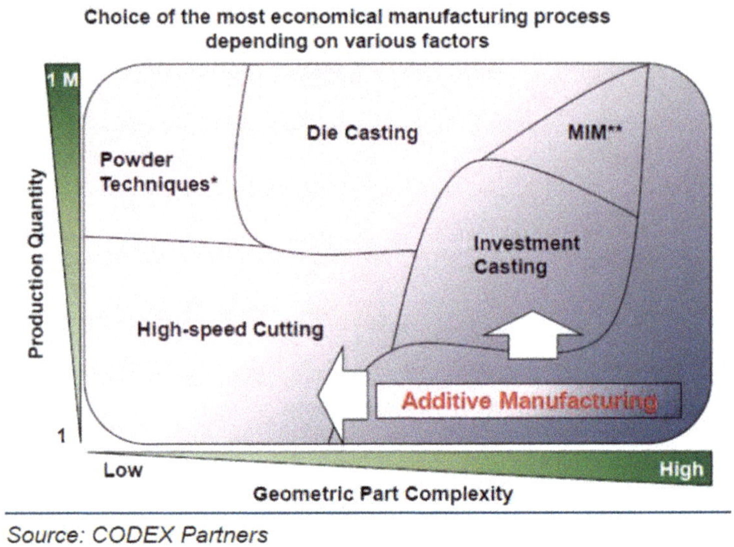

FIG. 1.1.38 – Production and geometric part complexity: AM opportunities.

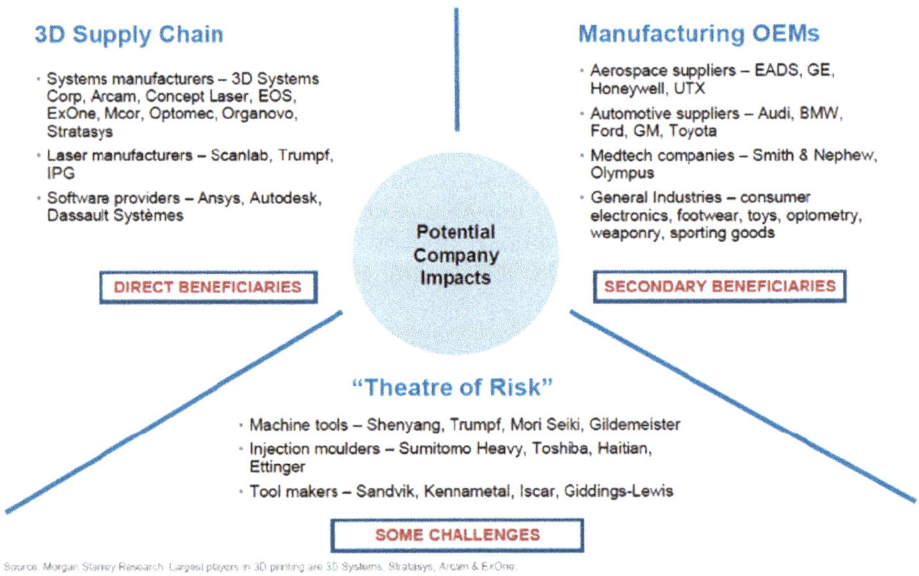

FIG. 1.1.39 – New scenario: what are the opportunities and risks of additive manufacturing [35].

Liberté • Égalité • Fraternité
RÉPUBLIQUE FRANÇAISE

MINISTÈRE
DE L'ENSEIGNEMENT SUPÉRIEUR
ET DE LA RECHERCHE

L'AVENIR DE LA METALLURGIE FRANCAISE

REFLEXIONS ET PISTES D'ACTIONS

Yves BRECHET

André PINEAU

Yves QUERE

FIG. 1.1.40 – The future of the french metallurgy: reflection and courses of action [36].

melting free form: EBFFF or EBM3) allowing the manufacture of large components. In 2011, NASA demonstrated the feasibility of manufacturing titanium support frames for the Lockheed Martin F-35 Lightning II aircraft. They also conducted flight manufacturing tests under "microgravity" – see figure 1.1.42.

In China, the Southern Fan Company announced the manufacture of a large laser projection machine for the production of 6 m × 6 m × 10 m parts weighing 300 tonnes. Beihang University was the first university to conduct research on the manufacture of titanium, nickel, intermetallic and refractory metal components (niobium, molybdenum, tantalum, tungsten and rhenium). In 2009, a university laboratory developed a "windshield" frame demonstrator for the Comac C919 aircraft with "laser engineered net shaping: LENS" technology. In 2010, this laboratory manufactured the central wing box of the same aircraft – see figure 1.1.43.

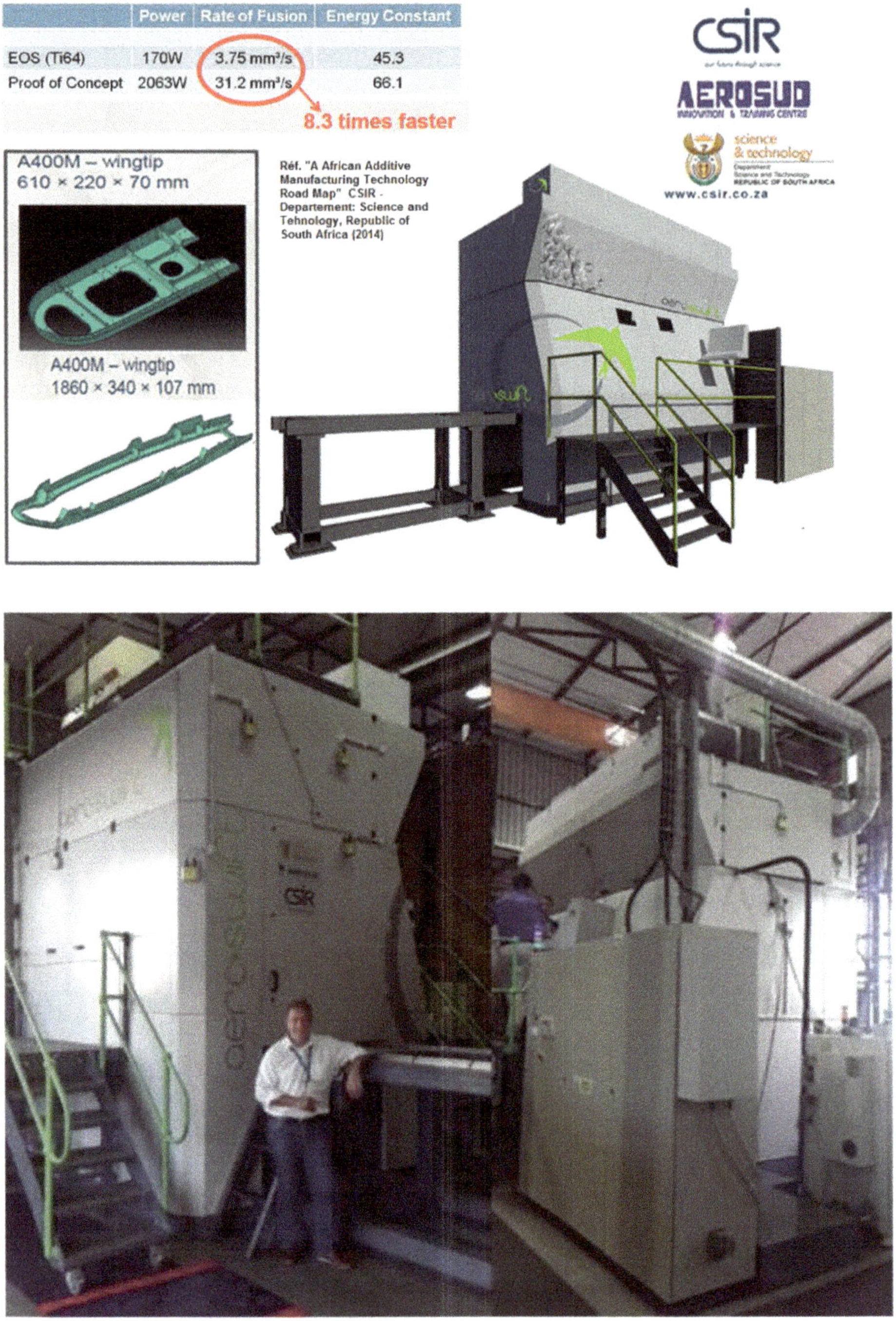

FIG. 1.1.41 – Machine developed as part of the "AEROSWIFT" project [38, 39].

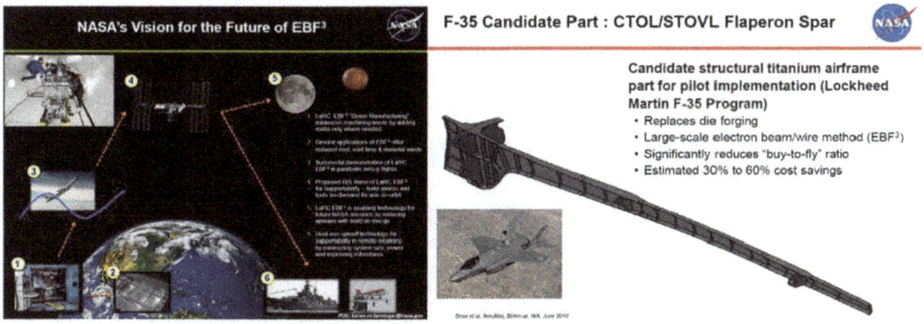

FIG. 1.1.42 – Freeform electron beam machine (EBF3) developed by NASA [40].

FIG. 1.1.43 – Large components manufacturing by direct metal deposition [41].

Northwestern Polytechnical University fabricated a COMAC 919 airliner titanium central wing frame by laser projection. This laboratory has good skills for making machines and in the metallurgy of titanium alloys, superalloys and stainless steels.

AVIC Laser Prototyping Manufacturing Co. start-up created by the University of Huazong, Wuhan Binhu Mechanical and Electrical and supported by AVIC heavy machinery group, produced the world's largest titanium component for the J-31 stealth fighter by a laser projection technique. According to AVIC laser, this technology is currently used to manufacture seven types of Y-20 transport aircraft, the J-15 fighter aircraft and the C919 airliner.

1.1.1.6 List of Various Industries Using AM

Currently, four rapidly growing application sectors have been identified in the industry:

- Medical/dental: industrial scale production of implants (Lima Corporate S.p.A).
- Tooling: small-scale production.
- Aeronautics: mass production (pilot state): qualification of components.
- The automobile: rapid prototyping in small series: motorsport/high range.

In figure 1.1.44, the technology readiness level (TRL) linked to each sector on a scale of 1–10 is shown.

1.1.1.7 Technical and Commercial Consideration of Each Process in AM

A second selection criterion is the expected evolution of technologies, because the potential is not identical for all processes – see figure 1.1.45. Metal additive manufacturing, in particular by EBM, is the most promising technology for defense applications.

The locks to be lifted to meet the needs of the industry are:

- Increase in capacity of manufacturing chambers (size).
- Speed and manufacturing strategies (several energy beams).
- Robustness of the process.
- Quality of interior surfaces: roughness of cavities.

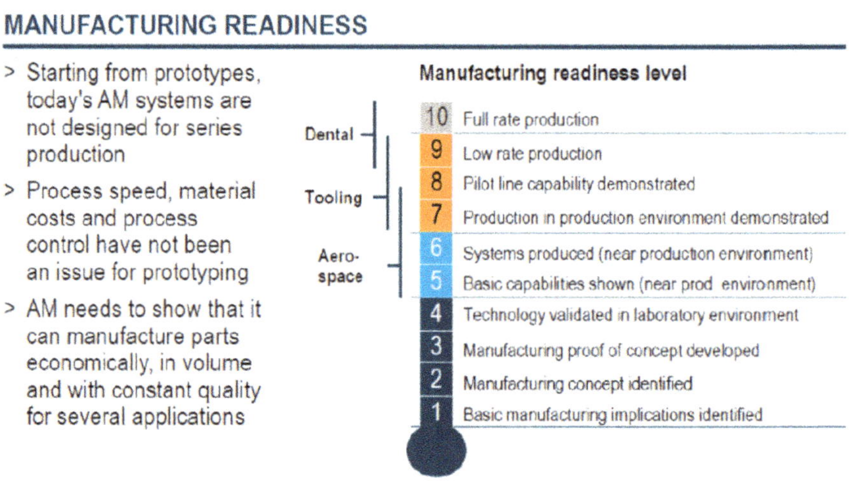

FIG. 1.1.44 – Differences in the of manufacturing readiness level of each industrial sector [42].

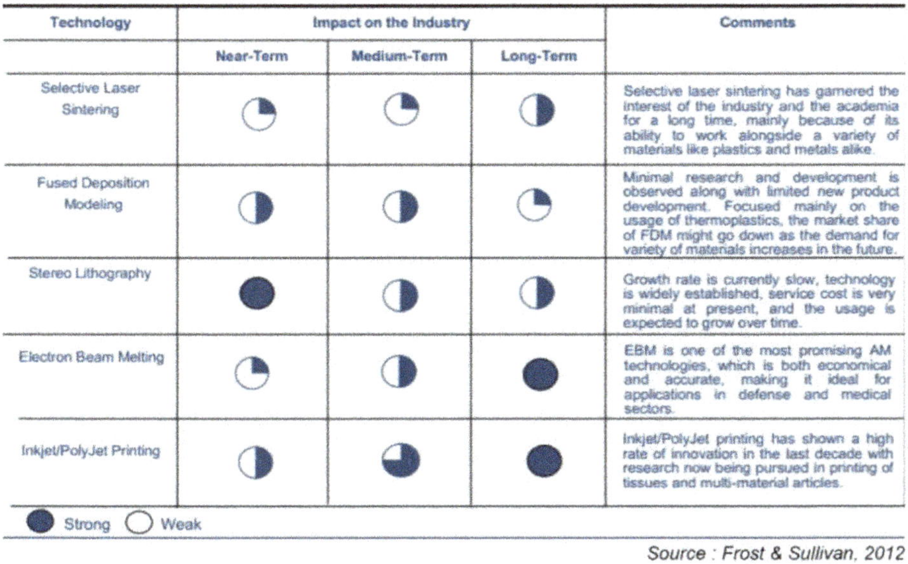

Technology	Impact on the Industry			Comments
	Near-Term	Medium-Term	Long-Term	
Selective Laser Sintering	◔	◔	◑	Selective laser sintering has garnered the interest of the industry and the academia for a long time, mainly because of its ability to work alongside a variety of materials like plastics and metals alike.
Fused Deposition Modeling	◑	◑	◔	Minimal research and development is observed along with limited new product development. Focused mainly on the usage of thermoplastics, the market share of FDM might go down as the demand for variety of materials increases in the future.
Stereo Lithography	●	◑	◑	Growth rate is currently slow, technology is widely established, service cost is very minimal at present, and the usage is expected to grow over time.
Electron Beam Melting	◔	◑	●	EBM is one of the most promising AM technologies, which is both economical and accurate, making it ideal for applications in defense and medical sectors.
Inkjet/PolyJet Printing	◑	◕	●	Inkjet/PolyJet printing has shown a high rate of innovation in the last decade with research now being pursued in printing of tissues and multi-material articles.

● Strong ○ Weak

Source : Frost & Sullivan, 2012

FIG. 1.1.45 – Potential of technologies.

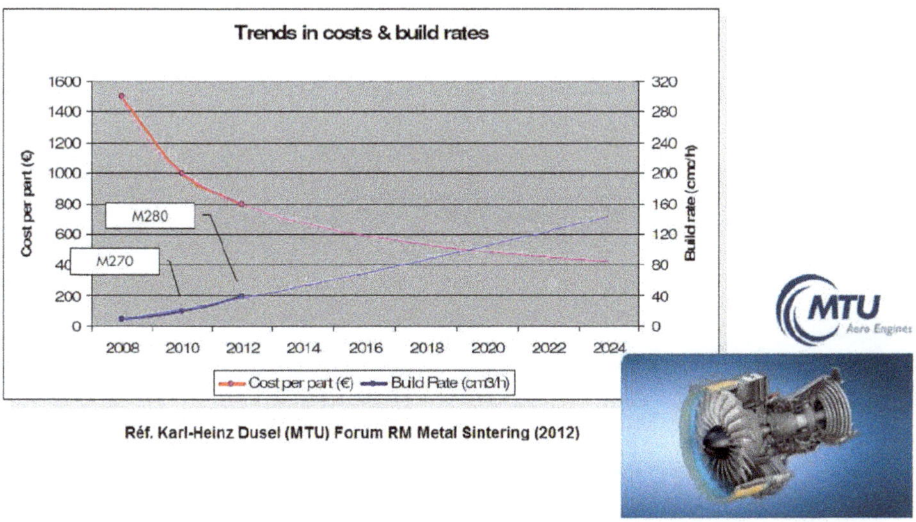

FIG. 1.1.46 – Tendency to reduce production costs according to technological evolution [43].

In particular, the improvement in manufacturing speeds will make it possible to reduce the price of components compared to traditional techniques. As shown in figure 1.1.46, the downward trend in the unit price of components as a function of the evolution of EOS machines (powder bed fusion) is strictly associated with the speed of manufacture.

1.1.1.8 Recycling Aspects: Circular Economy

This set of innovative technologies is overturning the concept of the traditional supply chain. It offers the advantage of a significant reduction in the material used, while minimizing rework in machining (subtractive manufacturing by removing material) and material loss. The image shows an example of an aircraft industry support – see figure 1.1.47.

A reduction of a factor of 5 in the "buy-to-fly" ratio often used in the aeronautical industry in terms of material for the manufacture of a support is obtained – the "buy-to-fly" measures the ratio between the mass of material used to make a part and the mass which actually flies. Furthermore, the topological optimization has made it possible to reduce the final weight of the part produced by additive manufacturing.

In addition, this family of technologies is economically suitable for the production of small components in large quantities or for the production of parts with great geometric complexity (cellular structures – lightening, reduction in assemblies/tools, personalization, internal channels, etc.).

In France, thanks to the creation of METAFENSCH in the Fensch Valley, the production and recycling of metal powders for the HIP (Hot Isostatic Pressing) and

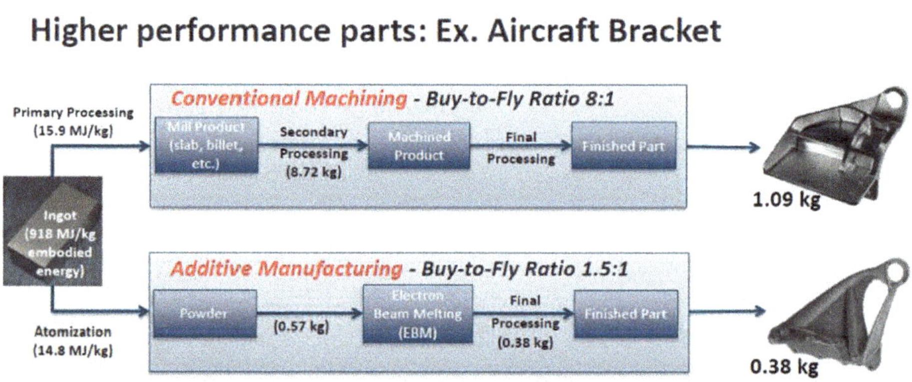

Process	Final part (kg)	Ingot consumed (kg)	Raw mat'l (MJ)	Manuf (MJ)	Transport (MJ)	Use phase (MJ)	Total energy per bracket (MJ)
Machining	1.09	8.72	8,003	952	41	217,949	226,945
EBM (Optimized)	0.38	0.57	525	115	14	76,282	76,937

FIG. 1.1.47 – Comparison between conventional machining technologies and additive manufacturing [44].

FIG. 1.1.48 – METAFENSCH platform [45].

new additive manufacturing processing routes are fully managed. This is a typical example of the whole control of the supply chain from powder production up to recycling by means of recycling and purifying strategic metals, in particular PACHR (Plasma Assisted Cold Hearth Remelting) and EBCHR (Electron Beam Cold Hearth Remelting). A vacuum induction melter (VIM) furnace and Electrode Induction Gas Atomization (EIGA) are also available in the platform – see figure 1.1.48.

References

[1] Quan Z., Wu A., Keefe M., Qin X., Yu J., Suhr J., Byun J-H., Kim B-S., Chou T-W. (2015) Additive manufacturing of multi-directional preforms for composites: opportunities and challenges, *Mater. Today* **19**, 503.

[2] Madeley D., Chaphalkar N. (2014) Additive and Hybrid Manufacturing. Think it. Model it. Make it, SIEMENS AG.

[3] ASTM-I F2792-12A (2012) Standard Technologies for AM Technologies.

[4] ISO-DIS 17296-2 (2015) Additive Manufacturing – General Principles – Part 2 Overview of Process Categories and Feedstock.

[5] Roland Berger study (2017) Additive Manufacturing in Aerospace and Defense.

[6] Bram A. (2010) Pulvermetallurgische Verfahren für die generative Fertingung, Fraunhofer IFAM Dresden.

[7] Industrial applications of laser cladding and equipement required, Trumpf Laser (2012).

[8] Peletet M. (2009) La fabrication directe chez Mecachrome, Mecachrome.

[9] Ghidini T. (2013) European Space Agency Perspective on Additive Manufacturing (AM), ESA.

[10] Behrendt U., Chinellato V. (2013) Additive Manufacturing for Aerospace, EOS.

[11] Tomlin M., Meyer J. (2011) Topology Optimization of an Additive Layer Manufactured (ALM) Aerospace Part, EADS Innovation Works.

[12] Morris G. (2014) Additive Manufacturing at GE, GE Aviation.

[13] Cotteleer M. (2014) 3D Opportunity: Additive manufacturing path to performance, innovation and growth, Deloitte Services.

[14] Varetti M. (2014) The achievements, the challenges, the future, Avio Aero au Arcam User's Group Meeting.

[15] La Fabrication Additive Enjeux et Attentes pour SNECMA, Safran Aircelles- division des Moteurs Spatiaux (2014).
[16] Waller J., Saulsberry R., Parker B., Hodges K., Burke E., Taminger K. (2014) Summary of NDE of Additive Manufacturing Efforts in NASA, National Aeronautics & Space Administration.
[17] Waller J., Parker B., Hodges K., Burke E., Walker J. (2014) NASA/ TM-2014-218560 – Nondestructive Evaluation of Additive Manufacturing: state-of-the-discipline report, National Aeronautics & Space Administration – Langley Research Center.
[18] Mathers D. (2014) UK Fuel Cycle R&D in Advanced Ceramic Fuels, National Nuclear Laboratory (2014).
[19] Anderson S. (2015) Final Technical Report on Laser Direct Manufacturing (LDM) for Nuclear Components, Lockheed Martin Corporation, DOE Award Number DE-NE0000542.
[20] Lyer J., Anderson S., Ramachadan G., Baca G., Meise S., Gardner S. (2014) Laser Direct Manufacturing of Nuclear components, Lockheed Martin – Quad city Manufacturing Laboratory.
[21] MST e-news, Newsletter of the Materials Science and Technology division 12-2014, Los Alamos National Laboratory (2014).
[22] Dryepondt S., Tatlock G., Jones A. (2014) Development of New Commercial ODS Alloys, Oak Ridge National Laboratory.
[23] Hunt R.M., Kramer K.J., El-Dasher B. (2015) Selective laser sintering of MA956 oxide dispersion strengthened steel, *J. Nucl. Mat.* **464**, 80.
[24] Gao R., Zeng L., Ding H., Zhang T., Wang X., Gang Q. (2016) Characterization of oxide dispersion strengthened ferritic steel fabricated by electron beam selective melting, *Mater. Design* **89**, 1171.
[25] Scudamore R. (2014) Additive Manufacturing – A European Perspective, The Welding Institute Ltd.
[26] Paul C.P., Singh B. (2015) Laser Additive Manufacturing of Specially Design Nuclear Components, Ramanna Centre for Advanced Technology, vol. 28, issue 2.
[27] China National Nuclear Corporation (CNNC) website: en.cnnc.com.cn/2016-01/12/c_49470.htm.
[28] Vasquez E., Giroux P.-F., Lomello F., Chniouel A., Maskrot H., Schuster F., Castany P. (2019) Elaboration of oxide dispersion strengthened Fe-14Cr stainless laser melting, *J. Mater. Process. Tech.* **267**, 403.
[29] Marggraff M. (2014) Science-based Additive Manufacturing for NNSA missions, Lawrence Livermore National Laboratory.
[30] Balazic M. (2013) Additive Manufacturing and 3D Printing LENS Technology, Efesto LLC.
[31] Dehoff R.R., Tallman C., Duty C.E., Peter W.H., Yamamoto Y., Chen W. (2013) Case study: additive manufacturing of aerospace brackets, *Adv. Mater. Process.* **171**, 19.
[32] Simonson M. (2013) Characterization Materials and Components at ORNL's, Neutron Sources Oak Ridge National Laboratory.
[33] Impellizzeri F. (2013) La fabrication additive: un procédé industriel innovant pour les produits de demain, Poly-Shape.
[34] Schaffer J., Mitchell J., Clarke C. (2014) Report: Additive Manufacturing, Credit Suisse.
[35] Uglow B., Davies R., Carrier L., Coe N., Ronkowitz D., Vig R., Ackerley J., Wood A., Mehra S., Jungling M., Wood P., Koike Y., Ibara Y., Luo K., Lamke L., Huberty K., Kogay N. (2013) Morgan Stanley Blue Paper: Capital Goods – 3D printing : don't believe (all) the hype, Morgan Stanley.
[36] Bréchet T., Pineau A., Quere Y. (2014) L'avenir de la métallurgie française: reflexion et pistes d'action, Ministère de l'Enseignement Supérieur et de la Recherche.
[37] Motibane L.P., Tshabalala L.C., Mathe N.R., Hoosain S., Knutsen R.D. (2019) Effect of powder bed preheating on distortion and mechanical properties in high speed selective laser melting, *Mater. Sci. Eng.* **655**, 012026.

[38] de Beer D., du Preez W., Greyling H., Prisloo F., Sciammarella F., Trollip N., Vermeulen M., Wohlers T. (2014) A South Africa Additive Manufacturing Technology Roadmap, Council for Scientific and Industrial Research.
[39] Sciammarella F. (2013) The 3-D Printing Advantage for A&D: a look into the past, present and future, Northern Illinois University.
[40] Brice C.A., Taminger K. (2011) Additive Manufacturing Workshop NASA-Commonwealth Scientific and Industrial Research Organisation, National Aeronautics & Space Administration – Langley Research Center.
[41] Castel A. (2014) Technologies duales: les imprimantes 3D en Chine, Russie, Inde et Singapour, Chlorodia pour la Délégation aux Affaires Stratégiques – Ministère de la Défense.
[42] Roland Berger Strategy Consultants (2013) Additive Manufacturing: a game changer for the manufacturing industry.
[43] Ferrari R. (2013) La tecnologia di microfusione laser selettiva di polveri metalliche: stato dell'arte, vantaggio competitivo e applicazioni, MET e-manufacturing company.
[44] Marshall B., Wright B., Lunt B. (2014) Unlocking the potential of additive manufacturing in the fuel cells industry, US Department of Energy.
[45] METAFENSCH website: https://www.metafensch.fr/.

1.2 Materials Challenges in Metal Additive Manufacturing

Matteo SEITA

School of Mechanical and Aerospace Engineering, Nanyang Technological University, Singapore

Abstract

Most metals used in engineering applications are polycrystalline – aggregates of crystal grains with different crystallographic orientation. Figure 1.2.1 shows an electron micrograph of a polycrystalline specimen of Inconel – a nickel-based advanced alloy that is extensively used in aerospace and deep hole applications for its remarkable performance and corrosion resistance at high-temperatures [14].

Depending on the size, shape, and orientation distribution of grains, polycrystalline materials may exhibit substantially different properties. For instance, the strength of polycrystalline metals is known to be inversely proportional to the grain size [16]. Materials with strong crystallographic texture – namely polycrystals in which most grains share a preferred crystallographic orientation – exhibit anisotropic mechanical properties, including ductility and toughness [17] (*i.e.*, the mechanical response of the material to a load differs depending on the direction the load is applied). In view of this important relationship between *microstructure* (*i.e.*, the geometry, orientation, and distribution of the constituent grains and phases) and *materials properties*, it is crucial to understand the processes that lead to the formation of solids and how they affect the resulting microstructure.

1.2.1 Structure-Property Variability in Metal AM

One of the main challenges associated with metal parts fabricated *via* fusion-based AM processes is the variability in the material's microstructure, which yields scatter in mechanical properties. This concept is exemplified in figure 1.2.2. The microstructure of an AM-produced Inconel sample is clearly heterogeneous; it consists of a very "patchy" distribution of grains with vastly different shape, size, and orientation (which is represented by the different colours in the map). The heterogeneity of this microstructure is obvious when compared to that shown in figure 1.2.1. Microstructure heterogeneity yields local differences in mechanical properties – such as yield and tensile strength – across the same sample. This variability is referred to as *intra-sample* microstructure variability. There is also an *inter-sample* microstructure variability, which refers to differences in microstructure – and thus in properties – across batches of otherwise identical samples. Because of these phenomena, it is difficult to incorporate AM-produced parts in industrial applications, since their behaviour, performance, and lifetime are difficult to predict. This is one of the main challenges in metal AM, and is the focus of this chapter.

DOI: 10.1051/978-2-7598-2446-5.c021

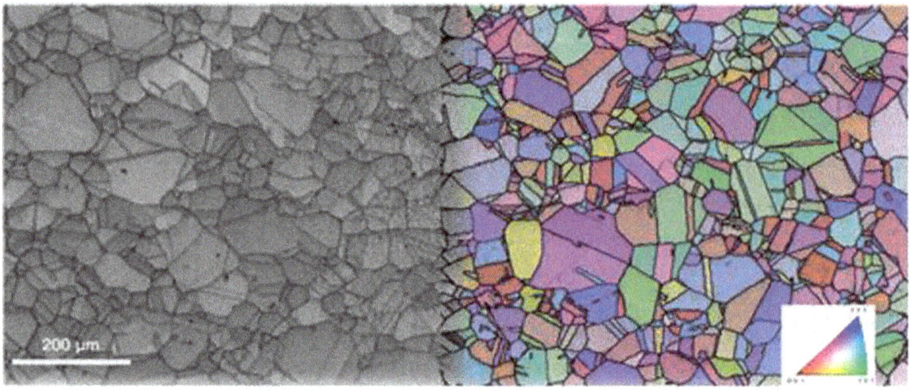

FIG. 1.2.1 – (Left) scanning electron (SE) micrograph with (right) overlaid electron backscatter diffraction (EBSD) grain map of an Inconel 725 specimen. The SE map shows the morphology of the material's constituent grains. Colours in the EBSD map indicate the crystallographic orientation of the grains, following the legend in the bottom right corner.

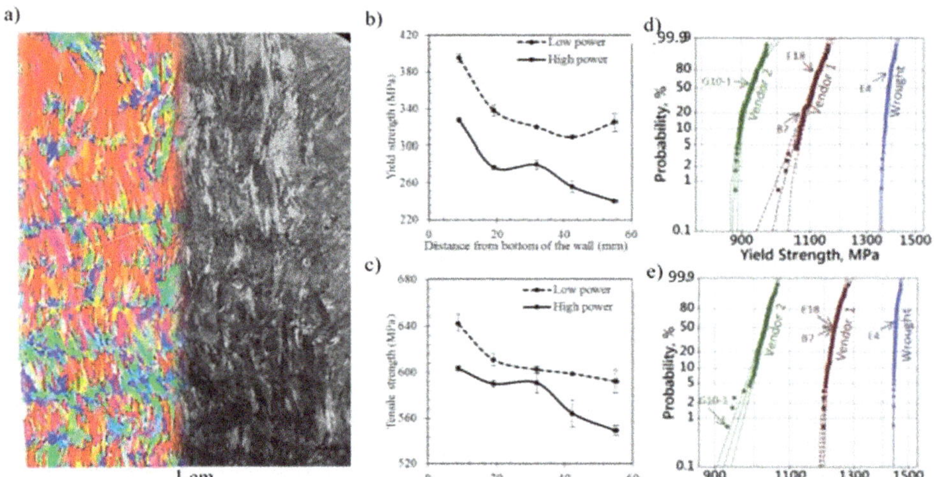

FIG. 1.2.2 – (a) (Left) EBSD grain map and (right) optical micrograph of an Inconel 718 sample produced *via* directed energy deposition (DED). The grain map shows a highly heterogeneous microstructure consisting of "bands" of grains with different size, morphology, and orientation (*i.e.* colour). (b, c) Quantify the intra-sample property variability in a stainless steel 304 sample produced by DED. The graph shows how the yield and tensile strength of the material change along the sample cross-section (*i.e.* from bottom to top along the build direction), respectively [9]. (d, e) Describe the inter-sample property variability in yield and tensile strength of two batches of stainless steel 316L produced by selective laser melting (SLM) compared to wrought samples [15]. It is remarkable to see scatter of properties on the order of 10% or 20%.

Section 1.2.2, will review the fundamentals of solidification, which will provide the basis to understand the origin of the microstructure heterogeneity resulting from fusion-based AM processes, described in section 1.2.3.

1.2.2 Solidification and Structure-Property Relationships in Metallic Materials[1]

Solidification

Most metals – including those produced by fusion-based additive manufacturing (AM) techniques – are formed *via* solidification of a melt. The solidification process consists of a phase transformation from liquid to solid and can be divided into two stages: *nucleation* and *growth* (see [1, 2]).

The kinetics of this transformation dictates the final microstructure and is determined by the amount of *undercooling* in the melt, as well as the *temperature gradient* across the solid/liquid (S/L) interface and its growth *rate* (or *velocity*).

Curvature Undercooling and Nucleation

Undercooling refers to the difference between the melting temperature of the material and the actual temperature of the melt: $\Delta T = T_m - T$. Thermodynamics dictates that solidification cannot occur unless $\Delta T > 0$. In other words, it is required to undercool the melt by a quantity ΔT in order to trigger the nucleation of a solid phase. The amount of undercooling required for nucleation depends on several factors, including the curvature of the S/L interface and the presence of a pre-existing solid in the melt.

Nucleation of a solid will occur if the energy change associated with the formation of an "embryo" or "cluster" of atoms in the melt (figure 1.2.3a) is negative – which means that nucleation of a solid is energetically favourable. The energy change associated with nucleation depends on two factors: the energy associated with the volume of the embryo, Δg_v, (which is negative if the temperature is lower than T_m) and that associated with its surface area, Δg_γ, (which is always positive). Since Δg_v and Δg_γ have a cubic and quadratic relationship to the embryo radius, respectively, a critical radius, r_0, exists above which the embryo will grow further into the liquid to form a stable nucleus and below which it will re-dissolve. Schematically, this energy balance is shown in figure 1.2.4.

[1]This section has been written following the class notes of Ali Miserez, MS2016 – Phase Transformation and Kinetics, NTU; [1] Kurz W., Fisher D.J. (1998), *Fundamentals of Solidification*. Trans Tech Publications Limited.; and [2] Stefanescu D.M., Ruxanda R. (2004) Fundamentals of solidification, in *Metallography and microstructures*. ASM International. p. 0.

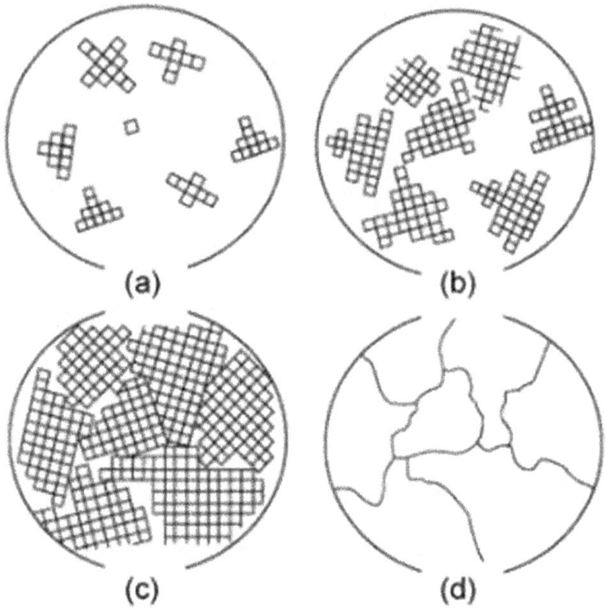

FIG. 1.2.3 – Time-resolved schematics of solidification. (a) Nucleation of solid clusters or embryos; (b) growth of nuclei; (c) impingement and coalescence of nuclei; and (d) microstructure formation. Adapted from MS2016 – Phase Transformation and Kinetics, NTU.

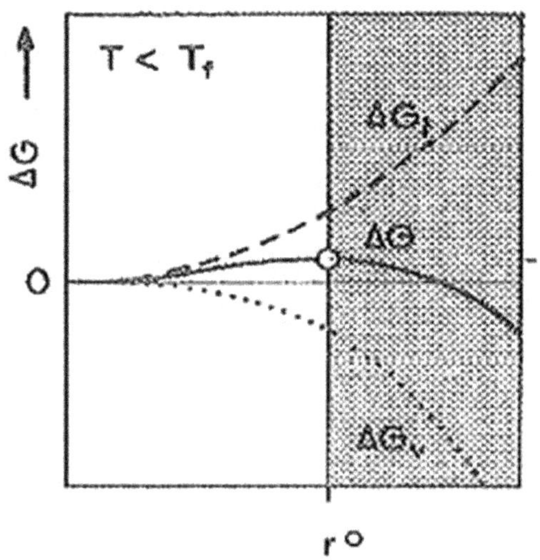

FIG. 1.2.4 – Change in energy associated with the creation of a solid nucleus in an undercooled melt. Adapted from reference [1].

Assuming a spherical embryo, we can write:

$$\Delta g_\gamma = \gamma_{s/l} \cdot 4\pi r^2$$

$$\Delta g_v = \frac{-\Delta s_f \Delta T_r \cdot 4\pi r^3}{3}$$

where $\gamma_{s/l}$ is the energy associated with the formation of the S/L interface, and Δs_f is the entropy of fusion. When the embryo reaches a critical radius, r_0, at $\Delta T > 0$ (*i.e.*, for a positive undercooling), the change in the total energy reaches a maximum:

$$\frac{d(\Delta g_\gamma + \Delta g_v)}{dr} = 0$$

which leads to the constitutive relationship for curvature undercooling:

$$\Delta T_r = \frac{2\gamma_{s/l}}{r_0 \cdot \Delta s_f}$$

In a practical sense, this relationship states that it is difficult to form a stable nucleus at small undercooling, ΔT_r, because that would require the nucleaus to have a large radius (which is unlikely). Conversely, the larger the undercooling the easier it is for nucleation to occur.

Nucleation Rate

It follows that the nucleation rate of solid in a melt, I, is the function of the amount of undercooling. For a small amount of undercooling, the energy barrier for nucleus formation is high and the nucleation rate is low. Low nucleation rate yields the formation of microstructures with relatively large grain size (the few, sparse solid nuclei in the melt grow larger and larger until they coalesce). As ΔT_r increases, there is a corresponding increase in I (the growth of these solid nuclei is limited by their high density in the melt and, as they impinge against each other, they form a microstructure consisting of fine grains). For even higher ΔT_r, the nucleation rate decreases again because atomic transport from the liquid to the S/L interface of an embryo becomes the limiting factor to forming a stable solid nucleus. Mathematically, the nucleation rate can be expressed as:

$$I = I_0 \cdot \exp\left(-\frac{\Delta g_n + \Delta g_D}{k_B T}\right)$$

where Δg_n is the activation energy for solid nucleation, which scales with $-1/(T \cdot \Delta T_r^2)$, and Δg_D corresponds to the activation energy for atomic migration, which scales with $-1/T$.

Heterogeneous Nucleation

The theory described in the foregoing considers a nucleation event in the melt. In other words, the solid nucleus is surrounded by a liquid phase. This nucleation is referred to as *homogeneous nucleation.* By contrast, *heterogeneous nucleation* occurs when the melt contains pre-existing solid particles or is in contact with a solid surface (*e.g.,* the surface of a mold). Heterogeneous nucleation is easier than homogeneous nucleation because the established solid/solid interface has lower energy than the S/L interface. In other words, the pre-existing solid lowers the activation energy for a solid to nucleate by decreasing its surface area.

Constitutional Undercooling, Thermal Gradients, and Growth Morphology

During solidification of alloys, solute atoms are rejected from the solid phase. This concept can be understood with reference to any binary phase diagram. At temperatures for which both solid and liquid phases co-exist, the lever rule dictates that the solute concentration in the solid is lower than that in the liquid (see figure 1.2.5).

The rejection of solute atoms during solidification results in the formation of a solute-enriched boundary layer at the S/L interface (see figure 1.2.6). The direct consequence of this phenomenon is a continuous change of the melting temperature across this boundary layer, which is lowest at the S/L interface and increases further away from it (top schematic in figure 1.2.6).

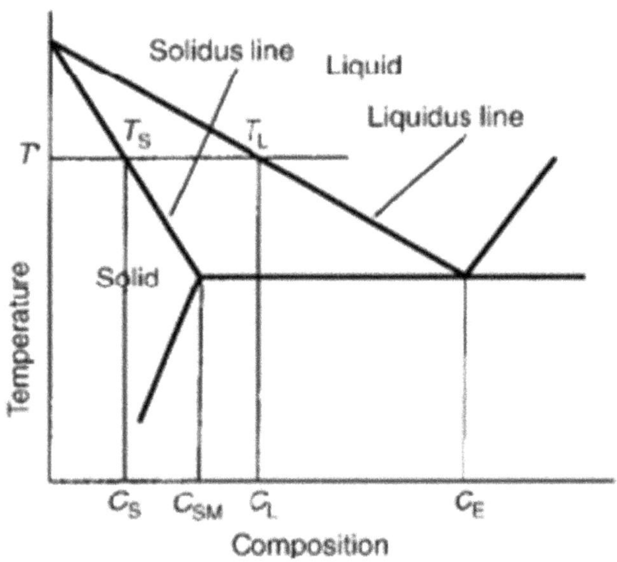

FIG. 1.2.5 – Exemplary phase diagram showing that, in a solid–liquid phase region and at a certain temperature, solute concentration in the liquid is larger than that in the solid ($C_L > C_S$). Adapted from [2].

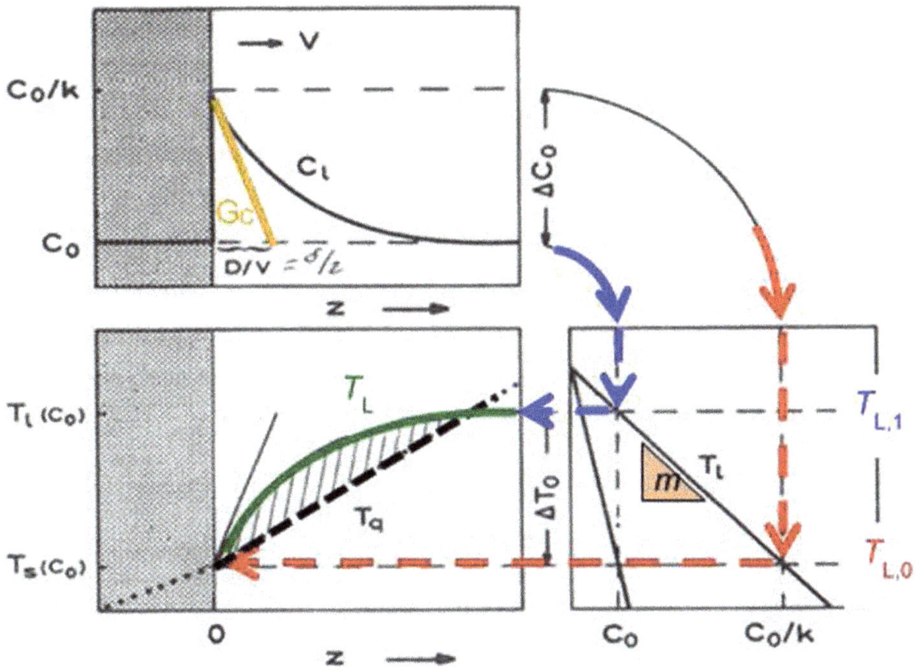

FIG. 1.2.6 – Constitutional undercooling. The top schematic shows the local change in composition as a result of solute rejection from the solid. The composition gradient C_L yields a change in the local melting temperature according to the phase diagram on the right hand side of the figure. High solute concentration near the S/L interface, C_0/k, lowers the melting temperature of the material facilitating solidification and thus advancement of the solid front along z (the red path in the figure). Noteworthy is that the thickness of the solute-enriched boundary layer is inversely proportional to the solidification velocity, V. Adapted from MS2016 – Phase Transformation and Kinetics, NTU and [2].

The heat flow that is established during solidification of a melt – for instance as a result of heat conduction and convection – yields the formation of a thermal gradient, $G_T = \frac{dT_q}{dz}$, in the material, where T_q is the temperature (dashed line in figure 1.2.6) and z is the solidification direction. The magnitude of G_T dictates the local temperature in the melt. If G_T is lower than the gradient of the melting temperature imposed by the undercooling, G_L (the tangent to the change of T_L across the boundary layer), then the melt is *constitutionally undercooled* (*i.e.* the temperature in the melt will be lower than the equilibrium melting temperature).

In addition to affecting solid nucleation (similarly to curvature undercooling), constitutional undercooling determines the morphology of the growing solid interface of an alloy. If the melt is constitutionally undercooled ($G_T < G_L$), the liquid at the S/L interface is at a lower temperature than that dictated by thermodynamic equilibrium and any instability in the front geometry will grow (consider an

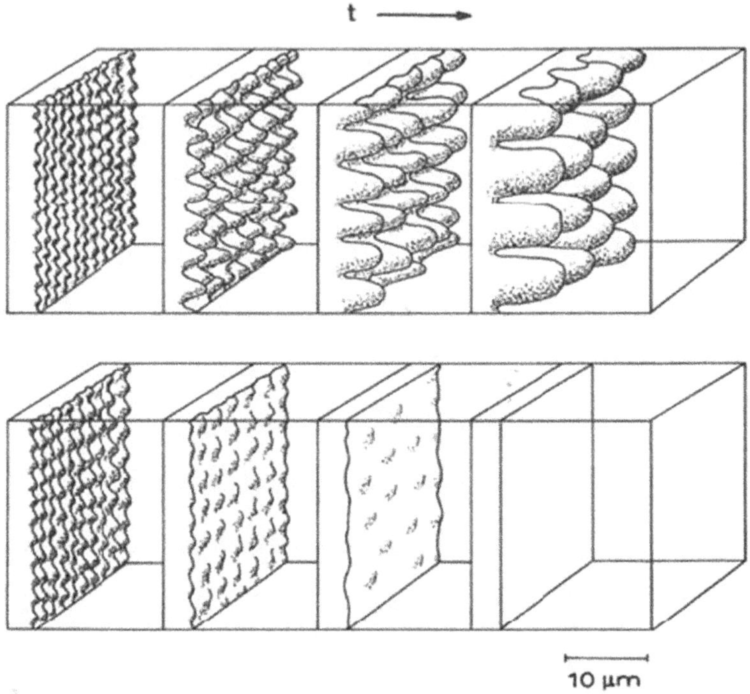

FIG. 1.2.7 – Schematic illustration of (top) unstable and (bottom) stable or planar interfaces during solidification. Adapted from reference [1].

instability as the random addition or removal of atoms at this interface, which is driven by local thermal fluctuations). In this case, the growing front is defined *unstable*. Local variations of temperature and composition across the S/L growing front will be enhanced and result into columnar or cellular growth (see top image in figure 1.2.7). On the other hand, if $G_T > G_L$ the instabilities will re-melted (because the temperature of the liquid at the interface is higher than T_L) and level off during growth. The resulting interface is referred to as *stable* and assumes a planar geometry (bottom image in figure 1.2.7).

For small constitutional undercooling, the instabilities will only grow parallel to the solidification direction. The resulting microstructure is referred to as *cellular*. As G_T decreases (and the amount of constitutional undercooling increases), the spacing between the cells increases, allowing for the establishment of constitutional undercooling along the direction perpendicular to the growth direction. Interface instabilities can then develop on the side walls of the cells, resulting in the formation of *dendrites*. The resulting growth is referred to as *dendritic*. Both cellular and dendritic growths are described as *columnar growth*, since they occur from the wall along a direction that is opposite to the heat flux. At greater constitutional undercooling,

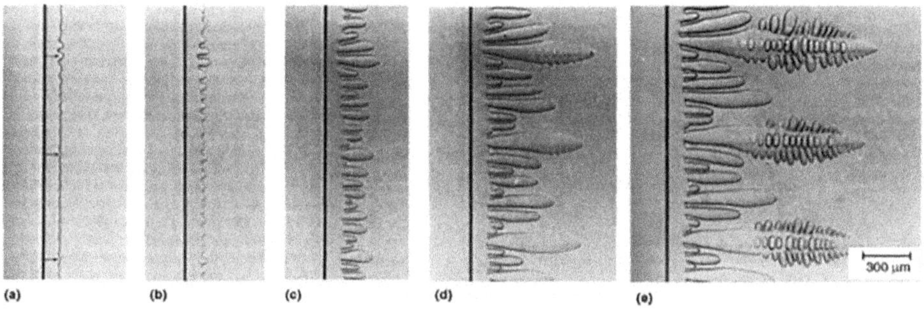

FIG. 1.2.8 – Change in interface morphology of a succinonitril-4% acetone solution when increasing the solidification velocity. Adapted from reference [2].

equiaxed grains can be nucleated in the liquid away from the interface. These morphological transitions can be visualized in figures 1.2.8 and 1.2.9.

All the above described morphological transitions are controlled by the initial composition of the alloy – which dictates the amount of constitutional undercooling – by the solidification velocity, V, and by the thermal gradient G_T.

To understand the effect of the solidification velocity, it is useful to consider that the rate at which solute atoms are rejected from the solid at the S/L interface is proportional to the growth rate of the solid into the liquid. So, the thinner the solute-enriched boundary layer (top schematic in figure 1.2.6), the steeper the concentration profile across the boundary layer, C_l, and the higher the growth rate (or solidification velocity).

The reason for this relationship stems from the fact that the concentration gradient in the boundary layer, $G_C = \frac{dC_l}{dz}$, is directly proportional the flux of solute atoms diffusing away from the S/L interface (following Fick's first law of diffusion). At steady-state, this flux of atoms escaping the S/L interface must be compensated by a flux of solute atoms that are rejected from the growing solid. So, the higher this flux, the faster is the solidification velocity, V. This phenomenon is translated into the following relationship:

$$G_c = -\left(\frac{V}{D}\right) m \cdot \Delta C_0$$

where D is the liquid diffusivity, m is the slope of the liquidus line in the phase diagram, and ΔC_0 is the difference between solute concentration at the S/L interface and that far away from it (see figure 1.2.6).

As previously mentioned, surface instabilities evolve if $G_T < G_L$. In alloys, G_L is function of G_c (because of constitutional undercooling). So we can derive a condition for the establishment of surface instabilities as:

$$\frac{G_T}{V} < \frac{\Delta T_c}{D}$$

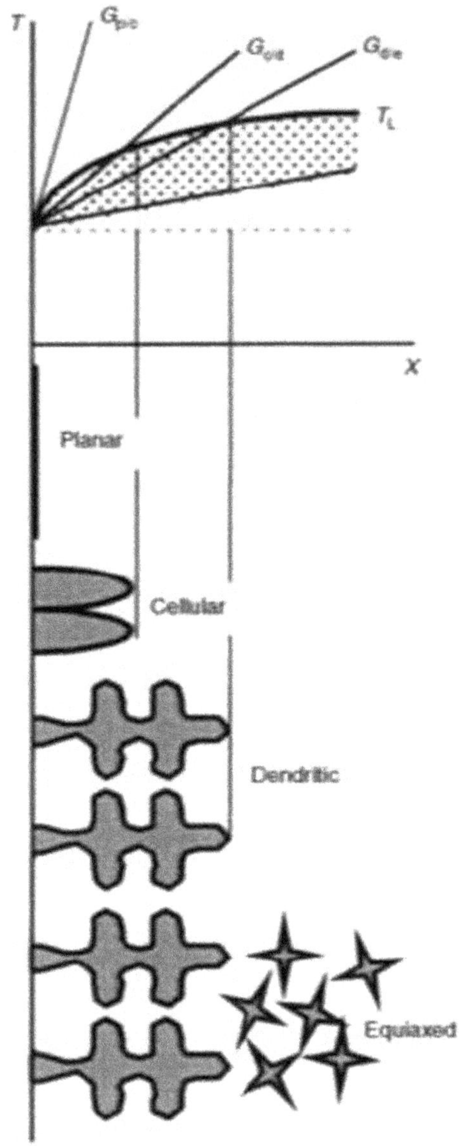

FIG. 1.2.9 – Correlation between the thermal gradient at the S/L interface and the interface morphology. Adapted from reference [2].

where ΔT_c is proportional to $-m \cdot \Delta C_0$. This relationship states that there is a critical constitutional undercooling ΔT_c (or a critical solute content in the liquid) for a given G_T/V ratio at which the interface will become unstable. In other words, surface stability is directly dependent upon the ratio between the external thermal gradient, G_T, and the solidification velocity, V.

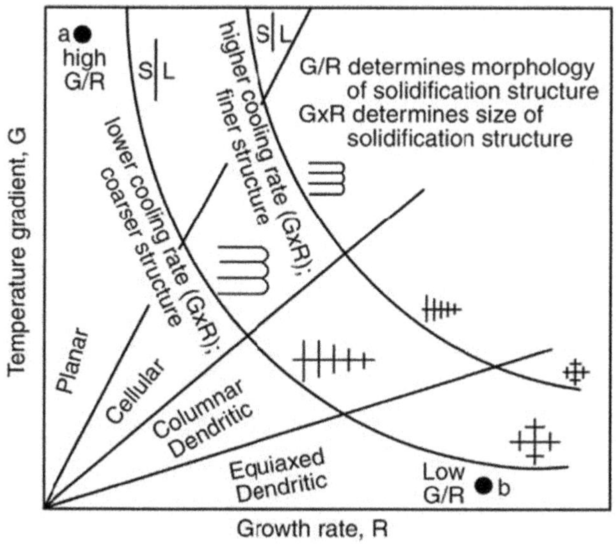

FIG. 1.2.10 – Exemplary solidification map, showing the effect of temperature gradient G_T and growth rate R (or solidification velocity V) on the morphology and size of solidification microstructure. Adapted from reference [6].

Based on this relationship, we can build solidification maps such as that in figure 1.2.10 (note that in figure 1.2.10 V is replaced by the solidification rate, R). These maps are extremely useful to predict the solidification microstructure as a function of the solidification process (*i.e.*, G_T and V). Noteworthy is that, while the morphology of the solidification microstructure is controlled by G_T/V, the scale of the microstructure formed (*i.e.*, the thickness of the cells/dendrites or the size of equiaxed grains) is governed by $G_T \cdot V$, which is proportional to the cooling rate of the material.

Solidification Microstructures

Following the map shown in figure 1.2.10, the solidification microstructure can be broadly categorized as:

- *Planar*: Planar growth of alloys can be achieved only in crystal growth furnaces at high-temperature gradients – of the order of $G_T = 100$ K/cm – and low solidification velocities – of the order of $V = 2\ \mu$m/s.
- *Cellular and dendritic*: When constitutional undercooling occurs, the S/L interface morphology becomes cellular or dendritic as $\frac{G_T}{V}$ becomes smaller and smaller compared to $\frac{\Delta T_c}{D}$. These structures grow mostly along one direction (and are thus called directionally solidified), which is antiparallel to the heat flux (or the direction of maximum G_T). This direction often corresponds to specific crystallographic directions in the material. This phenomenon stems from the different efficiency of atoms to attach to an interface. In cubic crystals, this

direction coincides with the <1 0 0> crystal axis. For this reason, directionally solidified microstructures often retain a strong crystallographic texture. Meaning that the constituent grains (*i.e.*, the solidifying dendrites) share the same <1 0 0> orientation along the direction of solidification. That also means that these microstructures exhibit columnar grains, with their axis aligned perpendicular to the walls of the mold (in the case of casting, for instance), since heat is extracted from the melt into the cold mold walls.

Within individual grains, it is often possible to observe a micro-segregation structure, which stems from the solute that is rejected from the growing solid into the regions between the cells and the secondary dendrites. These regions are the last portion of material to solidify. These segregation structures can be easily observed under the electron microscope after etching the alloy surface (figure 1.2.11).

- *Equiaxed*: For very high solidification velocities and low thermal gradients, equiaxed dendrites will nucleate directly in the melt. After nucleation, each equiaxed dendrite will grow independently until impingement on adjacent growing dendrites, forming individual grains. In contrast to columnar growth, during equiaxed growth heat is dissipated by the melt itself (which is undercooled) and the local G_T surrounding an equiaxed dendrite can be randomly oriented in space. For this reason, grains formed from the growth of equiaxed

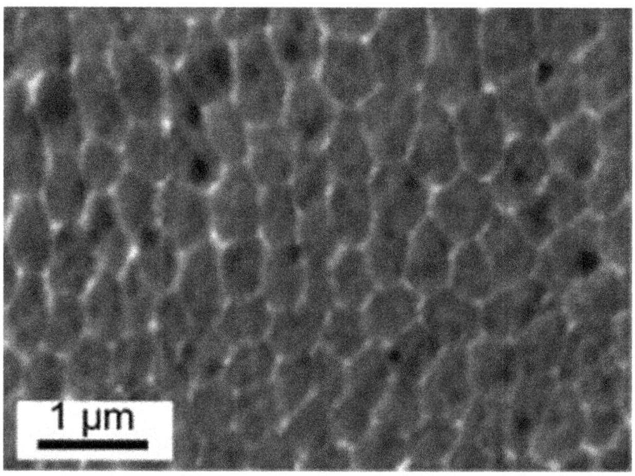

FIG. 1.2.11 – Micro-segregation structure in stainless steel produced by selective laser melting (SLM). The structure assumes the form of a "granular network" because the plane of view is perpendicular to the axis of secondary arms of the dendrites. In other words, the image shows the cross-section of secondary arms in a single grain. The bright contrast in the micrograph stems from differences in local composition deriving from solute segregation in between the dendritic arms.

dendrites have random crystallographic orientation (*i.e.* they are not textured) and assume globular geometry (as opposed to columnar).

1.2.3 Microstructure Variability in Fusion-Based Metal Additive Manufacturing[2]

AM is essentially a highly localized, rapid solidification process. Consequently, coarse macroscale microstructural features, such crystallographic texture and grain morphology, may be understood following the concepts described in section 1.2.2. Many experimental investigations, however, have demonstrated that the structures resulting from AM processes deviate from standard solidification microstructures and are often heterogeneous. These differences are difficult to control and predict and thus cast uncertainty over the performance and lifetime of different parts [15], hindering the adoption of AM components in industrial applications. In this section, some of the phenomena that give rise to this microstructure heterogeneity in AM will be discussed, including variable G_T directions, variable $\frac{G_T}{V}$, and variable $G_T \cdot V$.

Overview of Fusion-Based Metal AM Processes

During fusion-based AM processes – broadly categorized into powder bed fusion (PBF) or directed energy deposition (DED) – a feedstock material, such as powders or wires, are consolidated into a dense metallic part by melting and solidification with the aid of an energy source, such as a laser or an electron beam. At the beginning of the PBF process, a fresh layer of metallic powder is spread evenly using a rake or roller mechanism. The high-energy laser/electron beam then selectively melts the deposited layer of powder. After the layer is completed, the build platform is lowered and the cycle is repeated until the part is completed. In DED processes, the feedstock material is directly fed into the high-energy source *via* a nozzle and deposited "on-the-fly" on the build platform, layer by layer. Both processes are schematically shown in figure 1.2.12. Central to both processes is the creation of a *melt pool* of material (figure 1.2.13), which extends multiple layers in depth (to ensure metal bonding across different layers) and enables deposition of fresh material on the substrate. As the high-energy source moves, it carries the melt pool along line-by-line and layer-by-layer, creating a solid part. Melt pool geometry depends on thermal energy input (from the high-energy beam) and output (from the dissipation of heat through the underlying material) during the AM process. The most frequently used parameters to control the melt pool are beam power, P_B, beam velocity, V_B, and a term that estimates the melt pool volume, which depends on the AM process employed. In PBF, for instance, this term is $t \cdot \sigma$; where t is the thickness of the powder layer and σ is the overlap between consecutive melt pools.

[2]This section has been written following the review paper [6] DebRoy T., *et al.* (2018) Additive manufacturing of metallic components – process, structure and properties, *Prog. Mater. Sci.* 92, 112.

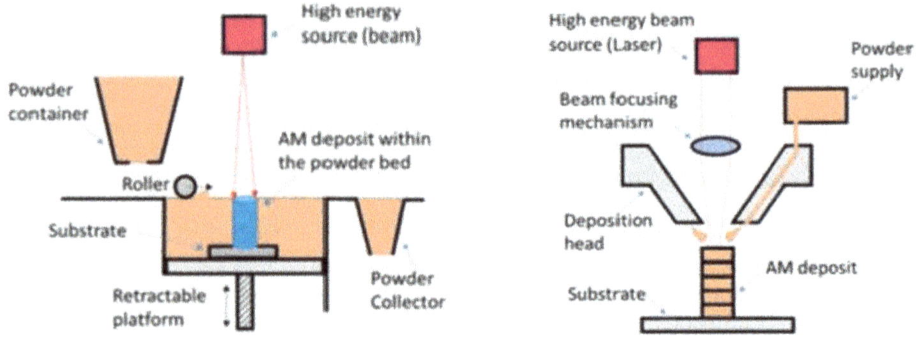

FIG. 1.2.12 – Schematic illustration of (left) PBF and (right) DED processes. Adapted from reference [4].

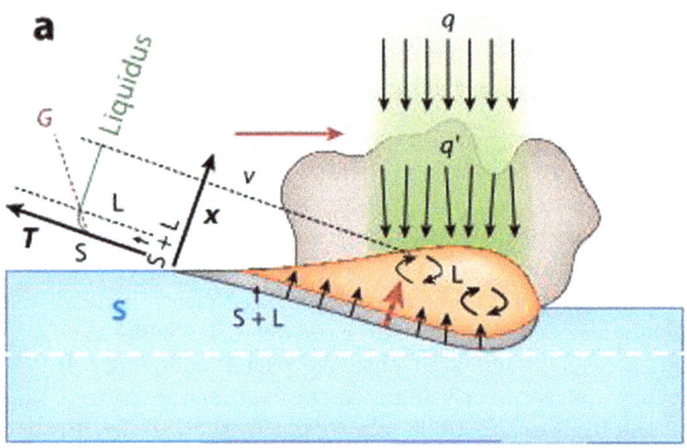

FIG. 1.2.13 – Schematic of a melt pool. Adapted from reference [6].

These parameters allow a coarse estimation of the total energy input per unit volume, which is referred to as the volumetric energy density, VED:

$$\text{VED} = \frac{P_B}{V_B \cdot t \cdot \sigma}$$

Melt Pool Shape

The grain structure and crystallographic texture of an AM component depend on the melting and solidification of the melt pool. Because melt pool solidification occurs on top of previously deposited (solid) layers, the solidification process in fusion-based AM is controlled by the *growth* of the underlying pre-existing grains, rather than by the *nucleation* of new ones. This phenomenon is generally referred to as *epitaxial growth*. The resulting solidification microstructure near the melt pool boundary is dominated by epitaxial growth of the grains in the previous layer. However, as melt pool solidification continues along the build direction (BD) – which corresponds to the normal to the material surface – the microstructure may evolve through *competitive growth.* Competitive growth occurs among dendrites with various crystallographic orientations in the solidifying polycrystalline melt pool. Dendrites whose easy-growth crystallographic axis is aligned with the direction of G_T will grow at the expense of others. In cubic systems, these crystallographic directions correspond to the <1 0 0> orientation. Therefore, as the melt pool solidifies along BD, grains with <1 0 0> orientation aligned to the direction of G_T will preferentially grow over other randomly oriented grains.

By contrast to standard directional solidification processes, however, the direction of G_T in fusion-based AM processes may not be uniform across the S/L interface (*i.e.*, the melt pool boundary). In general, the melt pool boundary is curved and its exact geometry depends on many factors, including VED, the beam shape, and heat dissipation. Consequently, the direction of G_T varies across the melt pool volume and may even change during the solidification of the melt pool (*i.e.*, along BD). These differences are schematically shown in figure 1.2.14. Thus, the resulting columnar growth and crystallographic texture may vary significantly as a function of the process parameters used. Therefore, the melt pool shape is a new

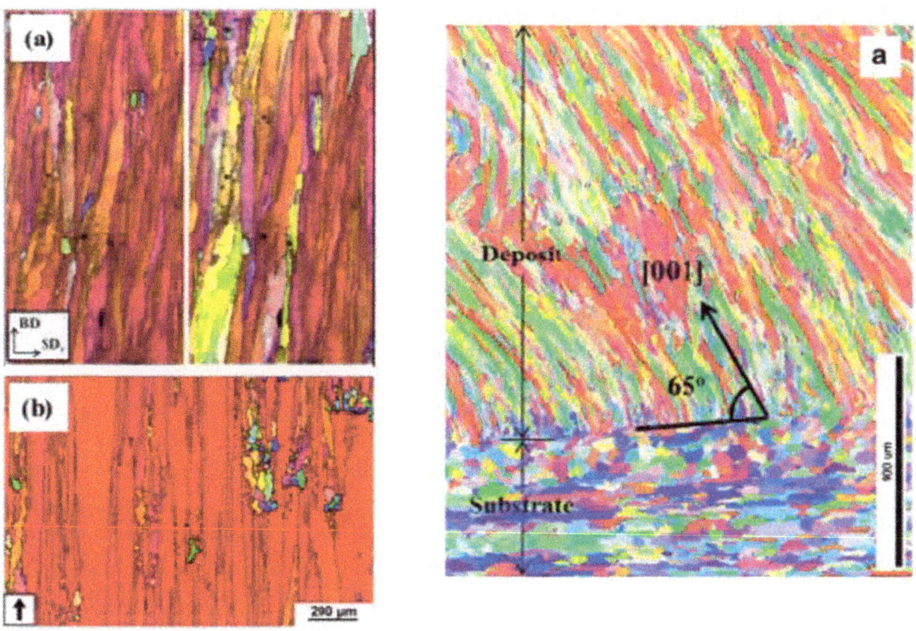

FIG. 1.2.14 – Schematic cross-sectional views of melt pool geometry for different process parameters. From (a) to (b) to (c) the melt pool becomes shorter and deeper. That, in turn, affects the local G_T direction (in red). Adapted from reference [6].

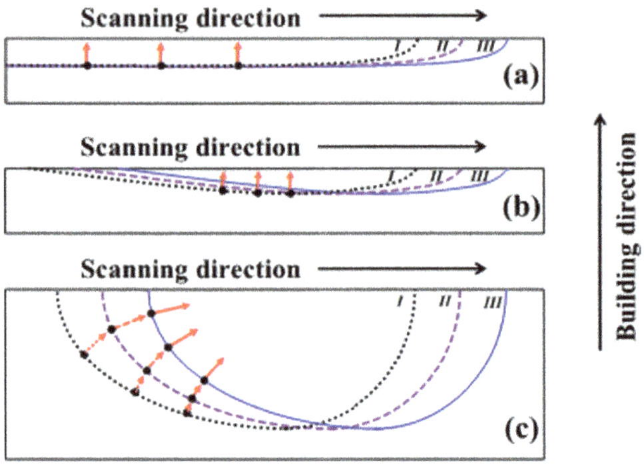

FIG. 1.2.15 – (Left) typical microstructures from PBF produced (a) silicon steel and (b) Inconel 718, consisting of highly elongated, textured grains aligned parallel to BD. (Right) DED-fabricated Ti-6Al-4 V sample consisting of columnar grains that are tilted away from BD. Adapted from reference [6].

important parameter to consider in order to predict and control the solidification microstructure of AM metals.

For fast scanning speeds, for instance, melt pools are long and shallow, giving rise to G_T directions that are aligned with BD (figures 1.2.15a and b). This type of microstructure is typical of PBF processes (which involve fast scanning speeds) and result in columnar growth with strong crystallographic texture along BD (left micrographs in figure 1.2.15).

Conversely, DED grain morphologies are often inclined with respect to the build direction as a result of slower scanning speeds and thus shorter and deeper melt pools (figure 1.2.15c). In these cases, the resulting <1 0 0> crystallographic textures are tilted away from the BD axis (right micrograph in figure 1.2.15).

Even more intriguing are some intermediate cases, in which the grain structure consists of a mixture of grains aligned parallel to the BD axis and grains that are tilted away from it within the same melt pool. These phenomena arise in melt pools with an oval cross-section, which is flat at the bottom, but grows steeper at the sides. The resulting microstructures consist of grains with alternating alignment and yields "bands" or "lamellae" (figure 1.2.16), which have been shown to affect the mechanical properties in non-obvious ways [10].

Keyhole-shaped melt pools are characterized by very elongated boundaries parallel to BD and which extend multiple layers across the build. In these extreme cases, grains may even grow parallel to the metal layer surface, resulting in completely different crystallographic textures [10] (figure 1.2.17).

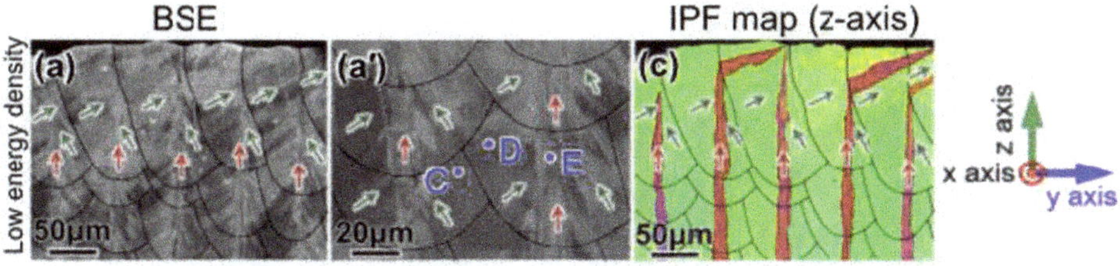

FIG. 1.2.16 – Cross-sectional views of a stainless steel 316L sample produced by SLM, characterized by melt pools consisting of different families of grains, growing along different directions (the green and red arrows in the electron micrographs). Because of epitaxial growth from the bottom layers, these grains can survive throughout multiple layers, giving rise to a unique lamellar microstructure. Adapted from reference [10].

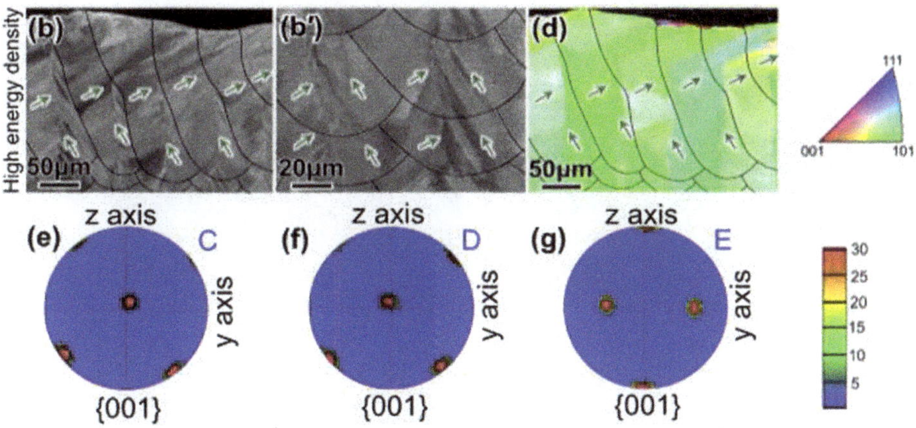

FIG. 1.2.17 – Keyhole-shaped melt pools give rise to highly tilted grains with respect to the BD, up to 45°. Because 45° is the exact angular distance between <1 0 0> and <1 1 0> crystal directions in cubic crystal systems, these melt pools give rise to a <1 1 0> crystallographic texture (as opposed to the more typical <1 0 0>) in a SLM stainless steel 316L. Adapted from reference [10].

Scanning Direction and Crystal Growth

Another interesting parameter that can be controlled in fusion-based AM processes is the "scanning strategy". The scanning strategy refers to how the high-energy beam is rastered across the portion of powder layer to be consolidated. Common scanning strategies are either mono-directional or bi-directional – meaning that the beam is scanned always along the same direction or back and forth across the layer, respectively. A comparison between the two strategies in a DED-produced Inconel 718 sample is shown in figure 1.2.18. As the EBSD micrographs visually suggest, bi-directional strategies are beneficial to make the solidification microstructure more uniform [18]. This result stems from the difference in cooling rate in between

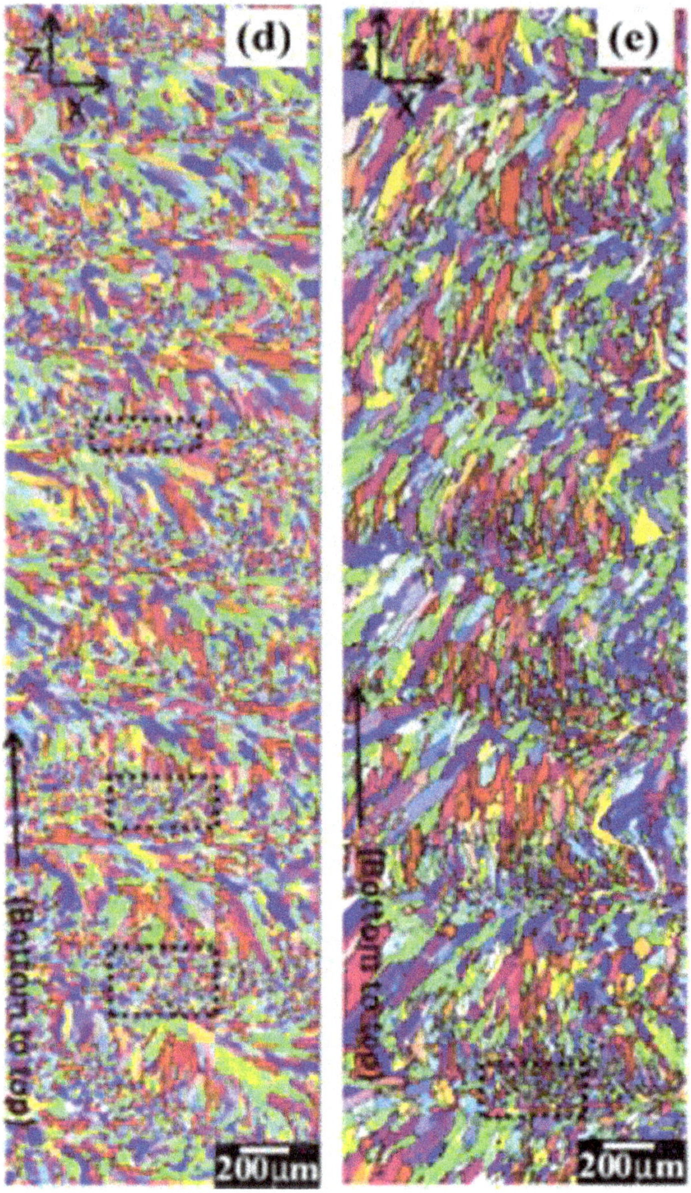

FIG. 1.2.18 – Effect of the scanning strategy on Inconel 718 produced by DED. Mono-directional strategy (left) induces more heterogeneities in the sample microstructure compared to a (right) bi-directional scanning strategy. Adapted from reference [6].

deposition passes when using different scanning strategies. In mono-directional scanning (figure 1.2.18d), the beam is turned off in between layers, allowing the material to cool down more than in the bi-directional scanning strategy sample

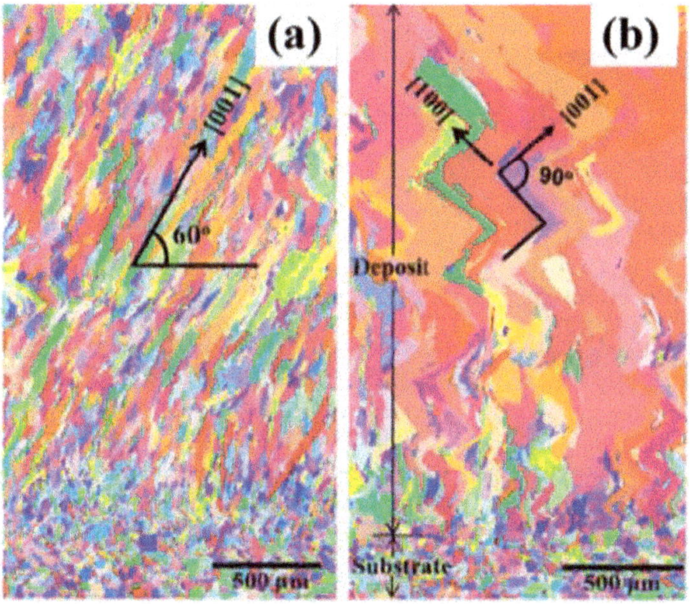

FIG. 1.2.19 – Different grain morphologies resulting from a rotation of the scanning direction across multiple layers in DED Inconel 718. Adapted from reference [12].

(figure 1.2.18e). The lower base temperature yields very high constitutional undercooling and drives nucleation of fine, equiaxed grains, which is not as prominent when using a bi-directional scanning strategy.

Irrespective of the scanning strategy used, the direction the beam is scanned across the layer can be varied (*i.e.*, rotated) throughout the different layers (*i.e.*, along BD). This approach is commonly employed to maximize build density and mitigate the establishment of residual stresses. However, rotating the scanning direction across layers has important implications on the solidification microstructure. A simple example is that depicted in figure 1.2.19, where the scanning direction was rotated by 180° every other layer. The resulting microstructure consists of "zig-zag" grains, whose growth direction follows the alternating G_T direction.

This result proves that rotating the scanning direction affects competitive growth and may be used as an effective means to tailor crystallographic textures in AM materials. The establishment of a texture during solidification of melt pools is the compromise between the alignment of easy growth directions of grains and the maximum direction of G_T at the melt pool boundary. Generally, growth is favoured when this maximum aligns closely with one of the easy-growth directions (<1 0 0> in cubic crystal systems). However, the actual direction of grain growth may deviate considerably from the local maximum G_T direction for certain scanning patterns. Figure 1.2.20 schematically shows possible primary dendrite growth scenarios to explain the results shown in figure 1.2.19 [11]. In the first scenario, the primary dendrite trunk (or growth axis) is assumed to be aligned with the local maximum

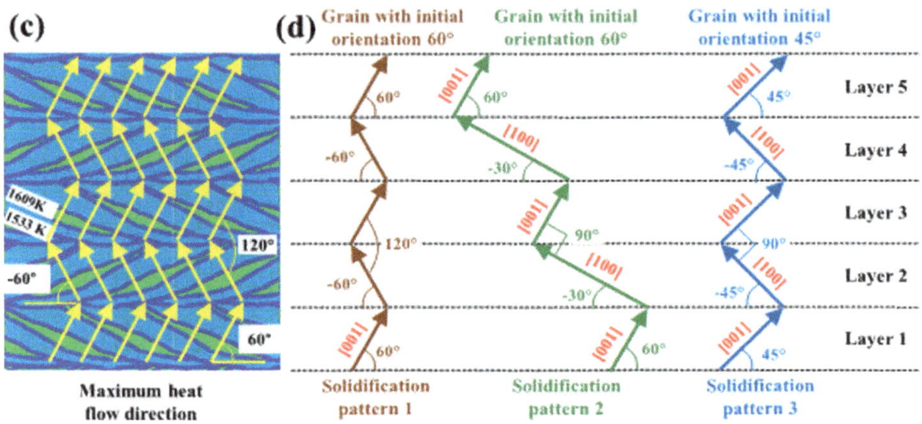

FIG. 1.2.20 – (c) Finite element model of the G_T direction across layers in the system shown in figure 1.2.19. (d) Of the three possible scenarios, the solidification pattern #3 is the only plausible one. Adapted from reference [11].

G_T direction throughout different layers, which is shown to alternate between +60° and −60° in the finite element simulation. This type of growth cannot occur because it would require nucleation of new dendrites aligned with the G_T direction at each layer. The driving force required for nucleation is much too high compared to that for grain growth. In the second scenario, the primary dendrites of successive layers are perpendicular to each other. However, in the second layer the maximum G_T direction deviates 30° with respect to the primary dendrite growth direction. This deviation is too large to be plausible. In other words, it is impossible to drive the growth of dendrites that are misaligned by 30° with respect to the maximum G_T direction. In scenario 3, the grain orientation in the first layer is at 45° with respect to the scanning direction. The primary dendrite in the second layer grows epitaxially from the secondary dendrites in the first layer and makes an angle of ~ 45° with the horizontal line. As a compromise, there is an acceptable deviation of only 15° between the maximum G_T direction and the primary dendrite growth direction in all the layers, resulting in the texture shown in figure 1.2.19.

Following the same principle, scan rotations of 60° have been shown to yield <1 1 1> (rather than the more conventional <1 0 0>) crystallographic textures along the BD in PBF-processed tantalum (see figure 1.2.21) [13].

Solidification Maps and Grain Morphology

The most widely observed solidification microstructures in AM components are columnar and equiaxed structures. Under certain solidification conditions, the growth of a columnar structure is terminated with the formation of an equiaxed structure. These instances occur when nucleation of sufficiently numerous equiaxed dendrites takes place in the constitutionally undercooled liquid adjacent to the columnar dendritic front (figure 1.2.9). Columnar grains are generally coarse because of their epitaxial growth across many layers (see figure 1.2.22) and

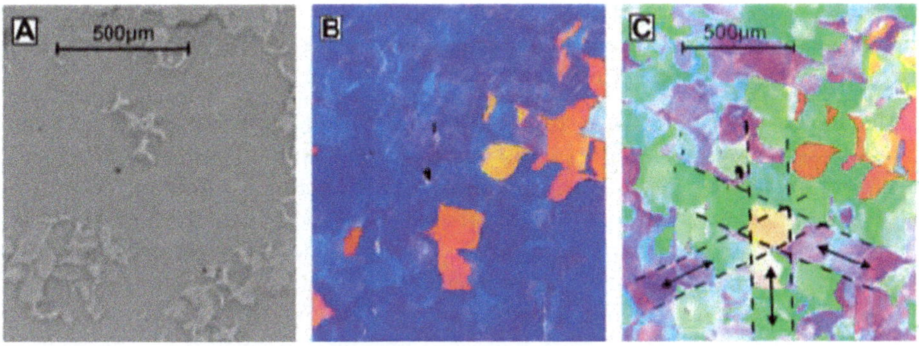

FIG. 1.2.21 – Tantalum sample produced by SLM. Using a scan rotation of 60° yields the formation of a <1 1 1> texture along the BD. Adapted from reference [13].

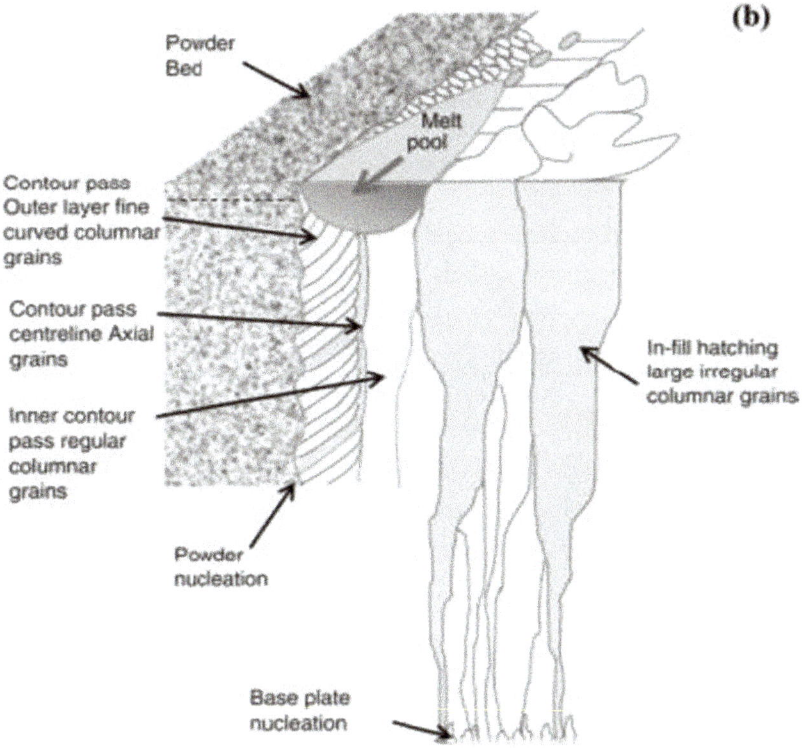

FIG. 1.2.22 – Schematic illustration showing progressive lateral growth of epitaxially growing grains throughout multiple layers in a PBF process. Adapted from reference [6].

characterized by anisotropic mechanical properties and prone to cracking [17]. In contrast, equiaxed grains are usually small because of the high cooling rates in AM processes, and characterized by more isotropic mechanical properties.

Mechanical behaviour anisotropy is detrimental for applications involving multidirectional stresses. Thus, attaining only aligned columnar grain structures is inappropriate for the fabrication of components that require isotropic mechanical properties. The anisotropy can be significantly reduced with the introduction of equiaxed grain structures, which improve the ductility and fracture toughness of metal parts. Equiaxed grains can be generated by promoting the nucleation of equiaxed dendrites in the melt pool, as predicted by the G_T/V ratio. G_T/V decreases as the scanning speed is increased and further away from the melt pool boundary. Furthermore, G_T/V decreases towards the upper layers of a build – since heat accumulation along the BD reduces the magnitude of G_T – which yields

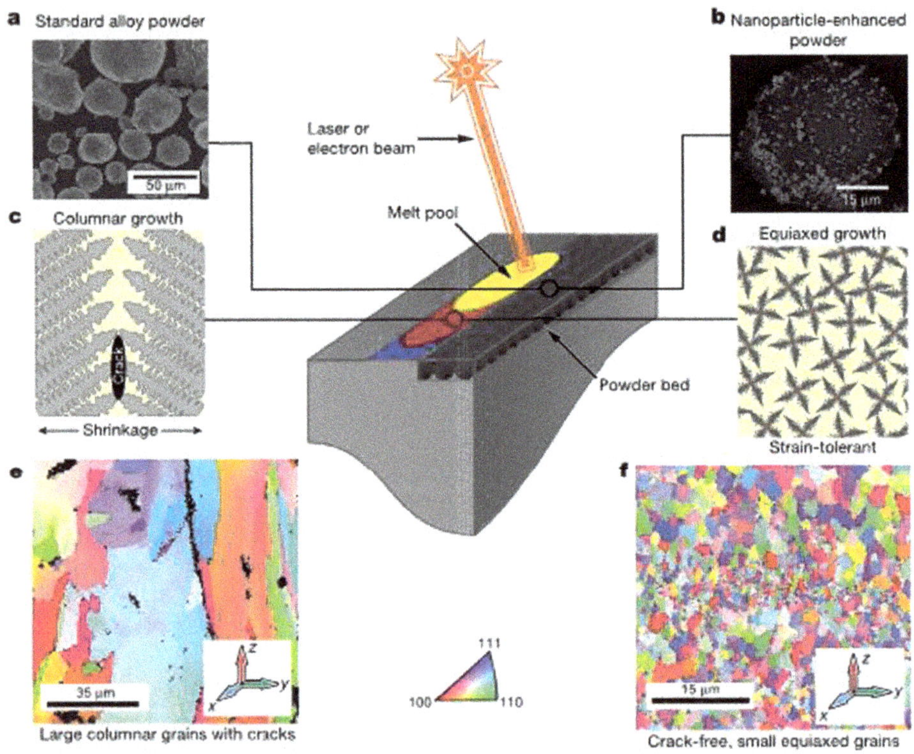

FIG. 1.2.23 – (a) Conventional Al7075 powder feedstock. (b) Al7075 powder functionalized with inoculants (*i.e.* nanoparticles). (c) Schematic representation of the directionally solidified dendritic structures in conventional Al7075 produced by SLM. (d) Inoculants induce heterogeneous nucleation and facilitate equiaxed grain growth. (e) Coarse grains resulting from competitive epitaxial growth. (f) Functionalizing the powder feedstock with nanoparticles produces fine microstructures and eliminates hot cracking. Adapted from reference [7].

increased instability of the solidification front (figure 1.2.18). Generally, equiaxed grains are rare in polycrystalline materials processed by PBF due to the extremely high G_T, which is of the order of 100 K/m. However, equiaxed grains may still form as a result of large constitutional undercooling near the S/L interface, or because of the presence of partially melted powders or added refractory particles – both of which act as inoculant and lower the activation energy for nucleation (see figure 1.2.23 and refer to "heterogeneous nucleation" in section 1.2.2).

Because G_T and V can be controlled separately – to some extent – in AM processes, it is possible to induce transitions in grain structures from columnar to equiaxed by changing VED; and specifically the beam power (which affects G_T) and scanning speed (which dictates V). Figure 1.2.24 shows examples of these controlled transitions in Inconel 718 specimens fabricated *via* SLM [5] and electron beam melting (EBM) [8]. In the case of SLM (top portion of figure 1.2.24) the different

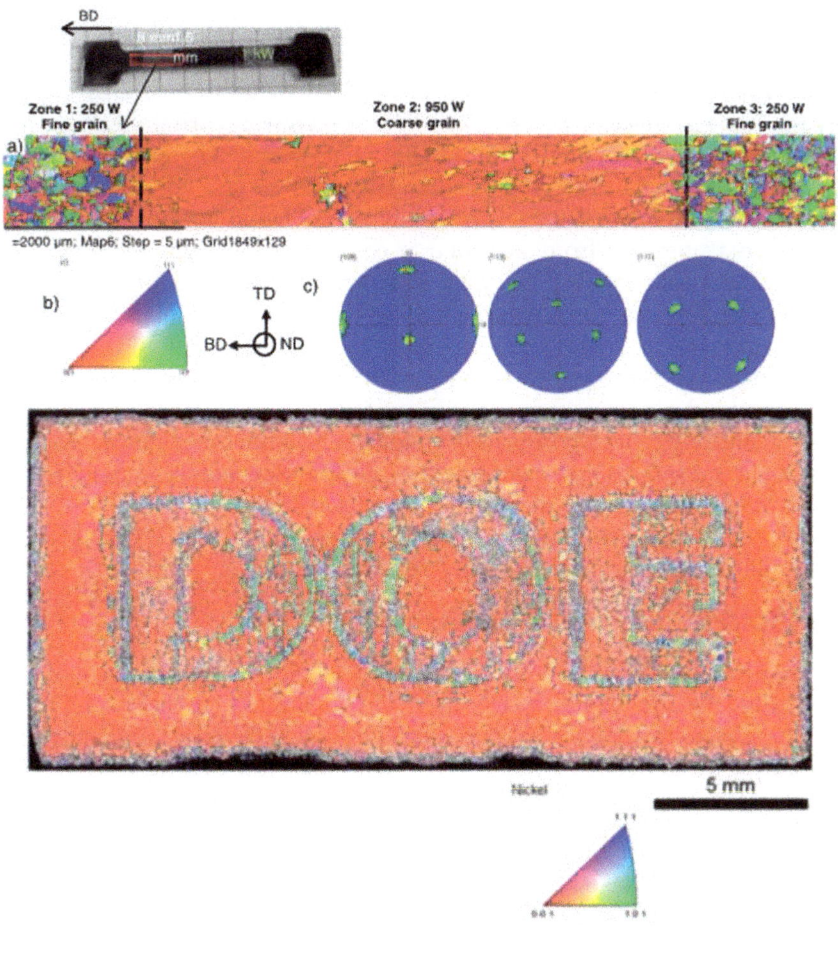

FIG. 1.2.24 – Examples of Inconel 718 samples with site-specific microstructures produced by (top) SLM [5] and (bottom) EBM [8].

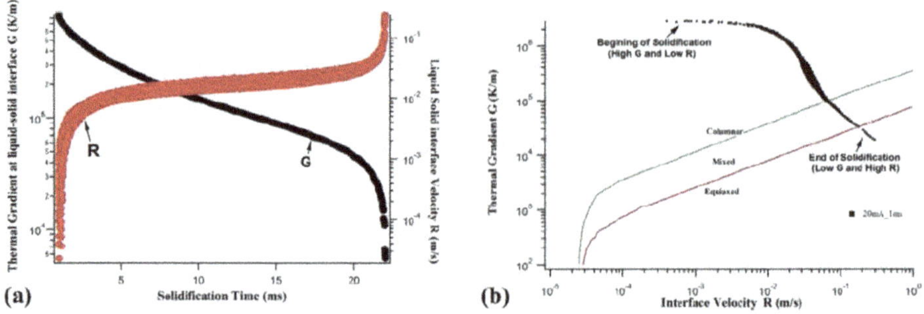

FIG. 1.2.25 – Calculated solidification parameters for grain morphology in Inconel 718 produced by EBM. (a) Variation of G_T and V with solidification time. (b) Solidification map with superimposed G_T and V indicating the formation of columnar and equiaxed grains. Adapted from reference [6].

microstructures were obtained using high and low energy densities by varying the laser-beam power from 950 to 250 W, respectively. High energy density microstructures exhibit highly textured columnar grains, whereas low energy density SLM processes yield randomly oriented equiaxed grains. The EBM experiments (bottom portion of figure 1.2.24) use a slightly different approach. In this case, microstructure transitions from columnar to equiaxed were promoted by varying the scanning strategy and speed. Columnar microstructures were induced using a line scan pattern and a scanning speed of 4530 mm/s, while equiaxed regions were produced using a spot strategy (which essentially utilizes a pulsed as opposed to a continuous beam of electrons) at a speed of 800 mm/s.

These results demonstrate the possibility of tailoring local microstructure transitions by identifying the appropriate regions in solidification maps. To this end, however, it is necessary to first calculate the material-dependent G_T and V values as a function of the AM process-dependent parameters used.

Figure 1.2.25 shows the superposition of these different processing conditions onto the solidification map for Inconel 718.

Cooling Rates and Part Geometry

As expected from standard solidification theory (figure 1.2.10), the size of grains in AM is dictated by the cooling rate $G_T \cdot V$. Finer grain structures solidify at high cooling rates because there is shorter time for grains to grow, and *vice versa*. In general, during fusion-based AM processes, the material cooling rate decreases as the layer height increases [3]. This phenomenon is exemplified in figure 1.2.26, where the grain structure, simulated melt pool peak temperature and size, and computed cooling rates are compared side by side for a 316L stainless steel wall produced by DED (figure 1.2.26). The thermal buildup along BD is the result of the limited thermal conductivity of the environment surrounding the build. In PBF processes, in fact, the solid material is embedded in a powder bath. Although heat transfer through powders is lower compared to that of a dense metal (due to the

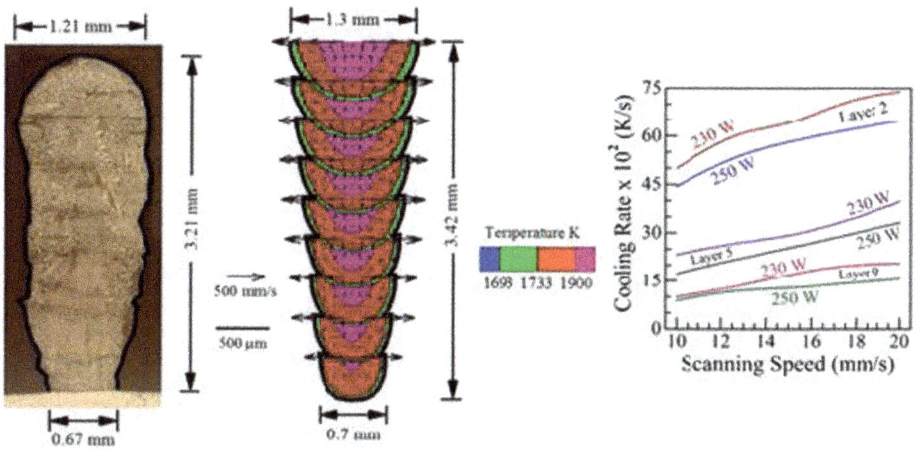

FIG. 1.2.26 – (From left to right) Optical micrograph of the grain structure, simulated melt pool geometry and peak temperature, and computed cooling rates (for different powers) of a 316L stainless steel sample produced by DED. The computed cooling rate values progressively decrease along the BD. Adapted from reference [3].

small contact points in between particles), thermal buildup in DED is more severe than in PBF owing to the lack of material surrounding the build. The only heat sink in DED processes is the base plate. However, as more and more material is added to the build, heat dissipation through the base plate becomes more difficult and the newly deposited material has less time to cool down before a new layer is added. Thus, grain structures become coarser in the upper layers due to lower local cooling rate compared to those in the lower layers (figure 1.2.27) [9].

1.2.4 Concluding Remarks

Metal AM is often regarded as a *new disruptive* technology because it enables the fabrication of parts that are otherwise impossible to make. However, the underpinning physics of metal AM is that of conventional solidification (section 1.2.2 in this chapter), which man started to uncover as early as in the Bronze Age. In spite of this solid foundation, materials produced by AM are difficult to control (section 1.2.3 in this chapter) and thus difficult to employ in many industrial applications.

In the author's perspective, these challenges stem in part from the small length scale at which solidification occurs and in part from the additional degrees of freedom that AM provides over conventional manufacturing processes. The highly localized energy input during fusion-based AM induces the establishment steep thermal gradients in the material, whose directions are process- and

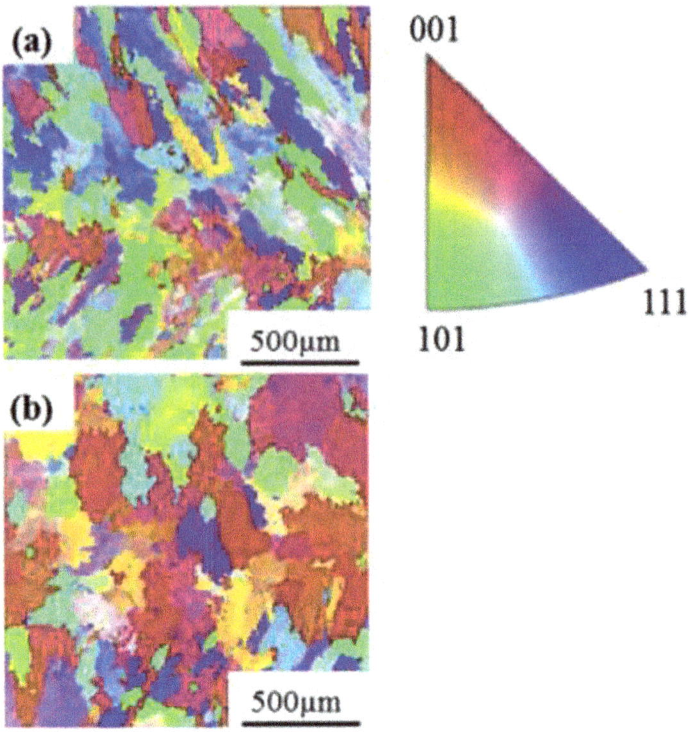

FIG. 1.2.27 – EBSD maps showing the grain structure at the (top) bottom and (bottom) top of a 70 mm tall 304 stainless steel wall produced by DED. The average grain size at the bottom of the build is ~1530 μm^2 but it reduces to ~1150 μm^2 at the top of the wall. Adapted from reference [9].

material-dependent. The relatively small melt pool volume created by this energy input – on the order of 10^{-4} mm^3 – is cooled over timescales that vary as a function of part geometry. Fluctuations in material feedstock or in the interaction of the high-energy source with it open the way to the incorporation of defects (such as porosity) across the build volume. Finally, the repeated thermal cycles each layer is subjected to during fusion-based AM processes have profound – and non-obvious – implications on the resulting parts (such as the presence of residual stresses).

Addressing all of the above-mentioned challenges in a single chapter would be too ambitious of an endeavour. Instead, the focus in this work has been set on describing and interpreting microstructure (and property) variability in metal AM from a solidification standpoint to explicitly point out the material-related challenges in this growing field. The implicit goal of this chapter, however, is to provoke thinking of these "challenges" as "opportunities" that are yet to be taken.

What if we embraced the microstructure variability offered by AM?

What if we capitalized on the unmatched capability of integrating dissimilar microstructures within the same part, such as those shown in figure 1.2.24?

What if we considered materials microstructure as an additional feature that can be *optimized* through AM, in analogue to material topology?

References

[1] Kurz W., Fisher D.J. (1998) *Fundamentals of solidification*. Trans Tech Publications Limited.

[2] Stefanescu D.M., Ruxanda R. (2004) Fundamentals of solidification, in *Metallography and microstructures*. ASM International, p. 0.

[3] Manvatkar V., De A., DebRoy T. (2015) Spatial variation of melt pool geometry, peak temperature and solidification parameters during laser assisted additive manufacturing process, *Mater. Sci. Technol.* **31**, 924.

[4] Kok Y., *et al.* (2018) Anisotropy and heterogeneity of microstructure and mechanical properties in metal additive manufacturing: a critical review, *Mater. Des.* **139**, 565.

[5] Popovich V.A., *et al.* (2017) Functionally graded Inconel 718 processed by additive manufacturing: crystallographic texture, anisotropy of microstructure and mechanical properties, *Mater. Des.* **114**, 441.

[6] DebRoy T., *et al.* (2018) Additive manufacturing of metallic components – process, structure and properties. *Prog. Mater. Sci.* **92**, 112.

[7] Martin J.H., *et al.* (2017) 3D printing of high-strength aluminium alloys, *Nature.* **549**, 365.

[8] Dehoff R.R., *et al.* (2015) Site specific control of crystallographic grain orientation through electron beam additive manufacturing, *Mater. Sci. Technol.* **31**, 931.

[9] Wang Z., Palmer T.A., Beese A.M. (2016) Effect of processing parameters on microstructure and tensile properties of austenitic stainless steel 304L made by directed energy deposition additive manufacturing, *Acta Mater.* **110**, 226.

[10] Sun S.-H., *et al.* (2019) Excellent mechanical and corrosion properties of austenitic stainless steel with a unique crystallographic lamellar microstructure via selective laser melting, *Scr. Mater.* **159**, 89.

[11] Wei H.L., Mazumder J., DebRoy T. (2015) Evolution of solidification texture during additive manufacturing, *Sci. Rep.* **5**.

[12] Dinda G.P., Dasgupta A.K., Mazumder J. (2012) Texture control during laser deposition of nickel-based super alloy, *Scr. Mater.* **67**, 503.

[13] Thijs L., *et al.* (2013) Strong morphological and crystallographic texture and resulting yield strength anisotropy in selective laser melted tantalum. *Acta Mater.* **61**, 4657.

[14] Seita M., *et al.* (2015) The dual role of coherent twin boundaries in hydrogen embrittlement, *Nat. Commun.* **6**.

[15] Salzbrenner B.C., *et al.* (2017) High-throughput stochastic tensile performance of additively manufactured stainless steel, *J. Mater. Process. Technol.* **241**, 1.

[16] Cordero Z.C., Knight B.E., Schuh C.A. (2016) Six decades of the Hall–Petch effect – a survey of grain-size strengthening studies on pure metals, *Int. Mater. Rev.* **61**, 495.

[17] Carroll B.E., Palmer T.A., Beese A.M. (2015) Anisotropic tensile behavior of Ti-6Al-4V components fabricated with directed energy deposition additive manufacturing, *Acta Mater.* **87**, 309.

[18] Parimi L.L., *et al.* (2014) Microstructural and texture development in direct laser fabricated IN718, *Mater. Charact.* **89**, 102.

1.3 Correlation Between Process Parameters and Properties of Use: Example of Corrosion

Fanny BALBAUD-CÉLÉRIER and Beatriz PUGA

Université Paris-Saclay, CEA, Service de la corrosion et du comportement des matériaux dans leur environnement, 91191 Gif-sur-Yvette, France

Abstract

Emerging fabrication processes such as additive manufacturing allow the elaboration of complex geometries, gradient materials, and functionalized materials. However, additive manufacturing processes such as selective laser melting (SLM) and direct laser deposition (DLD) also generate specific microstructures, which need to be qualified regarding the final use of the materials or components produced. Few publications are available on the link between the process parameters and the microstructure obtained with the corrosion behaviour, and most of them consider exclusively the corrosion behaviour in aqueous medium, mostly marine environment. In this chapter, we will consider essentially SLM and DLD metal additive manufacturing processes together with their impact on the generated microstructures and regarding corrosion processes we will focus on both aqueous corrosion processes and high temperature processes. In the first part, we will recall generalities on corrosion, then in the second part we will show the effect of the process parameters on the microstructure and then on the corrosion behaviour. Finally, we will show how these new manufacturing processes could bring potential solutions for the development of sustainable materials by the possibly of designing corrosion resistant materials, corrosion resistant components and by the emergence of corrosion protection solutions.

1.3.1 Introduction

With the increasing need in materials with improved properties (mechanical, environmental, thermal, wear...), longer lifetimes, complex geometries and optimised designs, massive development of innovative materials solutions, such as new manufacturing processes and emerging coating and thin films technologies is ongoing [1]. These new processes have to be lower carbon production processes than the conventional ones and have to minimize the use of scarce elements. Indeed, the global sustainability of the materials cycle, from the design to the end of life, has to be considered in a circular economy approach. Among new fabrication processes, additive manufacturing allows the elaboration of complex geometries, gradient materials, and functionalized materials. However, additive manufacturing also generates specific microstructures, which need to be qualified regarding the final use of the materials or components produced. Many researches have already been performed on the link between process parameters, microstructures obtained and

DOI: 10.1051/978-2-7598-2446-5.c031

mechanical properties [2–10] using various metal additive manufacturing processes, mainly selective laser melting (SLM) and direct laser deposition (DLD). However, fewer publications are available on the link between the process parameters and the microstructure obtained with the corrosion behaviour, and most of them consider exclusively the corrosion behaviour in aqueous medium, mostly marine environment [11, 12]. In this chapter, we will consider essentially SLM and DLD metal additive manufacturing processes together with their impact on the generated microstructures and regarding corrosion processes we will focus on both aqueous corrosion processes and high temperature processes. We will try to show how the main metallurgical parameters influencing corrosion processes can be connected to the process parameters showing the need for linking them intimately. In the first part we will recall generalities on corrosion, then in the second part we will show the effect of the process parameters on the microstructure and then on the corrosion behaviour. Finally, we will show how these new manufacturing processes could also bring solutions to corrosion concerns.

1.3.2 What is Corrosion? A Few Recalls

Corrosion is the deterioration of a material, usually a metal, by reaction with its environment. As a result of this process, changes in the properties of the metal can occur modifying its performances and lifetime. Corrosion process depends on the interfacial characteristics between the metal and its environment (an electrolyte for aqueous corrosion). The corrosion morphology for the same metal can be very different depending on its exposure to an aqueous or gaseous environment at room or at high temperature. For that reason, corrosion control is mainly achieved by recognizing and understanding the corrosion mechanisms at the material-environment interface. From a thermodynamic point of view, corrosion is an inevitable process. Metals evolve toward their natural state (oxides, sulfides, carbonates, etc.) in order to return to lower energy states. Consequently, using corrosion-resistant materials, designs, protective systems, devices and treatments is the only way to slow down this process.

The corrosion phenomena that take place when the metal is exposed at high temperatures or in an aqueous environment are described below.

1.3.2.1 Aqueous Corrosion: Electrochemical Corrosion

Electrochemical corrosion involves simultaneously the anodic (oxidation/dissolution of a metal) and the cathodic (reduction of an oxidizing agent) reactions on the metal surface, causing current flow from anodic to cathodic areas (that can be either the same global area or two distinct areas). Operating mode of this corrosion cell is shown in figure 1.3.1. The anode and cathode are connected through the solution by an ionic current path, and they are connected through the metal by an electronic path. The corrosion rate is controlled by the net balance among all of the components of the corrosion cell. Elimination of any of the four components (anode, cathode, ionic path or electronic path) stops the corrosion. For example, increasing

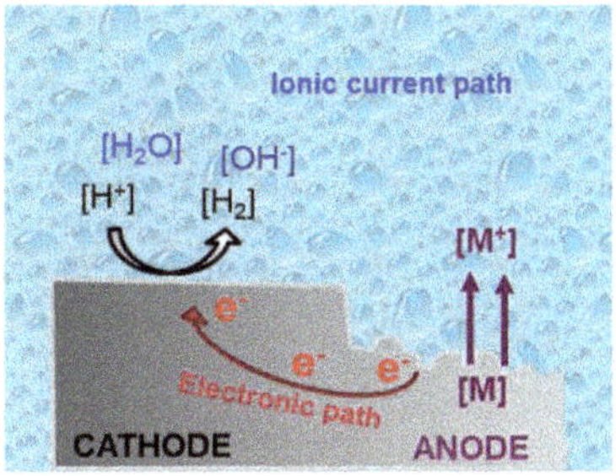

FIG. 1.3.1 – Components of an electrochemical corrosion cell. Electrons generated at the anode by oxidation reactions ($M \leftrightarrow M^{n+} + ne^-$) move to the cathode, where they are consumed by reduction reactions ($2H^+ + 2e^- \rightarrow H_2$, $O_2 + 2H_2O + 4e^- \rightarrow 4OH^-$). The movement of charged ions in the solution is the vehicle for current flow through this portion of the corrosion cell.

the resistance in the ionic current path by using a low-conductivity electrolyte limits the amount of current that can flow in the corrosion cell. Therefore, the corrosion behavior depends on the combination of the characteristics of the electrolyte (conductivity, pH, oxidizing power, degree of ionization, temperature, etc.) with the characteristics of the metal or alloy (metallurgical characteristics, inherent reactivity and tendency to form insoluble corrosion products). For instance, the metallurgical characteristics determine if the anodic and cathodic reactions are located in the same area or not. Different areas can be due to an inhomogeneous microstructure: existence of grains boundaries, different phases, chemical gradients and microstructural defects as presented in figure 1.3.2.

1.3.2.1.1 Morphology of Corrosion

Corrosion can be divided into two main categories according to appearance of the corrosion damage or mechanism of attack: uniform or general corrosion and localized corrosion.

- Uniform or general corrosion: the loss of thickness of sound material is uniform throughout the surface. The microcells are spread all over the metal surface. From a technical point of view, the life of the equipment that presents this type of corrosion can be accurately estimated by immersion tests.
- Localized corrosion: this type of corrosion occurs when one area of a metal becomes anodic with respect to the rest of the surface (figure 1.3.1). The attacks are isolated from each other on the surface. The detection of localized corrosion is therefore complex. It can cause failure by perforation while producing only a small weight loss on the metal. Pitting corrosion, crevice corrosion, galvanic corrosion, erosion-corrosion, intergranular corrosion, dealloying and stress-corrosion cracking are included in this category. Some causes that produce localized corrosion are summarized in table 1.3.1.

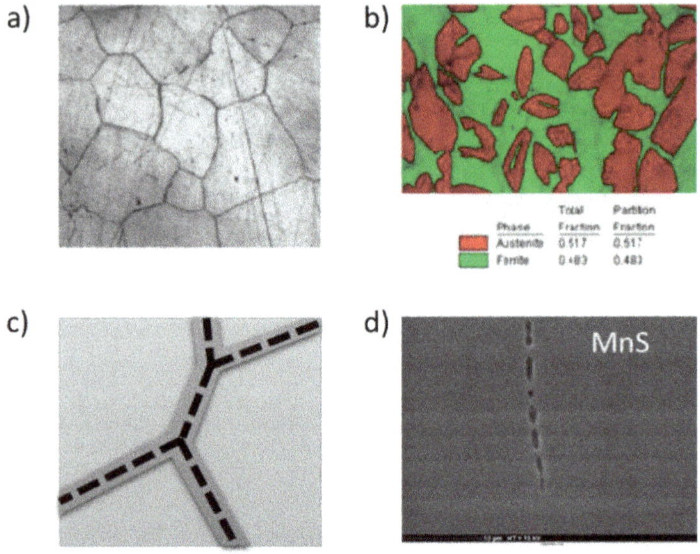

FIG. 1.3.2 – Some metallurgical characteristics that can influence the corrosion: (a) the microstructural and chemical differences of grains boundary with respect to the bulk grains, (b) multiphase materials, (c) precipitates of carbides (black particles) in the grain boundaries with depletion of chromium in regions surrounding the particles (grey) and (d) inclusions represented by a manganese sulfide particles.

Many corrosion problems are due to more than one form of corrosion acting simultaneously. Moreover, the same material can present different corrosion morphologies depending on the aggressive environment to which it is exposed.

1.3.2.1.2 Thermodynamic Approach

Thermodynamics is a powerful science to understand corrosion phenomena. It is used to predict the predominant species at equilibrium and when corrosion can occur. Gibbs energy difference (ΔG) can be used to predict the feasibility of reactions. The metal will react spontaneously in an aqueous environment to produce metal ions when $\Delta G < 0$. Another parameter such as thermodynamic potential gives us information about the boundaries between stable oxidized species and stable reduced species. It can be estimated using the Nernst equation 1.3.1:

$$E = E^0 - \frac{RT}{nF} \ln \frac{[\text{ox}]}{[\text{red}]} \tag{1.3.1}$$

where E is the thermodynamic potential, E^0 is the standard thermodynamic potential, R is the gas constant, T is the absolute temperature (in Kelvin), n is the number of electrons involved in the reaction, F is the Faraday constant, and [red] and [ox] are the concentrations of reduced and oxidized species (for instance a pure metal and its cation: Fe/Fe^{2+}), respectively. The thermodynamic potential can be

TAB. 1.3.1 – Main causes that produce localized corrosion.

Localized corrosion	Causes of corrosion	Morphology of corrosion
Pitting corrosion	Local inhomogeneity on the metal surface such as manganese sulfide inclusions (MnS), local break down of protective oxide film, mechanical ruptures in protective coatings by contact with electrolyte that contains halide ions, particularly chlorides	Reconstructed image from 3D microscopy of a steel (18% Cr) in NaCl + HNO_3. 
Crevice corrosion	Occluded regions by the presence of joints, rubber O-rings, deposits, accumulations or biological materials, etc. where oxygen diffusion is restricted. A difference aeration cell tends to be set up between crevice and bulk environment	Optic image of a stainless steel under a deposit containing graphite 
Galvanic corrosion	Potential difference between dissimilar metals or alloys. Presence of impurities, multiphase materials	Scheme of galvanic corrosion process 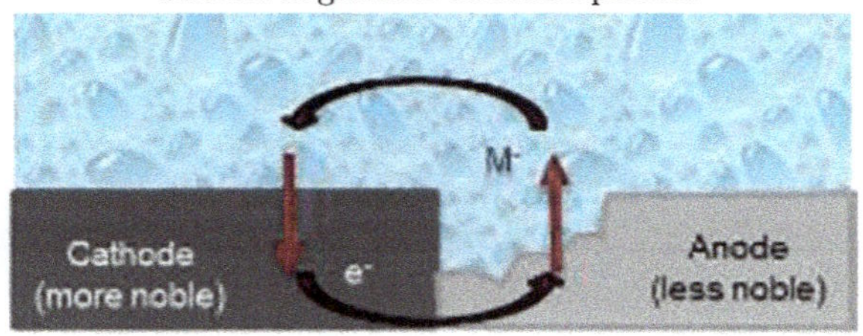

TAB. 1.3.1 – (continued).

Localized corrosion	Causes of corrosion	Morphology of corrosion
Intergranular corrosion	Potential difference between the regions adjacent to the grain boundary and the precipitates, intermetallic phases or impurities, which are formed at grain boundary	SEM image of cross section of stainless steel in 8 M HNO_3 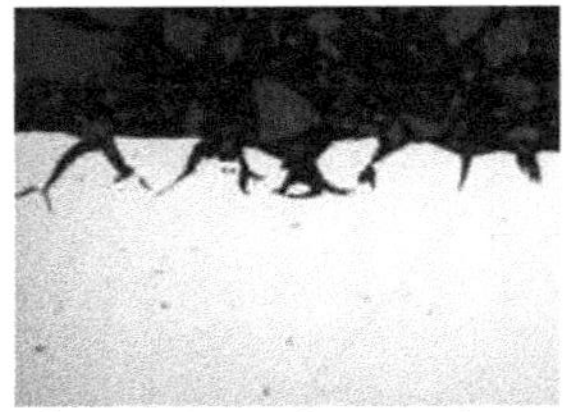
Erosion-corrosion	Mechanical wear or abrasive contributions in combination with corrosive environments	Optic image of zirconium after erosion-corrosion test 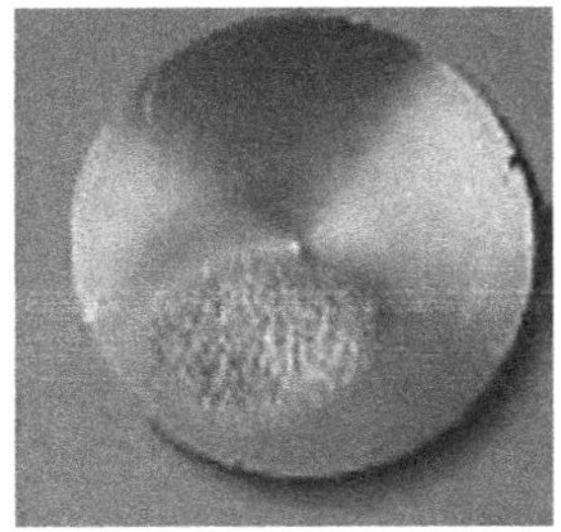
Stress-corrosion cracking (SCC)	Joint action of a surface tensile stress and the presence of a specific corrosive environment. Alloy composition and microstructural aspects such as the presence of phases in the matrix and the grain boundaries, the composition of phases, the grain size and orientation and grain boundary segregation can influence this corrosion	SEM image of U-bend specimen in Na_2CO_3 + NaOH + NaCl at boiling temperature 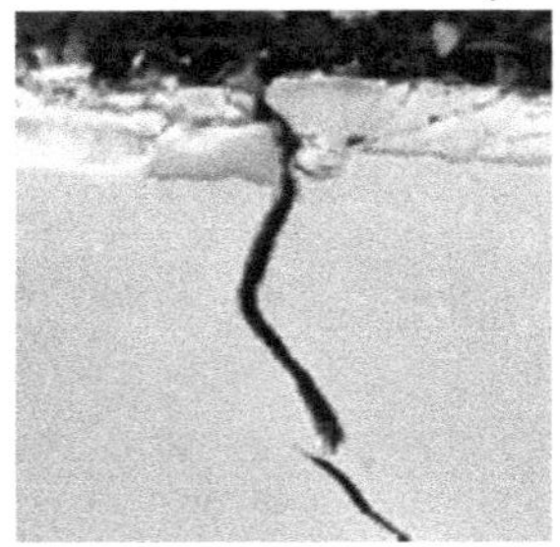

measured by an electrochemical system constituted of two electrodes immersed in the electrolytic environment containing the cation, ox. One electrode is the material composed of the pure metal, red and the other is a reference electrode. In that case, the potential difference $E_{el} - E_{ref}$ is measured. E^0 is the standard electromotive force (emf) measured when a metal/metal ion electrode is coupled to a hydrogen electrode under standard conditions (1 M concentration and 25 °C). The standard thermodynamic potential of the electrochemical reaction which is linked to the standard Gibbs energy of the reaction by $\Delta G^0 = -nF\Delta E^0$ is higher for less oxidizing metals (noble metals) and lower for more oxidizing metals. Then measuring E^0 of series of metals is used to generate the emf series where the more noble metals are at the more positive potentials and the more active metals, those prone to corrosion, are at the more negative potentials. For example, when graphite is coupled to steel in seawater, graphite (more noble) becomes the cathode (site of the reduction reactions) and the steel (more negative potential than the graphite) becomes the anode, thus corrosion of steel occurs (figure 1.3.3).

Potential-pH diagrams, also called Pourbaix diagrams (figure 1.3.4), are another widely used thermodynamic tool to predict the behavior of a metal. Considering oxido-reduction reactions of a metal and water (the environment), the thermodynamic potentials E of all the electrochemical reactions involving metal and water can be calculated as a function of pH in comparison with the hydrogen electrode. These diagrams allow to identify the E-pH domains for which [13]:

- The metal is stable under its metallic form (immunity domain), in that case no corrosion occurs.
- The metal is stable as a cation under solid oxide form and, if this oxide is continuous, forming a protective oxide layer, in that case the corrosion rate is very low (passive domain).
- The metal is stable as a soluble oxidized species (corrosion domain). In that case, the corrosion rate can be very high.

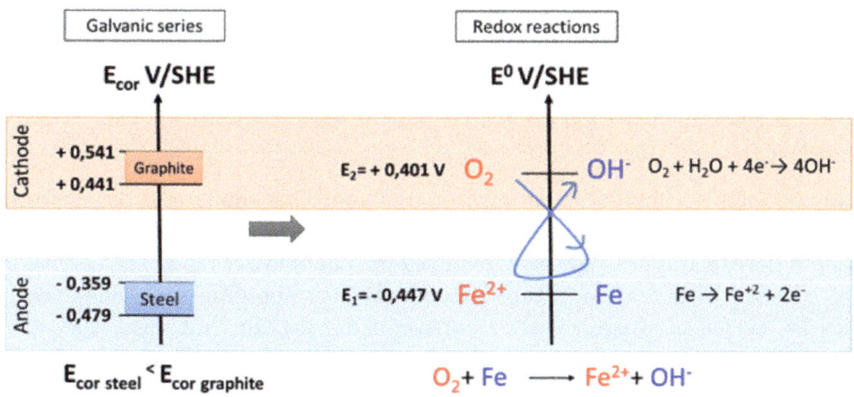

FIG. 1.3.3 – Corrosion potentials of graphite and steel in seawater (galvanic series) and prediction of the direction of a redox reaction using the gamma rule.

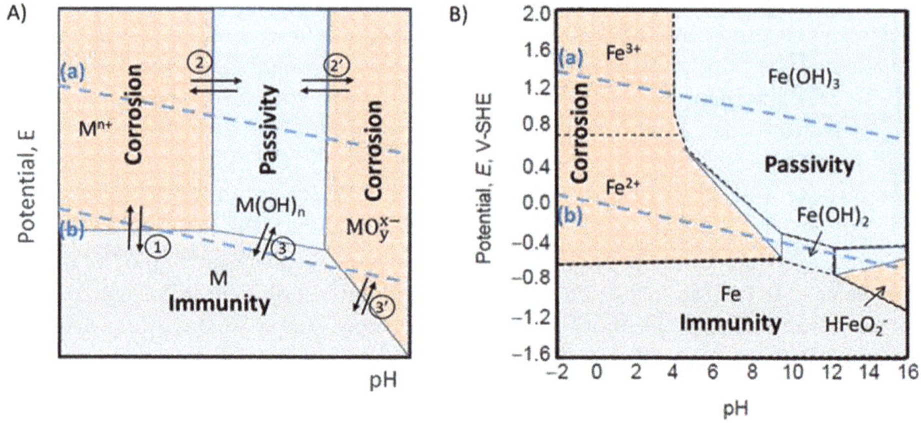

FIG. 1.3.4 – (A) general Pourbaix diagram: horizontal lines (1) depict pure redox reactions, not pH-dependent, vertical lines (2) and (2′) represent pure acid–base reactions, sloped lines (3) and (3′) give both reactions (redox and acid–base) and the broken lines, (a) $O_2 + 4H^+ + 4e^- \rightarrow 2H_2O$ and (b) $2H^+ + 2e^- \rightarrow H_2$, indicate the domain of thermodynamic stability of water. For potentials and pH conditions above line (a), water is thermodynamic unstable with respect to the evolution of oxygen and for any value of potential and pH below line (b), water is thermodynamically unstable with respect to generation of hydrogen gas. (B) Pourbaix diagram for iron at 25 °C showing reaction/corrosion products.

However, if thermodynamics can forecast the nature and the composition of the corrosion products (soluble cation, cation constitutive of a solid), it does not provide any information about kinetics reactions or corrosion rate.

1.3.2.1.3 Kinetic Approach

In the case of a uniform corrosion process, the corrosion rate can be expressed as:

- The mass of metal changed into corrosion products per unit area of surface per unit of time (weight loss in mg/dm^2.day).
- Or an increase in corrosion depth per unit of time (penetration rate in mm/year or μm/year).

Immersion tests are the most commonly used corrosion tests for determining corrosion rates. It has to be noted that when the notion of "corrosion rate" is considered here, it implies that, as long as all parameters acting on the corrosion rate remain constant, the corrosion rate is constant corresponding to a linear corrosion kinetics (in terms of weight loss or corrosion depth). In that case, the corrosion process will be uniform and generalized corrosion occurs. The case of localized corrosion will not be treated in this chapter on kinetics, as it is more complex and specific and cannot be analyzed in a general way.

The weight loss of a metal exposed in an aggressive medium is measured for a given time period. Assuming linear kinetics, and thus a uniform corrosion process, the corrosion rate is calculated from 1.3.2:

$$v_{\text{cor}} = \frac{\Delta m}{\rho S \Delta t} \tag{1.3.2}$$

where Δm is the weight loss variation in g, ρ is the density in g/cm^3, S is the area of exposure in cm^2 and Δt is the time of exposure in h.

Although immersion tests are easy to perform and are representative of the conditions commonly found in service, it is not possible to get direct information from these tests on the corrosion mechanism and the exposure times can also be long to reach significant corrosion rates depending on the environment. Electrochemical tests are presented as a good alternative given that they allow in a short time determining the corrosion rate as well as they allow acquiring information on the corrosion mechanism.

The corrosion rate can be calculated using electrochemical methods considering the mixed-potential theory and Faraday's law [14, 15]. Mixed-potential theory indicates that electrochemical corrosion cell works at a mixed potential between anodic and cathodic reactions. This potential, called the corrosion potential (E_{cor}), is acquired spontaneously when the metal is in contact with the electrolyte. At E_{cor}, the net current density (i), the algebraic sum of the anodic (i_a) and cathodic (i_c) partial current densities, is equal to zero. The value of either the anodic or cathodic current densities at E_{cor} is called the corrosion current density, i_{cor}. This means that 1.3.3:

$$|i_{\text{cor}}| = |i_a| = |i_c| \tag{1.3.3}$$

The metal dissolution rate (v_{cor}) at corrosion potential (E_{cor}) is related to the corrosion current using Faraday's law 1.3.4:

$$v_{\text{cor}} = \frac{i_{\text{cor}}}{nF} \tag{1.3.4}$$

where n indicates the number of electrons exchanged in the dissolution reaction and F is the Faraday constant, $F = 96\ 485$ C/mol.

If the anodic and the cathodic reactions are controlled by the charge transfer, the Butler-Volmer equation describes the current–potential relationship of a corrosion system according to 1.3.5:

$$i = i_{\text{cor}} \left\{ \exp\left(\frac{2.3(E - E_{\text{cor}})}{b_a} \right) - \exp\left(\frac{2.3(E - E_{\text{cor}})}{b_c} \right) \right\} \tag{1.3.5}$$

where i is the measured current from the cell, E is the electrode potential, b_a is the anodic Tafel constant and b_c is the cathodic Tafel constant.

Unfortunately, i_{cor} cannot be measured directly. However, it can be determined by a variety of electrochemical techniques such as current–voltage techniques and the electrochemical impedance spectroscopy (EIS). These techniques are based on measuring the current (or potential) response, when a potential (or current) is applied. Figure 1.3.5 shows two examples of polarization curves. The log(i)-E diagrams or Evans diagrams depict the anodic and cathodic current densities (figure 1.3.5a). i_{cor} can be obtained from Evans diagrams by extrapolating the anodic or the cathodic partial current density at the corrosion potential.

The anodic polarization curve for a passive metal is shown in figure 1.3.5b. Three behaviors are represented as a function of the potential and of the current density:

- Active: the material dissolves according to the process $M \rightarrow M^{n+} + ne^{-}$. The products of this reaction are soluble in the solution (corresponding to the corrosion domain of the Pourbaix diagram). Active corrosion is characterized by a sharp increase in current density as a function of potential. Consequently, high corrosion rates are obtained in this domain.
- Passive: the current density suddenly decreases to a very low value, and then remains stable with the potential due to the formation a protective oxide layer at the surface of the metal (corresponding to the passive domain of the Pourbaix diagram). This film is very thin, in the range of a few nanometers. In the case of stainless steels, this layer is mainly composed of Cr_2O_3.
- Transpassive: at more positive potentials, the passive film breaks down and the current density increases rapidly again. The chemical species such as Cr^{+3} or Fe^{+2} turning into higher valence and soluble products are formed (corresponding to the corrosion domain of the Pourbaix diagram). Higher rates of dissolution are achieved in this region.

As shown in figure 1.3.5b, passive metals show an active, passive or transpassive behaviour depending on the oxidizing power of electrolyte. It should be emphasized that factors such as temperature and alloying elements can modify the anodic behavior. The corrosion resistance of many alloys is increased by adding elements that enhance the passive region due to the formation of stable oxide film on surface of metal. This is the case for chromium. Samples with higher Cr content (>11%)

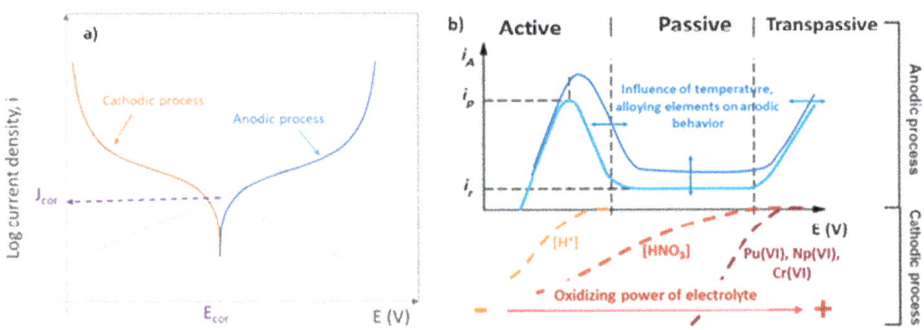

FIG. 1.3.5 – Schematic diagrams of polarization curves: (a) log(i)-E diagram of a mixed electrode ($E < E_{cor}$: cathodic process, $E > E_{cor}$: anodic process) showing the intersection of the partial current densities (E_{cor}, i_{cor}) and (b) anodic current–potential curve for a passive metal (*e.g.* stainless steel).

show a larger passive domain and a lower current density in the active domain (lower corrosion peak). This behavior is due to the formation of mixed surface oxides (Fe, $Cr)_2O_3$ and $Cr(OH)_3nH_2O$.

1.3.2.2 High Temperature Corrosion

As in the case of aqueous corrosion, high temperature corrosion is a physicochemical interaction between a metal and an environment leading to a modification of the properties of the metal and often to a functional deterioration of the metal itself, its environment or even of the global system.

Unlike aqueous corrosion, high temperature corrosion occurs in the absence of an aqueous electrolyte, at temperatures at which diffusion in solids becomes important. Even if there is no strict temperature limit for high temperature corrosion, we will consider phenomena occurring above approximately 500 °C [16].

Oxidation is the most common form of high temperature corrosion – almost all metals and alloys will oxidize above a certain temperature, leading to scaling, loss of material and changes in physical properties. However, gaseous attack is not limited to oxygen, with sulphur-bearing gases, carbon oxides, nitrous oxides, hydrogen, and halogens...

Furthermore, high temperature corrosion is not restricted to the gaseous phase – solid ash and salt deposits contribute to the corrosive effect, with associated erosion and removal of scale. In the liquid phase, molten metals, molten salts and molten glasses cause specific and complex corrosion phenomena.

Industrially, high temperature corrosion is a significant issue. Any component exposed to a high temperature in a non-inert environment is potentially at risk. This includes aerospace, power, metal processing, automotive, waste incineration and chemical processing industries among others.

High temperature corrosion corresponds to a heterogeneous chemical reaction between two different phases:

$$\text{Solid}_1 + \text{Gas/Liquid} = \text{Solid}_2 \tag{1.3.6}$$

In most cases, an oxidation–reduction process occurs; the metal is oxidized and the environment is reduced.

In the following, we will consider that oxidation of the metal leads to formation of a solid oxide only, even if it has to be kept in mind that many other processes occur in high temperature corrosion.

We will thus consider the following reaction:

$$\alpha\text{Metal} + \beta\text{Oxygen} = \gamma\text{Metal oxide}(s) \tag{1.3.7}$$

1.3.2.2.1 Thermodynamic Approach

As said previously, except gold, which is stable in air, all usual metals exist on earth combined with oxygen, carbon, phosphorus, sulphur or nitrogen. Metals themselves are not stable and they will therefore react to form stable compounds (oxides, sulfides, chlorides, phosphates, nitrates...) leading to the corrosion of the metal.

Knowing the stable form of the metal in a given environment (composition, temperature, and pressure) is a first key to the knowledge of a corrosion process. This knowledge of course does not give any insight into the kinetics of the process, but it is a first step to understand a corrosion mechanism.

Thermodynamics allows determining which corrosion products should form in given conditions, supposed to be at equilibrium.

In the case of an oxidation process leading to the formation of an oxide 1.3.7, the spontaneous tendency of a metal to form an oxide will be evaluated with the Gibbs energy difference (ΔG) of this oxidation reaction.

If this variation is positive, then the metal will not be attacked by oxygen, if it is negative the metal will be oxidized. The more negative will this variation be, the more stable will the oxide be. The comparison of these values for different metals gives a first idea of what chemical species could form.

When reaction 1.3.7 is at equilibrium, $\Delta G = 0$ with:

$$\Delta G = \Delta G^0 + RT\ln \frac{a^{\gamma}_{\text{Metal oxide}}}{a^{\alpha}_{\text{Metal}} P^{\beta}_{O_2}} \tag{1.3.8}$$

where ΔG^0 is the standard Gibbs energy of reaction, a_i the activities of the metal oxide and of the metal, α, β, γ the stoichiometric coefficients of reaction 1.3.7.

For a pure metal and a pure oxide, $a_{\text{Metal oxide}} = 1$ and $a_{\text{Metal}} = 1$.

Then equation 1.3.8 gives:

$$\Delta G^0 = RT\ln P_{O_2} \tag{1.3.9}$$

where R: gas constant (J.mol^{-1}.K^{-1}), T: temperature (K), P_{O_2}: oxygen partial pressure (bar).

So, for a given metal/oxide system, ΔG^0 varies linearly with temperature. Stability diagrams of oxides can be established as a function of temperature and oxygen partial pressure (diagram constructed for one mole of O_2 in reaction 1.3.7). The most well-known diagrams are those of Ellingham (figure 1.3.6) [17]. Such diagrams can help to visualize the relative stability of metals and their oxidized products. The values of ΔG^0 on an Ellingham diagram are expressed as kJ per mole O_2 to normalize the scale and be able to compare the stability of these oxides directly, *i.e.* the lower the position of the line on the diagram the more stable is the oxide.

These diagrams allow knowing for a given temperature, the oxygen partial pressure for the metal/oxide equilibrium. These diagrams are thus a tool to know in given conditions (T, P_{O_2}) if a metal is going to be oxidized or not.

Similar diagrams can be established for any other kind of species (carbides, nitrides, sulphides...).

Of course, even if these diagrams give a view of how the system metal/oxide should evolve, they do not give any idea of the kinetics of this evolution. This information is indeed essential in the study of a corrosion process and requires a kinetic analysis of the oxidation process.

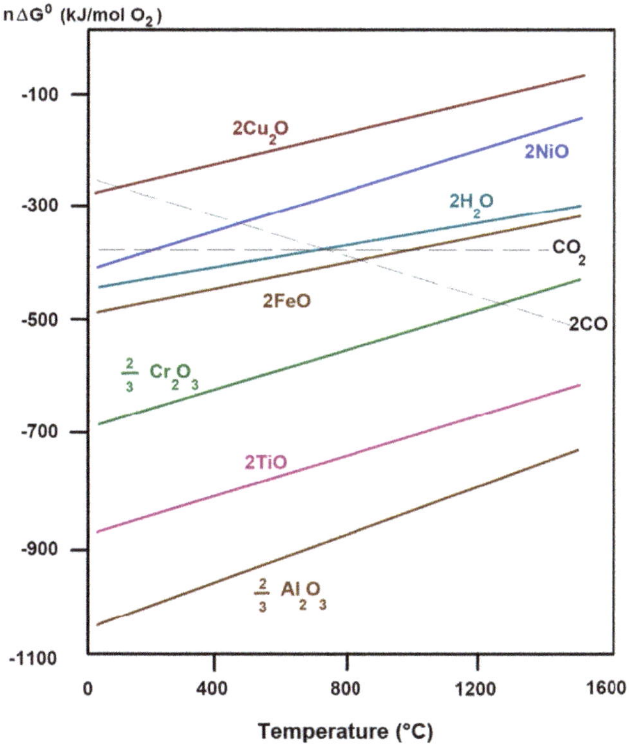

FIG. 1.3.6 – Ellingham diagram for various metallic oxides.

1.3.2.2.2 Kinetic Approach

In order to get the information on the oxidation kinetics, it is necessary to follow the growth of the oxide layer. The knowledge of the oxidation kinetics combined with the thermodynamics data available and with all the information obtained from the experimental data (corrosion facies, chemical/microstructural evolution of the alloy and environment...) will allow the establishment of the oxidation mechanism.

The determination of the oxidation mechanism is essential:

- to be able to find solutions to slow the oxidation process if it is too fast,
- to model the corrosion process in order to predict the corrosion rate for the whole lifetime of a component,
- to determine the time necessary before inspection, cleaning or change of a component.

To determine the oxidation kinetics, the weight variation of the samples exposed to the oxidizing atmosphere has to be followed with time; this can be performed with classical oxidation tests performed in oven with controlled atmosphere or with oxidations test performed in thermobalance.

Oxidation kinetics laws are of the type [16]:

$$y^n = kt \qquad (1.3.10)$$

where y is the thickness of the oxide layer formed for a given time, t, k the kinetics constant. The reaction kinetics depends on temperature according to the Arrhenius law:

$$k = k^0 \exp\left(-\frac{Q}{RT}\right) \qquad (1.3.11)$$

where Q: activation energy, R: gas constant, T: temperature (K).

In most of the cases, n is equal to 1 or 2 or 3 (figure 1.3.7).

In practice, the most frequent law is, for thick layers, the parabolic law ($n = 2$ in equation 1.3.10). In that case and considering the Wagner theory, diffusion in the oxide layer is the controlling step of the oxidation process. With the increase in the oxidation duration, the oxidation rate decreases as long as the oxide layer remains dense, adherent and compact.

The linear law occurs for metals with porous or cracked oxide films with no efficient protection of the metal by the oxide layer. There is always a contact between the metal and the surrounding environment.

For the logarithmic law, the oxidation process is controlled by an interfacial reaction (sorption, metal/oxide interfacial reaction).

Logarithmic laws are characteristic of the oxidation of numerous metals at low temperatures (below 673 K). The oxidation rate initially very high (corresponding to the formation of a very thin oxide layer, approximately few tens of nanometres), becomes very low and even equal to zero.

Another parameter influencing the protective ability of the oxide layer are the mechanical constraints inside the oxide layer. The Pilling-Bedworth ratio, P, which

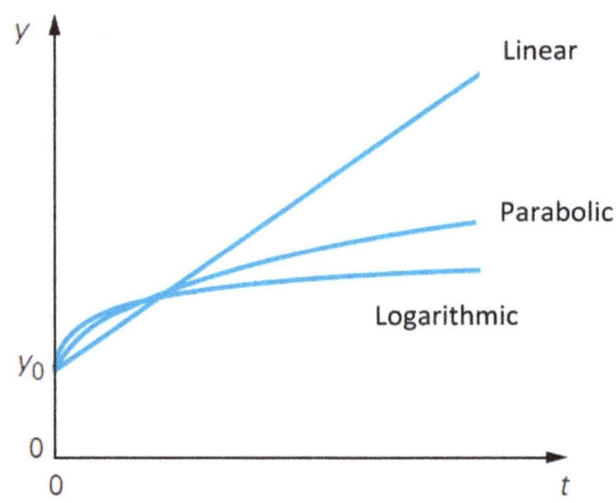

FIG. 1.3.7 – Graphic representation of the oxidation laws.

expresses the expansion coefficient of the metal oxide is defined as the ratio of the volume of the metal oxide, which is produced by the reaction of metal and oxygen, to the consumed metal volume [16]:

$$P = \frac{M_m(\text{oxide}) x \rho_{\text{metal}}}{M_{\text{at}}(\text{metal}) \rho_{\text{oxide}}} \tag{1.3.12}$$

where M_m(oxide): molecular weight of the metal oxide and ρ_{oxide}: density of the metal oxide and, M_{at}(metal): atomic weight of the metal and ρ_{metal}: density of the metal.

- When P is less than 1, the metal oxide has a lower volume than the consumed metal. The oxide cannot cover the whole metal surface. It is therefore not protective and the kinetic law will be linear. This is the case of alkaline and alkaline-earth metals, Na, Ca, Li, Ba and of metals forming liquid or volatile oxides such as Mo, V, W...
- When P is higher than 1, the oxide covers completely the surface and acts as a barrier. However, for excessively high values of P (>2.4), large compressive stresses are likely to exist in metal oxide, leading to buckling and spalling. This can occur in the case of the oxidation of Nb or Ta.
- For $1 < P < 2.4$; the oxide layer is compact, dense and adhesive. It is protective. In that case, the growth of the oxide is often controlled by the diffusion of species in the layer. The oxidation kinetics is parabolic resulting from Wagner's theory. It is generally observed for Fe, Ni, Cu, and Co. Other phenomena can also be observed with successive phases of oxidation and brutal weight losses leading to a global linear oxidation law. This phenomenon is called "breakaway" and is observed for Zr or other metallic materials.

Considering parameters influencing corrosion, the temperature is one of the major parameters as the kinetic constants follow an Arrhenius law. Most of the time, the higher the temperature is, the faster the oxidation process will be. However, in some cases an opposite behavior can be observed, for instance when a temperature increase leads to the formation of a more protective oxide. Another parameter influencing the corrosion process is the oxygen partial pressure. However, the variation of the corrosion kinetics as a function of the oxygen partial pressure can be complex to understand and analyse.

1.3.2.3 Characterisation Methods

In order to determine the corrosion kinetics and the corrosion mechanism, experiments have to be performed in the laboratory. Once the kinetics has been obtained by various methods, for instance, thermogravimetry for high temperature gaseous environment and Tafel method for aqueous medium, post mortem analyses of the corroded alloy are realized.

Post mortem methods allow to get information on the morphology of corrosion and on the nature and the quantity of the corrosion products after tests. These methods cover both imaging techniques, and analytical methods.

For very thin oxide layers (<10 nm), spectroscopic methods such as X-ray photoelectron spectroscopy (XPS), Auger Electron Spectroscopy (AES) are well adapted. They can allow the acquisition of both qualitative and quantitative data concerning the composition of the layer and the valence of the elements.

For thicker oxide layers (between 100 and 1000 nm) other techniques such as Secondary-ion mass spectrometry (SIMS) or Glow Discharge Mass Spectrometry (GDMS) are well adapted, they allow the acquisition of both qualitative and quantitative data on the composition of the layer. These techniques can also be used also for thinner and thicker layers.

For layers thicker than 1 μm, more classical techniques of imaging, covering optical microscopy for the lower resolution (few μm), Scanning Electron Microscopy (SEM) (hundreds of nm) and Transmission Electron Microscopy (TEM) (few nm). These imaging techniques can be coupled to analytical techniques like Energy Dispersive Spectroscopy (EDS) for SEM and TEM or Wavelength-Dispersive X-Ray Spectroscopy (WDS) to get qualitative and quantitative information of the composition of the oxide layer.

X ray diffraction is also used to identify the phases formed as well as Raman spectroscopy.

Other more specific experiments can be used to study the oxidation mechanisms involving tracers such as oxygen-18 to identify the growth interfaces in a complex oxide layer.

1.3.3 Influence of Metallurgical Parameters on Corrosion

1.3.3.1 Effect of Process Parameters on the Microstructure

Because of the rapid heating and fast cooling rates and directional solidification, metals produced by additive manufacturing have specific structures, microstructures, and architectures that differ from cast and wrought materials [2, 10, 11]. Fine grains, anisotropic microstructures with elongated grains, non-equilibrium microstructures, dislocation density, residual stresses and metallurgical defects, such as porosity due to unmelted powders and gas entrapment, are among the common microstructural features of metals elaborated by additive manufacturing (AM). The surfaces of metallic components fabricated by AM exhibit characteristic features, which can include weld tracks, unmelted powder particles or ejected molten droplets.

All these specificities of microstructure and metallurgical state can have an impact on the corrosion behavior of materials. It is thus essential to understand the link between the process parameters, the microstructures obtained and the corrosion behavior [2–10].

The main process parameters are the laser power, the scanning speed, the scan spacing which is the separation between two consecutive laser beams (also called hatch distance) and the layer thickness [11].

In this paragraph, a synthesis will be made of the known effects of process parameters on the microstructures produced. Mainly the case of steels will be considered. But it has to be noted that depending on the composition of the materials processed (steels, Ti alloys, Ni alloys...), the effects of the process parameters can be different.

1.3.3.1.1 Surface Roughness

Concerning the surface roughness obtained after processing, it is generally high for components elaborated by SLM (from 5 to 50 μm) or Electron Beam Melting (EBM) (from 10 to 60 μm) and even higher for components elaborated by DLD (from 10 to 200 μm). Surface roughness is attributed to two phenomena, the first one is evaporation during powder melting. Gas expansion causes instability and irregularity of the melt pool, which leads to an increase in the surface roughness and in porosity. In that case, thinning the powder layers could help decrease surface roughness. The second cause of surface roughness is improper powder melting and balling, balling is the formation of metallic droplets in opposition to a desired uniform liquid metal surface. In both cases, a link with energy density has been shown [2, 11, 12].

The laser energy density E (J.m^{-3}) represents the applied energy per unit of volume during a powder bed fusion process and can be expressed according to:

$$E = \frac{P}{Vht} \tag{1.3.13}$$

where P: laser power (W), V: scan speed (m.s^{-1}), h: hatch spacing/scan spacing (m), t: layer thickness (m).

When the energy density delivered is insufficient to completely melt the powder particles, solid particles remain at the surface leading to a high surface roughness. In that case, increasing the energy density should allow decreasing the surface roughness. However, a too high heat input can also be detrimental to the surface roughness possibly due to thermal distortions and excessive melting. However, the clear link between the process parameters and the surface roughness still needs to be investigated.

What has to be noted concerning surface roughness is that post processing of the fabricated components will be an essential step of the global elaboration process. Optimisation of this post processing step also requires also intensive work.

1.3.3.1.2 Porosity

There are two kinds of pores observed in SLM elaborated materials: one type is associated to unmelted powder and the other type is associated to gas trapped inside the powder occurring during gas atomisation. Process parameters influencing the porosity are laser power, scanning speed, scanning direction.

The link between specimen porosity in 316L, laser power and laser scanning speed was studied by different authors [18–20]. It was observed that porosities were more numerous with increasing the scanning speed [18, 19].

The effect of the energy density was also studied. At low energy density, a high porosity (around 8.8%) was observed attributed to the lack of melting of the metal powder and lack of sintering between the layers. At higher energy density, the porosity was still high (around 6.5%), it was suggested that the larger melt pools

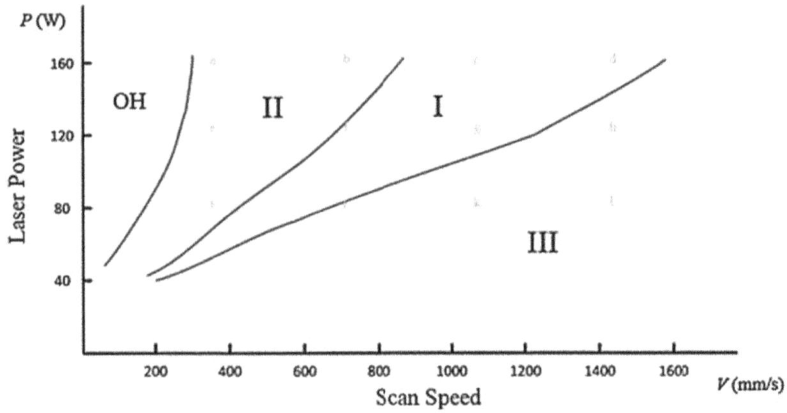

FIG. 1.3.8 – Process window for SLM Raymor Ti-6Al-4 V powder [22]. Copyright Elsevier 2020. OH: overheating zone – I: fully dense – II: over melting – III: incomplete melting.

formed for this high energy density could be susceptible to solidification micro shrinkage porosity. Moreover, the formation of gas voids could also result in the formation of spherical pores.

In fact, porosity can form due to insufficient energy input or due to the use of excessive energy [21]. In the case of Ti-6Al-4 V, a representation of the process window for a given layer thickness and hatch spacing was proposed (figure 1.3.8) [22].

In zone I, the sample is nearly free from porosities. In zones II and III, there is detectable porosity. In zone II ((overmelting zone), the porosities are due to excessive energy while those of zone III are due to insufficient energy. Finally, an overheating zone is observed for very low scan speed and high laser power [22]. In this zone, Ti-6Al-4 V samples cannot be built due to excess heating resulting in accumulated strain energy and serious deformation. Similar process windows can be established for various kinds of materials.

Again, the thorough understanding of the link between the laser parameters and the specimen porosity has still to be investigated. This understanding is required to be able to control this porosity towards a given application. In that case, the use of post treatments such as heat treatments including hot isostatic pressing should allow reducing porosities.

1.3.3.1.3 Residual Stress

The non-uniform heat distribution associated with the fast cooling rate results in high level of residual stress after the elaboration. It was observed that in the case of 316L elaborated by SLM, the residual stress was mainly compressive in the bottom part of the build and tensile in the upper part but with addition of new layers on the previously scanned layers, the tensile stress converts into compressive, and the compressive stress is enhanced due to subsequent thermal cycling [23]. Moreover, residual stress parallel to the scanning direction is much larger than that perpendicular to the scanning direction. Lower energy input and shorter line length induce smaller residual stress in SLM parts. Therefore, optimization of the scanning strategy can allow reducing residual stress efficiently. Finally, the peak values of residual stress occur at the onset of scanning track, which indicates that both ends of the part are easy to warp and crack. Therefore, stronger support structures are needed at both ends of SLM part, to prevent warps and cracks, during the SLM process [23]. Residual stress can also be lowered thanks to post process treatment such as stress relief heat treatment performed after component production.

1.3.3.1.4 Inhomogeneous Microstructure and Anisotropy

The rapid heating and cooling rates during elaboration result in the formation of non-equilibrium microstructures. In the case of 316L, MnS inclusions classically observed were replaced by inclusions containing Si and O and also Mn, Al and N preferentially formed at high temperatures and retained in the microstructure due to the rapid cooling [26]. A high laser power and high scanning speed were found to be responsible for laser cladded 420 martensitic stainless steel of a higher proportion of non-equilibrium phases such as retained austenite, ferrite, metastable carbides... compared to specimens fabricated with lower laser power and scanning speed [24].

The anisotropy is also caused by the thermal history of the fabrication process. Different solidification rates are obtained in different directions and heat conductivity is faster in the building direction than in the two other spatial directions. This usually results in the case of stainless steels, Co alloys and Ti alloys in a microstructural anisotropy with elongated grains in the building direction and equiaxial grains in the other directions [2, 25]. In the case of Ti alloys, it was determined that the formation of fully columnar or fully equiaxed microstructure depended on the cooling rate of the liquid metal pool and on the thermal gradient of this metal pool [2]. The liquid metal pool temperature and solidification rate depend on the laser power and on the other process parameters.

A fully columnar structure is observed for low values of cooling rate and high value of thermal gradient whereas a fully equiaxed structure is observed in the opposite case.

Therefore, being able to know and to control the thermal history of the liquid metal pool is essential to control the final microstructure. This research work performed on the link between process parameters and elaborated microstructures is a huge work and a lot still has to be performed. It is also strongly dependent on the nature of the materials elaborated and on the technologies used to fabricate the components. *In situ* instrumentation of the process is required to get more precise information and robust modeling will also be a key issue to progress on the understanding between process parameters and elaborated materials.

Finally, work on post process treatments is also needed to master the whole elaboration chain.

1.3.3.2 Influence of the Elaborated Microstructure on the Corrosion Behavior

Despite the influence of the unique and complex microstructure formed after AM, the corrosion behavior of these SLM/DLD elaborated materials has not been much studied, analyzed and understood yet. Indeed, studies performed reveal that the corrosion behavior can be affected by AM-prepared alloys, in a benefic or detrimental way depending on the conditions. Consequently, AM can play a significant role in corrosion performance and thus the control of its parameters will also be essential to be able to control and optimize the corrosion behavior.

For example, in the case of localized corrosion, it was shown in section 1.3.2.1 that it can occur due to the presence of different phases in the alloy (precipitates and matrix for example). The chemistry of these phases is also of major importance in the corrosion process. As presented in section 1.3.3.1.4, the SLM/DLD processes can induce different precipitates chemistry in comparison to that obtained in classical processes. Moreover, localized corrosion is also influenced by the surface state and the alloy microstructure. The influence of some of these parameters on the corrosion behavior is illustrated below.

The effect of the nature of inclusions on pitting corrosion of SLM 316L specimens in 0.6 M NaCl at 25 °C was studied by Chao *et al.* [26]. The results revealed that pitting potential (E_{pit}) values were notably higher in the printed SLM 316L than in wrought alloy (figure 1.3.9). This information is very important since the E_{pit} is a criterion for determining the pitting corrosion resistance of a stainless steel. When the corrosion potential (E_{cor}) is higher than E_{pit}, a stable pitting will appear and when E_{cor} is lower than E_{pit}, the stainless steels will show a passive behavior. The most resistant steels will be those that have a higher pitting potential, that is, the steels that show a larger passive domain.

As presented in table 1.3.1, MnS inclusions are the main cause that induces pitting corrosion in wrought alloys. The formation of Mn–Si oxides in the SLM 316L specimens, which replace the MnS inclusions in wrought alloys, could enhance resistance to pit initiation. However, this behavior seems to change as a function of porosity. Sun *et al.* showed that the resistance to pit initiation diminished for

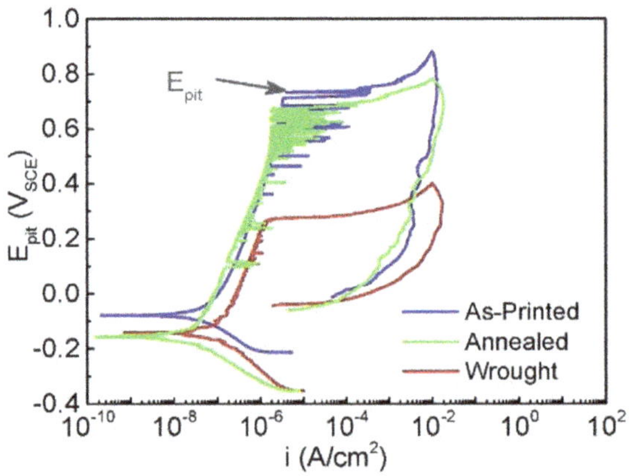

FIG. 1.3.9 – Cyclic potentiodynamic polarization curves for the as-printed (SLM), SLM annealed at 1100 °C for 5 min and wrought 316 stainless steels in 0.6 M NaCl [26]. Copyright Elsevier 2020.

porosities above 1%. The accumulation of chloride ions inside the pores could increase the probability of oxide layer breakdown [27]. Instead, Sander *et al.*, suggested that lower porosity (<0.4%) only affected the pit repassivation of SLM 316L [20]. However, how porosity detrimentally influences the repassivation characteristics of SLM 316L is still not well understood. Porosity was also identified as the main cause of the decreased corrosion resistance of LPBF 17-4PH in the same conditions [28]. The authors remarked the importance of the geometric factor on corrosion behavior. Samples with larger pores exhibited a more reduced passive region and higher corrosion rate compared to the smaller pores. A recent investigation provided evidence that irregular-shaped pores (lack of melting porosity), formed as a result of incomplete powder melting when inappropriate SLM process parameters are selected, have a detrimental effect on both pit initiation and propagation [29]. These pores act as occluded regions limiting the access of oxygen and acidifying the solution, therefore more severe environments for corrosion attack.

Electrochemical techniques, SEM and scanning kelvin probe force were used to understand the local corrosion mechanism in the interior of the melt pool border of AM AlSi10Mg specimens prepared by SLM in 0.1 M NaCl at room temperature [30]. The results showed the formation of galvanic cell due to the electropotential difference between the Al phase (less noble) and SiMg segregation (more noble). Consequently, preferential dissolution of Al phase occurred. A much higher driving force for an accelerated attack was present in the melt pool border where larger microstructures are generally found. Taking into account that galvanic corrosion depends on the ratio of the areas of the two coupled materials, this could explain why the melt pool borders are more susceptible to corrode than the other regions of the surface where the second phase is more refined. In that case, the same kind of behavior could be found for the cellular structures in SLM/DLD 316L, where the boundaries of the cells are rich in Cr and Mo [4]. The influence of the cellular structures and the associated solute segregation on the corrosion properties has still to be well clarified for any alloy systems.

The presence of multiple phases was shown to have a detrimental impact in the case of DLD FeCrNiMoTi in oxidizing medium at boiling temperature. EDX analyses revealed the preferential corrosion of Ni_3Ti phase compared to σ-phase (figure 1.3.10). The formation of phases of different composition could increase the galvanic corrosion susceptibility.

Other parameters as surface roughness, grain structures and residual stress could influence the corrosion behavior. The authors suggested that the localized corrosion increased as a function of the surface roughness due to stagnation of the electrolyte in the concave areas [31, 32]. Leon *et al.* observed that SLM AlSi10Mg unpolished samples in 3.5% NaCl solution showed a more intensive localized corrosion attack at the surface than the polished samples [31]. This behavior can be related to the difficulty of reforming a passive layer on rough surfaces. It was also observed that the surface roughness led to an acceleration of the corrosion fatigue failure. The large amount of cavities and other surface defects induce a large number of crack initiation sites and then, crack propagation [31]. In the case of general corrosion or high temperature corrosion, the surface roughness must be taken into account to estimate the surface area used to calculate the corrosion rate. Hence, an under-estimation of

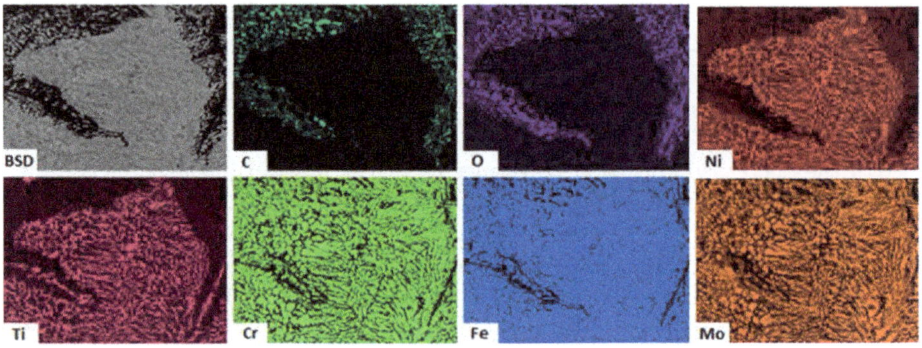

FIG. 1.3.10 – Scanning electron microscopy (SEM) and energy dispersive spectroscopy (EDS) of DLD FeCrNiMoTi in oxidizing solution (5 M nitric acid, 4 mM V(V)) at boiling temperature.

real surface area and consequently, an over-estimation of corrosion rate could take place [33].

It is important to highlight the effect of the build orientation on the corrosion process. Dai *et al.* observed that XY-plane (build plane) exhibited better corrosion resistance than XZ-plane (build direction) of SLM Ti–6Al–4 V in 1 M HCl solution. This behavior was attributed to the presence of different amounts de α′-martensite and β-Ti phase in microstructure [34]. The grain structure could also be different as a function of build orientation, consequently affect the corrosion susceptibility. Elongated grain structure or equiaxed microstructure could have a detrimental or beneficial effect depending on the aggressive environment. For example, the SCC growth rates increased in the case of SLM/DLD 316L with the elongated grain structure [35]. However, fine grain size and equiaxed microstructures were reported to have a positive effect on corrosion fatigue [36]. In fact, microstructural anisotropy can be eliminated after high-temperature recrystallization annealing. The results revealed that the annealed SLM 316L showed a slower crack growth rate in high temperature water compared to stress-relieved counterpart [36].

Moreover, Laleh *et al.* showed that the precipitation behavior during the sensitization heat-treatment (typical welding conditions) depends on the grain size and shape, as well as the grain boundary character of the SLM-produced specimen [37]. Fine precipitates enriched in Mo and Si were detected as well as a high frequency of twin boundaries and low-angle grain boundaries along with fine grains, which led to the avoidance of localized Cr depletion, leading to a lower intergranular corrosion susceptibility. This fine and unique distribution of alloying elements could influence the structure and the stability of the native oxide layer, and consequently the corrosion behavior of these materials.

For conventionally manufactured materials, externally introduced compressive stresses are beneficial in terms of corrosion resistance [38]. Furthermore, it was observed that stainless steel specimens under compressive stress exhibited lower passive current densities due to the formation of more compact passive film compared to specimens under tensile stress. In the case of SLM samples, the compressive residual stress also improved pitting resistance. The E_{pit} increased and the passive current density decreased with increasing compressive residual stresses in 316L specimens [26, 39]. The compressive residual stresses decreased the kinetics of oxide growth and the defect concentration, such as oxygen vacancies, in the passive film. The presence of a lower concentration of oxygen vacancies prevents Cl^- adsorption/ingress into passive film. However, these samples showed a poorer repassivation behavior due to the sluggish kinetics of the passive film growth [39].

All these studies demonstrate the essential role of the passive film on the corrosion behavior. Therefore, a more detailed analysis and characterization of the link between the native oxide film and the corrosion mechanism could contribute to a better understanding of the effect of microstructure on the corrosion behavior of AM-prepared alloys.

Further investigations should be performed to understand the influence of the microstructure on the corrosion mechanisms. Important studies are needed for the development and optimization of new materials. The global link between the process parameters, the generated microstructure and the final corrosion behavior should be studied in depth to be understood and controlled towards the final application of the material.

1.3.4 Advanced Manufacturing Processes as Solutions to Corrosion Concerns

Even if these new processes generate specific microstructures that have to be characterized, understood and controlled towards the final application, they also offer opportunities. First, the control of the feedstock powder is essential: the particle morphology, composition, structure have to be controlled to have a satisfying produced material and this also includes the way the powder is elaborated. Mastering the level of impurities in the powder (S, P) could thus allow reducing the corrosion behavior in certain conditions.

The manufacturing process itself by the understanding of the link between the process parameters and the final microstructure can become a tool to optimize the corrosion resistance for example by controlling the nature of the inclusions formed during the elaboration process [24, 26].

Moreover, these fabrication processes by allowing the direct construction of some components (within the size limitation of the actual equipments) can allow to avoid welds or assemblies which are generally parts that are more susceptible to corrosion than massive material. Conception of components has to be thought in a completely different way taking into account the new opportunities offered by these manufacturing processes.

Materials themselves can be imagined and designed in a completely different way and these manufacturing techniques should accelerate both the discovery of new materials as well as their ultimate qualification.

For example, in the field of innovative materials, high entropy alloys (HEA) are defined as alloys that have at least five elements in equimolar/equiatomic compositions and the concentration of each element is between 5 at% and 35 at%. The high configurational entropy of these alloys stabilizes simple solid solutions such as body-centered cubic, face-centered cubic and hexagonal close-packed [2, 40]. These alloys can have specific interesting properties, such as high mechanical resistance, good ductility, and also higher corrosion resistance [41–43]. These materials can be elaborated easily with AM techniques either as bulk materials or as claddings. Indeed, DLD seems to be particularly attractive to elaborate such materials as it allows to avoid pre-alloy powders by the use of elemental powders in different containers and the possibility to elaborate *in situ* the final homogeneous alloy with optimized process parameters [2]. Development of HEA claddings is also interesting using DLD, thick coatings (above 1 mm) can be produced using this technique and using different powder containers. The powder is then fed in a gas stream or as a wire to be melted by the laser beam with a thin layer of the substrate. DLD is generally used to produce coatings to protect materials exposed to severe environments and increase their resistance to wear, corrosion, abrasion... Usual materials deposited by DLD are steels, Ni-, Co-, Ti-, Al- based materials. In the last few years, developments on HEA alloys have increased to improve wear resistance as well as corrosion resistance [44–46].

As it was said, DLD is also largely used for the functionalization of materials by adding protective layers. Indeed, this technology allows the deposition of a thick protective layer on the surface of a massive material which can add a corrosion protection functionality to this material. This is for example the case for the development of cobalt free hardfacing materials for wear resistance in sodium fast reactors [47]. In the reactor, some parts can be submitted to sliding contact between each other. On these parts, the contact areas usually need a hardfacing coating. The standard hardfacing alloy is a cobalt-base alloy (as, for example Stellite®6). Unfortunately, in the primary coolant circuit and on wear conditions, cobalt can be released. Under neutron flux, the 59Co, stable, can be transmuted into 60Co by neutrons irradiation and, therefore, can contaminate the primary circuit and will be an issue for dismantling. Therefore, it is desired to replace this cobalt based hardfacing alloy by a cobalt-free one. Laser cladding was one of the options selected to deposit this coating. Several materials were selected for the application, mainly Ni based alloys. Process parameters were optimized to get a sound coating, with minimized porosities, no cracks, low dilution and low deformation [47]. Then, tribological tests were performed under argon. However, a deeper investigation is in progress in order to verify the proposed hypothesis on wear mechanisms. Tribological tests in sodium are planned.

Other interesting developments regarding corrosion concern the use of DLD for the development of functionally graded materials (FGM) [48]. FGM is a class of material of which the composition and microstructure change progressively in a predetermined way from one side to the other, resulting in a corresponding variation of properties. It is thus possible to manufacture a material with a progressive transition from a ferritic steel to an austenitic steel avoiding welding of two massive materials. Other examples concern the elaboration of gradient materials with a progressive composition transition from a nickel base alloy to an austenitic steel allowing the elaboration of a material which could combine high strength, excellent toughness, and high corrosion resistance at high temperatures, and could be used to resist oxidation and thermal shock [48].

Finally, DLD is also used for the repair of components directly on industrial sites [49]. Repair can concern corrosion defects and studies are ongoing to optimize the repair strategy and qualify this repair regarding the corrosion process in the case of marine environments.

1.3.5 Conclusions

The AM processes likes SLM or DLD involve very high temperature gradients, which generate specific microstructures different from those of cast or wrought alloys. These microstructures can have a strong influence on the corrosion behavior. Studies have started to link the process parameters to the microstructure and the corrosion behavior but they are still scarce and do not cover in depth all the parameters influencing corrosion and all the corrosion phenomena which can be extremely different from a material/environment system to another. Indeed, the corrosion behavior can take multiple forms. Depending on the temperature, the corrosive environment and the material studied, different corrosion processes can occur. General corrosion is observed in some cases and localized corrosion in some other cases. The understanding of the link: process/microstructure/corrosion behavior is required to master and optimize these new manufacturing processes; this understanding should then be combined to modelling to accelerate the developments. Instrumentation of the processes should also help this understanding. Post treatments covering heat treatments, heat treatments combining hot isostatic pressing, but also surface treatments, which can be mechanical, chemical, electrochemical…, are also an essential part of this work.

However, these new processes open also a wide range of opportunities for the development of sustainable materials by the possibly of designing corrosion resistant materials, of building in one step corrosion resistant components and by the emergence of highly efficient corrosion protection solutions. Indeed, these manufacturing techniques should accelerate both the discovery of new materials as well as their qualification by integrating combinatorial materials fabrication methods, high-throughput characterisation techniques, material modelling, and data analysis with the potential of incorporating machine learning (ML) and artificial intelligence (AI) schemes. The final objective of this "inverse design process" would be to be able to design the "perfect" material regarding a given application.

Acknowledgements

The authors want to thank the CEA Cross-Cutting Program on Materials and Processes Skills, for giving them the opportunity to participate in the AMETIS.

References

[1] Billard A., Maury F., Aubry P., Balbaud-Celerier F., Bernard B., Lomello F., Maskrot H., Meillot E., Michau A., Schuster F. (2018) *C.R. Physique* **19**, 755.

[2] Gorsse S., Hutchinson Ch., Gouné M., Banerjee R. (2017) *Sci. Technol. Adv. Mater.* **18**, 584.

[3] Gu D.D., Meiners W., Wissenbach K., Poprawe R. (2012) *Int. Mater. Rev.* **57**, 133.

[4] Barkia B., Aubry P., Haghi-Ashtiani P., Auger Th., Gosmain L., Schuster F., Maskrot H. (2020) *J. Mater. Sci. Technol.* **41**, 209.

[5] Sun Z., Tan X., Tor S., Chua C.K. (2018) *NPG Asia Mater.* **10**, 127.

[6] Alsalla H., Smith C., Hao L. (2018) *Rapid Prototyping J.* **24**, 9.

[7] Kurzynowski T., Gruber K., Stopyra W., Kuznicka B., Chlebus E. (2018) *Mater. Sci. Eng. A* **718**, 64.

[8] Yadollahi A., Shamsaei N., Thompson S.M., Seely D.W. (2015) *Mater. Sci. Eng. A* **644**, 171.

[9] Guo P., Zou B., Huang C., Gao H. (2017) *J. Mater. Process. Technol.* **240**, 12.

[10] Liverani E., Toschi S., Ceschini L., Fortunato A. (2017) *J. Mater. Process. Tech.* **249**, 255.

[11] Sander G., Tan J., Balan P., Gharbi O., Feenstra D.R., Singer L., Thomas S., Kelly R.G., Scully J.R., Birbilis N. (2018) *Corrosion* **74**, 1318.

[12] Kong D., Dong C., Ni X., Li X. (2019) *NPJ Mater. Degrad.* **24**, 1.

[13] Pourbaix M. (1963) Atlas d'équilibres électrochimiques à 25 °C, Publication du centre belge d'étude de la corrosion CEBELCOR.

[14] Stern M., Geary A.L. (1957) *J. Electrochem. Soc.* **104**, 56.

[15] Wagner C., Traud W. (1938) *Z. Electrochem.* **44**, 391.

[16] Armanet F., Béranger G., Moulin G. (2012) Techniques de l'Ingénieur, COR 378V1, pp. 1–32.

[17] Ellingham H.J.T. (1944) *J. Soc. Chem. Ind.* **63**, 125.

[18] Shuang Y., Yuan Y., Li D., Li Y., Chen J. (2017) *Int. J. Adv. Manuf. Technol.* **92**, 4379.

[19] Li R., Liu J., Shi Y., *et al.* (2010) *J. Mater. Eng. Perform.* **19**, 666.

[20] Sander G., Thomas S., Cruz V., Jurg M., Birbilis N., Gao X., Brameld M., Hutchinson C.R. (2017) *J. Electrochem. Soc.* **164**, 250.

[21] Gong H., Rafi K., Gu H., Janaki Ram G.D., Starr T., Stucker B. (2015) *Mater. Des.* **86**, 545.

[22] Gong H., Rafi K., Gu H., Starr T., Stucker B. (2014) *Additive Manuf.* **1–4**, 87.

[23] Liu Y., Yang Y., Wang D. (2016) *Int. J. Adv. Manuf. Technol.* **87**, 647.

[24] Alam M.K., Edrisy A., Urbanic J., Pineault J. (2017) *J. Mater. Eng. Perform.* **26**, 1076.

[25] Kok Y., Tan X.P., Wang P., Nai M.L.S., Loh N.H., Liu E., Tor S.B. (2018) *Mater. Des.* **139**, 565.

[26] Chao Q., Cruz V., Thomas S., Birbilis N., Collins P., Taylor A., Hodgson P.D., Fabijanic D. (2017) *Scrip. Mater.* **141**, 94.

[27] Sun Y., Moroz A., Alrbaey K. (2014) *J. Mater. Eng. Perform.* **23**, 518.

[28] Schaller R.F., Taylor J.M., Rodelas J., Shindelholz E.J. (2017) *Corrosion Sci.* **73**, 796.

[29] Laleh M., Hughes A.E., Yang S., Li J., Xu W., Gibson I., Tan M.Y. (2020) *Corrosion Sci.* **165**, 108394.

[30] Revilla R.I., Liang J., Godet S., De Graeve I. (2017) *J. Electrochem. Soc.* **164**, C27.

[31] Leon A., Aghion E. (2017) *Mater. Charact.* **131**, 188.

[32] Schaller R.F., Mishra A., Rodelas J.M., Taylor J.M., Schindelholz E.J. (2018) *J. Electrochem. Soc.* **165**, C234.

[33] Samviemvongsak T., Monceau D., Macquaire B. (2018) Corrosion Sci. **141**, 127.

[34] Daia N., Zhangb L.C., Zhanga J., Zhanga X., Nia Q., Chena Y., Wua M., Yang C. (2016) *Corrosion Sci.* **111**, 703.

[35] Lou X., Song M., Emigh P.W., Othon M.A., Andresen P.L. (2017) *Corrosion Sci.* **128**, 140.

[36] Lou X., Othon M.A., Rebak R.B. (2017) *Corrosion Sci.* **127**, 120.

[37] Laleh M., Hughes A.E., Xu W., Haghdadi N., Wang K., Cizek P., Gibson I., Tan M.Y. (2019) *Corrosion Sci.* **161**, 108189.

[38] Takakuma O., Soyama H. (2015) *Adv. Chem. Eng. Sci.* **5**, 62.

[39] Cruz V., Chao Q., Birbilis N., Fabijanic D., Hodgson P.D., Thomas S. (2020) *Corrosion Sci.* **164**, 108314.

[40] Wang S. *et al.* (2013) *Entropy* **15**, 5536.

[41] Nair R.B., *et al.* (2018) *Ultrasonics Sonochem.* **41**, 252.

[42] Ayyagari A., *et al.* (2018) *Mater. Chem. Phys., High-Entropy Mater.* **210**, 162.

[43] Zhao Y.L., *et al.* (2018) *Scripta Materialia* **148**, 51.

[44] Huang C., *et al.* (2012) *Mater. Des.* **41,** 338.
[45] Huang C., *et al.* (2011) *Surf. Coat. Technol.* **206**, 1389.
[46] Chao Q., *et al.* (2017) *Surf. Coat. Technol.* **335**, 440.
[47] Aubry P., *et al.* (2017) Proceedings of ICAPP 2017, April 24-28, Fukui and Kyoto, Japan.
[48] Su Y., *et al.* (2020) *J. Mater. Proc. Tech.* **283**, 116702.
[49] Yao J.H., Zhang Q.L., Kong F.Z. (2012) Laser remanufacturing to improve erosion & corrosion resistance, *Woodhead Publishing Series in Metals and Surface Engineering*, Chap. 9, pp. 320–354.

1.4 Additive Manufacturing, from Powder to *In Situ* Nanocomposites

Laurent CHAFFRON[1], Elodie VASQUEZ[3], Pierre-François GIROUX[1], Fernando LOMELLO[2] and Hicham MASKROT[2]

[1] *DES-Service de Recherches Métallurgiques Appliquées, CEA, Université Paris-Saclay, F-91191, Gif-sur-Yvette, France*
[2] *DES-Service d'Etudes Analytiques et de Réactivité des Surfaces, CEA, Université Paris-Saclay, F-91191, Gif-sur-Yvette, France*
[3] *Now at Thales Global Services, IPTISAS, 19/21 av. Morane Saulnier, 78140 Vélizy-Villacoublay, France*

Abstract

This chapter deals with the fabrication of nanocomposite materials by laser power bed fusion. It introduces first the presentation of the laser power bed fusion process and the main requirements that must meet the powder. Then, the different kinds of nanoparticles will be described and the way they can enter the microstructure of the grain to form the nanocomposite (NC) powder. Finally, example of a representative NC material produced from a nanostructured powder will be described.

1.4.1 Introduction

Nanocomposite materials, *i.e.* materials reinforced with nanoparticles, are of prime interest worldwide for some years, thanks to their promising properties suitable for a large number of functional and structural applications, including optical, magnetic biomedical, and nuclear industries... Exponential development of powder additive manufacturing processes has opened new opportunities for matching news challenges in materials sciences.

In this framework, combination of laser powder bed fusion (LPBF) process with nanocomposite powders should be one of the new breakthroughs for the next years.

A nanocomposite is defined as multicomponent materials comprising multiple different phase domains in which at least one type of phase domain is a continuous phase and in which at least one of the phases has at least one dimension smaller than one hundred nanometers [1].

The nature of the matrix as well as that of nanoparticles is very wide, going from metal to polymer. Very different combinations can been found depending on the final application. Even though the topic of this paper is focussing on metallurgy applications, let us know that one of the most recent and hopeful examples concerns the biomedical application where the association of a hydrogel and of hematite nanoparticles leads to successful therapy against cancer. Indeed, this hydrogel nanocomposite can be remotely heated and controlled by alternating electromagnetic field, leading to kill targeted cells [2]. Giant magnetoresistance (GMR) is also a domain where nanocomposites play a major role in connection with the physical

DOI: 10.1051/978-2-7598-2446-5.c041

properties of the nanoparticles. GMR properties are driven by the size and the distribution of the superparamagnetic nanoparticles embedded in a non-magnetic metallic matrix [3]. In the same way, optical properties such as light absorption are depending on the size of inorganic particles embedded in the polymer matrix [4].

Now, in the metallurgy field and especially for structural applications, the use of nanocomposite sounds differently. Nanocomposites are of prime interest for automotive and aerospace domains where increasing the ratio σ/ρ is the main objective. It means that increasing the mechanical properties while keeping the density low is the target coveted by R&D centres. The way to meet this objective is to disperse individually and uniformly the population of nanoparticles and in the meantime to guarantee a very good interphase between the particles and the matrix. Among the different strengthening mechanism contributions existing, Taylor, Hall Petch and Orowan strengthenings seem to play the major role for the finest microstructure, *i.e.* when the mean grain size of the matrix is lower than 100 nm [5].

Different matrices, predominantly classified as light metals, have been used for the production of composites reinforced by nano-ceramic particles such as carbides, nitrides, oxides as well as carbon nanotubes more recently.

The main issue to be faced in the production of these nanocomposites is the low wettability of ceramic nano-particles with the molten metal matrix, excluding generally conventional casting processes.

Several alternative routes have been proposed to overcome this concern. These processes are divided into two families: the *ex situ* route (i) and the *in situ* route (ii).

(i) The *ex situ* route consists of adding existing nanoparticles to the liquid or powdered metal matrix; the main advantage is the apparent facility to fabricate the nanocomposite. In fact, drawbacks are quite similar to those of the casting route, *i.e.* dispersing uniformly the nanoparticles, preventing secondary reactions between particles and matrix...

(ii) At the opposite, the *in situ* route refers to methods leading to the generation of a new nano phase induced by chemical reaction during processing, for example using reactive gases, solid state reaction or solid liquid reactions.

For the both routes, one of the key parameters is the holding time at high temperatures leading either to the densification step to occur or to the nanoparticles to form prior densification.

By the way, laser powder bed fusion process that brings at liquid state a tiny volume of matter in a very short time appears as a reliable alternative method to produce nanocomposite materials thanks to the very swift stay of the grains at liquid state.

This papers aims at presenting how additive manufacturing process can help to produce nanocomposite materials thanks to an appropriate choice of the powder and thanks to the extreme rapidity of the melting and solidification steps.

1.4.2 Additive Manufacturing

From a normative point of view, additive manufacturing (AM) defines all of the processes allowing the manufacturing of a part layer by layer, by adding material, from a 3D digital model [NF EN ISO/ASTM 52900 2017]. Three types of use of additive manufacturing have been distinguished: rapid prototyping, rapid tooling, for example for the production of molds, and direct manufacturing, which make it possible to directly obtain a part that meets functional characteristics. Nowadays, the challenge is to mass-produce functional parts bringing an improvement in terms of speed and finishes [6].

In what follows, we will focus more specifically on the laser powder bed fusion (LPBF), process that appeared in the early 1990s and remains the most commonly used.

1.4.2.1 Principle of the Laser Powder Bed Fusion Process

This process consists of building a part, layer by layer, by laser fusion of a bed of metal powders, as shown in figure 1.4.1.

Practically, the first step is to model in 3D the desired part by computer-aided design (CAD). Manufacturing supports (grid or "lattice" structure) are generated by the software in order to secure the parts to the manufacturing platform, dissipate heat, and reduce stresses. These supports must be removed later by machining. The part is then cut into slices, thus generating a succession of 2D plans stacked

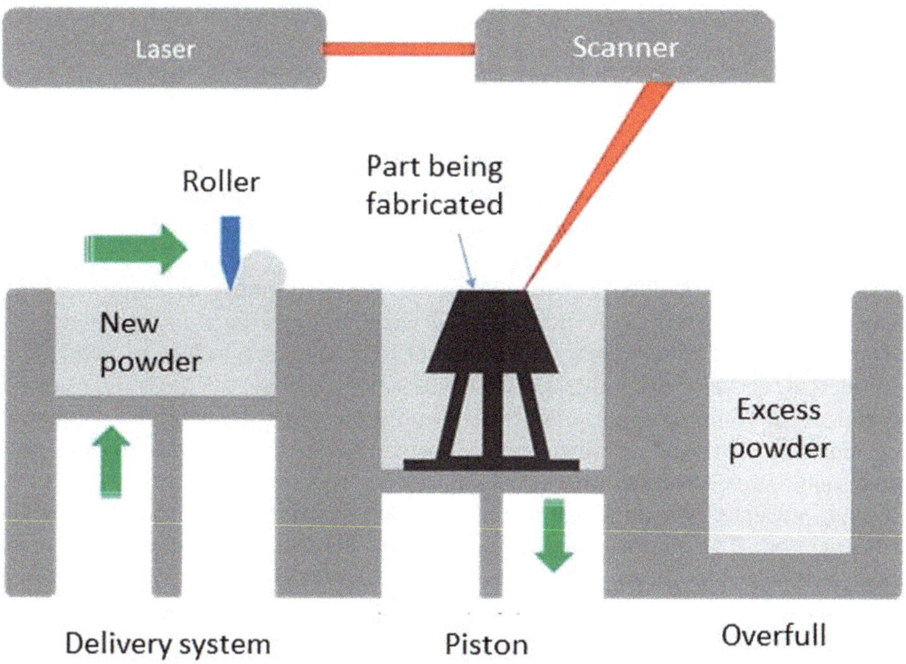

FIG. 1.4.1 – Functional diagram of the laser powder bed fusion process.

according to the direction of production. A STL (Standard Tessellation Language) file is generated.

The production then takes place under a controlled atmosphere to prevent oxidation of the part. At each layer, a roller spreads a fine thickness of powder; then, the laser selectively scans the surface of the powder according to the information provided by the STL file. Once the layer is finished, the production plate moves down from the height of a layer thickness in order to allow the deposition of a new layer of powder and so on. Generally, the mean grain size for metals is around 50 μm with a layer thickness close to 100 μm.

The lasers used are generally Nd: YAG fiber lasers with a wavelength of 1064 nm, with a Gaussian-type beam and used in continuous mode (or more rarely in pulsed mode). The powers of these lasers are between 50 and 1000 watts. The diameters of the laser beams are between 30 and 100 μm. The movement of the laser beam on the stage is ensured by two rotating mirrors and can reach speeds up to 5 $m.s^{-1}$ [7].

The range of materials that can be produced by LPBF is quite wide: stainless steels, maraging steels, nickel-based superalloys, cobalt chromium alloys, titanium alloys, and aluminium alloys. Attention must be paid to the absorptivity of the material and to the thermal conductivity of the grains. Recent work has shown that changing the laser wavelength allows using copper powders without any preparation/modification of the surface [8].

1.4.2.2 Advantages and Limits of Laser Powder Bed Fusion Process

Like the other AM processes, the LPBF process has many advantages:

Design freedom: As the parts are designed layer by layer, it is possible to access geometries that cannot be produced by other methods, to limit the machining of printed parts, to rethink the design of parts to lighten the structures (figure 1.4.2) and imagine new functionalities [9]. In addition, parts previously requiring several assemblies can be produced at one time. This saves manufacturing time but above all avoids assembly problems (breakage in welds, etc....).

Flexibility: The AM processes allows to produce parts in small series, on demand, quickly, without specific tools and without the need to store them.

Development of tailored microstructure: LPBF makes possible to produce tailored microstructures with controlled porosities. By the way, it is possible to develop materials with composition gradients [10] and gain access to new microstructures, often finer than those obtained by conventional methods, which leads to an improvement in mechanical properties [11].

Use of powders: Not all alloys are available in powder form and not all of them can be printed. The major problem lies in the variability that can exist between several batches of powders or suppliers of the same grade in terms of chemical composition and morphology [12]. The handling of powders also requires the installation of specific safety equipment.

Energy efficiency: LPBF allows to generate less waste by avoiding the machining of parts, to realize optimized geometries allowing to lighten structures and to save

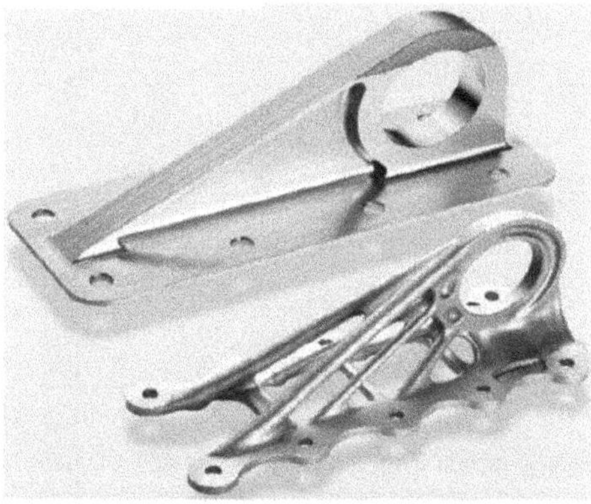

FIG. 1.4.2 – Example of construction with topological optimization: Airbus A350 XWB wing support (before and after optimization) (®Airbus Operations).

material [13]. In addition, to some extent, unprocessed powders can be re-used several times [14].

Although parts produced in additive manufacturing are already industrialized and marketed, the development of AM must face certain limitations to allow the mass production of functional parts:

Surface finish: The surface finish of parts at the end of manufacturing is often rougher than that obtained with conventional processes [15]. The roughness of the parts however tends to decrease considerably and does not always require post-treatment according to the process used.

Post-treatments: At the end of the elaboration step, it is necessary to detach the part from the production plate, to remove the production supports, to repackage the powder (sieving, steaming), and to carry out possible heat treatments of stress relieving or hot isostatic pressing to eliminate residual porosity. All of these steps are time consuming and can represent up to 20%–30% of production time. The different post-treatments represent a significant cost on the final price of the produced parts, as shown in figure 1.4.3.

Productivity: LPBF is not a process suitable for large mass production. The manufacturing speed is not yet sufficient to meet industrial needs. It takes several hours to several days to print a part. Nevertheless, progress in development of the technology in the last years is impressive. Different large LPBF machines have build rates over 100 cm^3/h, when these of the lab Trumpf True Print 1000 reach 20 cm^3/h [17]. The maximum volume of parts is also still limited. However, manufacturers are developing new machines equipped with several lasers with larger processing chambers [6].

Certification: The improvements of AM process over the last 10 years now allow considering seriously these production processes in the nuclear context, in order to propose components of suitable quality with improved design, and thus delivering more safety. The AM technology would also minimize or eliminate weakened areas generated by conventional welding processes. AM represents a promising solution to

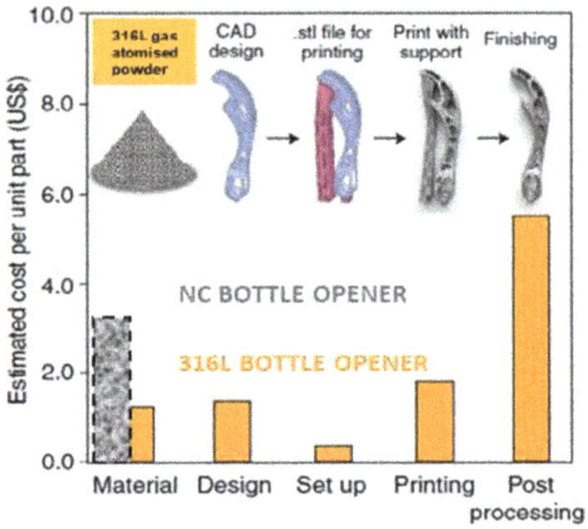

FIG. 1.4.3 – Estimation of the costs of each stage required for the production of a bottle opener either in 316L steel or in nanocomposite (NC) by LPBF [from [16], modified].

tackle those unsolved issues; however, research and development on materials fabricated by AM is required to ensure that they fit nuclear design code requirements and provide a sound alternative as compared to current qualified ones.

In AM, certification of the final product is not enough. It is necessary to certify a powder/machine pair. However, machines are constantly evolving, just like software, and it becomes difficult, if not impossible now, to certify a part over the long term [6].

1.4.3 From Powder to Nanocomposite Powder

1.4.3.1 Prerequisite for Powder Use in LPBF

It is now clearly established that the final properties of a component manufactured by LPBF mainly depend on the quality of the powder used, regardless of the processing parameters chosen [18]. It is also agreed that the main characteristics of interest for a powder dedicated to AM are flowability, particle size distribution (PSD), morphology and powder layer density (PLD). Some of these items can interact in a conflict manner.

1.4.3.1.1 Flowability

Flowability is a generic term that includes different specific tests in order to characterize the capability of a powder to flow. It covers few different elementary tests which are standardized. These are mainly: (i) Hall flowmeter funnel, (ii) Slope angle, (iii) Flodex test method, (iv) Hausner ratio and (v) Avalanche angle. The latter requires to determine the apparent density and the tapped density.

(i) The Hall flowmeter funnel method determines the flow velocity from 50 g of powder flowing through a funnel Hall having a calibrated orifice.

(ii) The measure of the angle of repose consists in flowing a powder on a cup through a funnel to distribute it evenly. The average of the angle measurements in the four cardinal directions gives the angle of repose.
(iii) The Flodex method determines the ability of a powder to flow by gravity through a hole of a defined diameter (Flodex diameter) [19].
(iv) The Hausner ratio (HR) represents the state of interparticle friction; the closer to 1, the better the flowability.
(v) The measurement of the avalanche angle consists of a rotating, transparent drum, partially filled with powder and a camera in front of a backlight. The latter records pictures of the powder free surface and the cross-sectional area of powder inside the drum, giving access to the value of an avalanche angle [20]. This is a kind a dynamic measurement of the static angle of repose technique. According to the authors, this measurement will be more a representative test for LPBF process.

1.4.3.1.2 Particle Size Distribution

The particle size distribution of powder is a complex parameter for LPBF. A narrow PSD will guarantee a good flowability of the powder but a low powder layer density that implies an incomplete densification of the powder. On the contrary, a wide PSD will be detrimental to flowability but beneficial for a good densification. Nevertheless, as the thickness of the layer of powder is limited to around 100 μm, large particles have to be excluded. In fact, several studies pointed that a bimodal distribution was a good compromise in terms of flowability and final part density [20].

1.4.3.1.3 Morphology

The role of morphology of grain powder on their capability to be used by LPBF is quite complex. Few years ago, it was assumed that sphericity of the grain was the key parameter [21] and many attempts have been carried out in this way, using plasma inductive torch *e.g.* [22]. In fact, it appears that general statements are not possible [23]. For coarse grains, there is no effect of the roughness of the surface of the grain. It means that ball-milled powders can be used [24]. A contrario, it has been demonstrated that small grains even with perfect sphericity could be detrimental for AM applications. It has been shown that it was due to interparticular forces (Van der Walls, mainly); the finest the particles size, the strongest the effect of the cohesive forces [25]. This behaviour can be minimized in case of rough surface small grains. Moreover, it has been observed that whatever the diameter is, elliptic grain improves flowability in general [20].

1.4.3.1.4 Powder Layer Density

The powder layer density (PLD) also called powder bed apparent density corresponds to an intermediate value between tapped and apparent density, due to the effect of the roller on the powder during processing. To determine the PLD, the roller delivers the powder cross the powder bed, but only the walls of the cube are melted, leaving the raw powder within the cube. Knowing the mass of the raw powder and

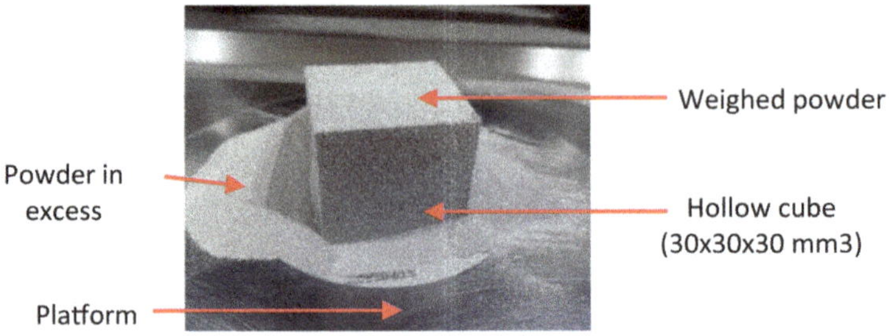

FIG. 1.4.4 – Picture of a hollow cube filled with powder that is used to estimate the bulk density of the powder bed [24].

the inner volume of the cube allows to calculate the PLD (figure 1.4.4). Of course, since the higher the powder bed density, the higher the part density, it is recommended to use a wider grain size distribution rather than a narrow one, or a bimodal distribution with both a small particle population and coarse particle ones [26]. Moreover, the use of small particles will be also beneficial to the decrease in the roughness.

1.4.3.2 Synthesis of Nanocomposite Powder

As previously reported, the preparation of nanocomposite powders can be done according to two approaches: an *ex-situ* route which consists in mixing/assembling the phases, one of which is at least nanosized; and an *in-situ* route that will generate the nanoscale phase after heat treatment generally, either produced before or during the application of the LPBF process.

In all cases, the powder or its precursor must meet the prerequisites of the LPBF process with an additional challenge: no uncontrolled reaction has to take place during the very short time in the liquid state. This may constitute an issue towards the deployment of parts made of nanocomposite material produced by this process compared to other standard powder metallurgy processes, such as hot isostatic pressing or spark plasma sintering.

1.4.3.2.1 Ex Situ Route

The synthesis of nanocomposite powder *via* the *ex situ* route implies generally the use of atomized metal powder, the mother phase, and of nanosized/nanostructured second phase powder, the nanoparticles (NP).

NP can be synthesized among various methods as summarized in figure 1.4.5 (for a review, see [27]). One representative bottom-up approach is laser pyrolysis that allows producing significant quantities of high quality nanopowders [28]. An example of nanoSiC powder is presented in figure 1.4.6a. The latter can be provided by the French company Nanomakers [29].

Another bottom process that is not (yet) reported by Kahn is mechanochemistry. That route, inspired from ball milling, allows to promote the formation of

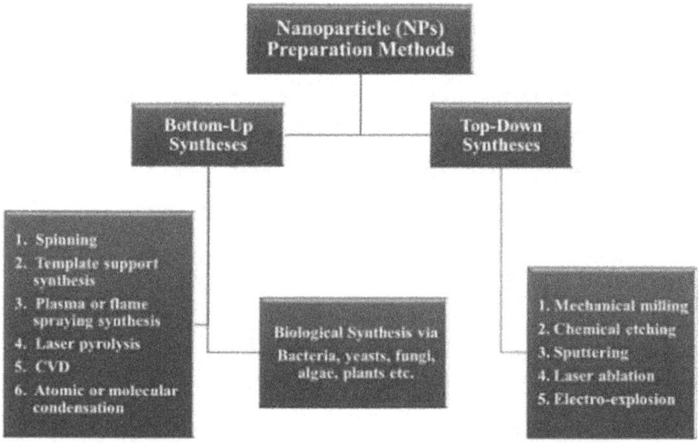

FIG. 1.4.5 – Typical synthetic methods for NPs for the (a) top-down and (b) bottom-up approaches [27].

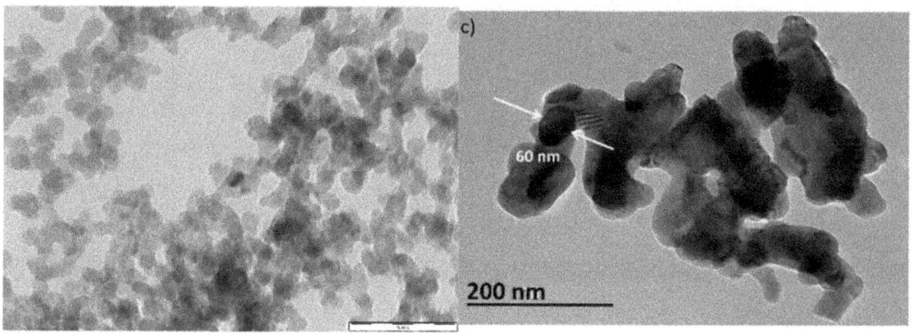

FIG. 1.4.6 – TEM micrographs of SiC nanopowders produced par laser pyrolysis.

nanostructured powders, *i.e.* agglomerates of NP, from metallic or MX compounds powders that are ball milled under reactive atmosphere [30, 31]. An example of nanostructured pyrochlore is shown in figure 1.4.6b. The size of the elementary crystallite is estimated to be roughly 60 nm.

More recently, mechanochemistry has been successfully used for the synthesis of individual metallic NPs [32]. Figure 1.4.7 reports the different steps leading to the formation of NP *via* the bottom-up approach by comparison to the top-down one.

The nanocomposite powder can be prepared in different ways.

Electrostatic deposition can be used for uniform deposition of NPs onto metallic powders [33]. The same result can be roughly obtained using low energy ball milling or a stirring device [34]. In this latter case, there is no fracturation of the mother phase grains, the NPs recovers progressively the surface of the metallic grains.

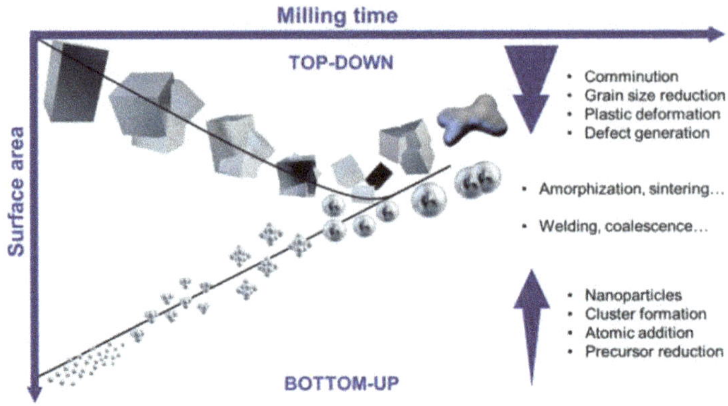

FIG. 1.4.7 – Mechanochemical synthesis of nanoparticles: top-down and bottom-up approaches [32].

FIG. 1.4.8 – Fe-14Cr powders ball-milled in an attritor (left); without NPs; (right) with NP [36].

On the contrary, high energy ball-milling (HEBM) allows to incorporate very finely the NPs inside the gains, due to the successive deformations, fracturations and weldings submitted by the metallic grains [35].

In addition of the incorporation of NPs inside the grains, HEBM allows to reduce significantly the mean size of the initial size of the metallic grains. When the NPS are gradually incorporated into the grains, the mechanical behaviour of the grain changes. The material becomes more brittle so that the sequence of fracturation becomes preponderant. This point is illustrated in figure 1.4.8 showing ball milled powders (Fe-14Cr) in the same conditions with and without NPs [36].

1.4.3.2.2 In-Situ Route

Contrary to the *ex-situ* approach where no reaction occurs inside the grains, the *in-situ* route implies chemical reaction during the process leading to the formation of NPs within the metal.

The *in-situ* process for direct synthesis of NC powders is mainly the reactive ball-milling (RBM). It was widely studied in the 1990s to develop advanced material first and to explore the limits of the process [35]. It was demonstrated that, even RBM cannot induce in one step the desired NC powder, it became a key process for the preparation of smart precursors prior fully transformation by thermal treatment (flash sintering, plasma torch, LPBF or DED) [24, 37].

1.4.3.2.3 Interest of Induction Plasma Treatment for NC Powders

Induction plasma is a major technology in the development of nanomaterials [38]. Since few years, its interest for AM applications is growing more and more. This is now a reliable process for big scale production, compatible with the harsh needs in powder metallurgy. Regarding specifications required for LPBF process, induction plasma can induce partial or fully melting of the powder that improves the sphericity of particles produced by gas or water atomization; it can also fully smooth the angular shape of particles produced by mechanical milling. Figure 1.4.9 clearly demonstrates these features.

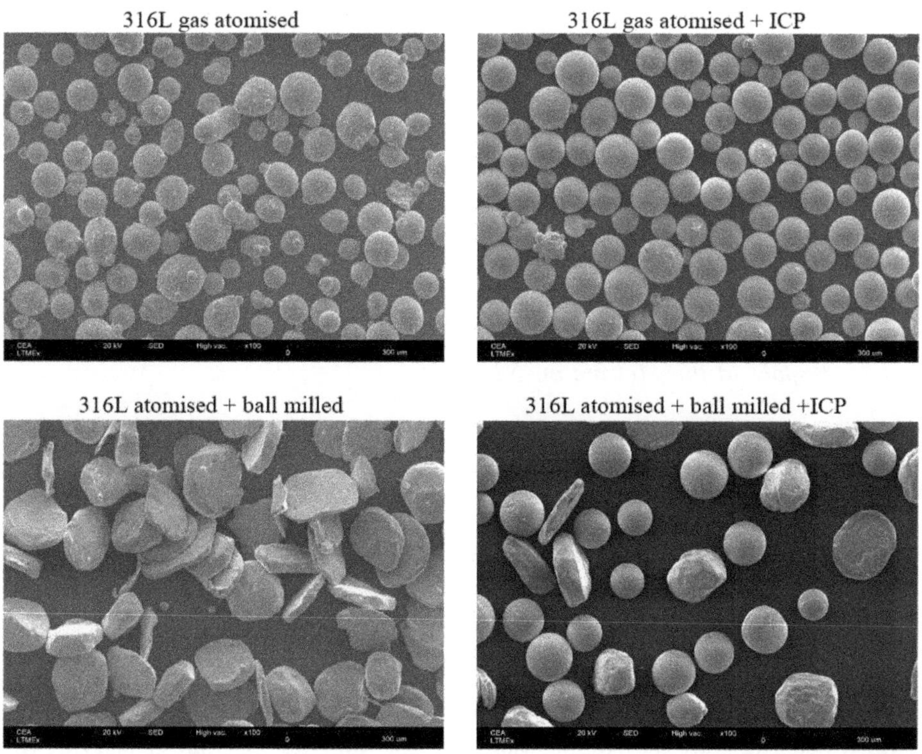

FIG. 1.4.9 – Effect of induction plasma on 316L stainless steel gas atomised grains (top) and on a 316L ball milled powder (bottom).

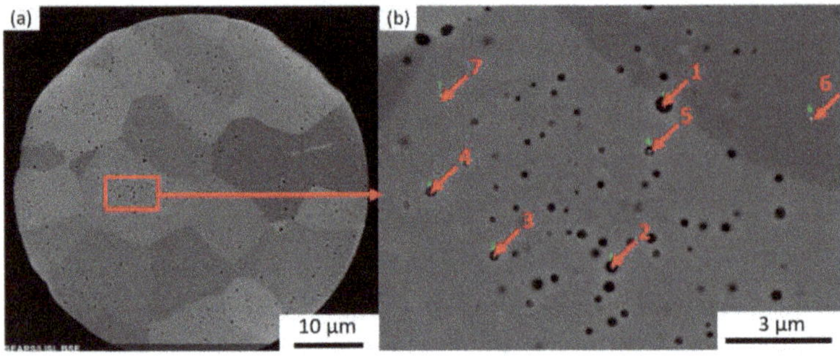

%m	O	Ti	Cr	Fe	Y
1	9	6	11	59	15
2	9	6	11	59	15
3	4	3	12	73	7
4	6	4	12	66	10
5	6	4	12	67	10
6	/	/	13	85	/
7	/	/	14	85	/

FIG. 1.4.10 – (a) SEM image of a polished grain after thermal treatment by induction plasma. The dark spot have been identified as pyrochlore oxide ($Y_2Ti_2O_7$) (b) [24].

More recently, it has been shown that an induction plasma treatment could induce chemical reactions within the powder grains without damages. To illustrate this point, the starting powder is an Fe-14Cr-1W ferritic steel powder which was ball milled together with yttrium oxide (Y_2O_3) and titanium hydride (TiH_2) powders. HEBM produces a homogeneous powder in which all the elements are intimately mixed without inducing nucleation of new places. Figure 1.4.10 shows a polished section of a ball milled grain after induction plasma torch. EDX analyses give evidence of oxide nanoparticles that are labelled on the picture. Indexation of electronic diffraction patterns allow to identify the oxide as pyrochlore compounds ($Y_2Ti_2O_7$) [24].

In addition to the nearly perfect sphericity of the grain, the remarkable feature is the good distribution of nanoxides that have been trapped uniformly in the whole grain. There is no specific segregation of NP at the grain boundaries.

1.4.4 Nanocomposite Material Produced by LPBF

1.4.4.1 Impact of the Melting on NPs

Previously, emphasis has been put on the properties or characteristics that should have powders for a good implementation in the LPBF process. Now, if we focus on the final product that was prepared with a powder with a good flowability and with a good powder layer density, we can observe that its properties depend on the effect of the LPBF parameters such as the laser energy density or the scanning speed. The molten state, occuring during the process, is one of the phenomenom which can affect significantly the material properties. The characteristics of this state depend on the powder chemical composition of course but also on the microstructure of the initial particles. In particular, when using powders with NPs, located within the grain or on the surface of the grain, the viscosity of the melt pool can be drastically modified according to the composition, the size and the volumic density of these NPs [37].

A contrario, the molten state can be beneficial to promote reaction or displacement reaction in order to form new phase by nucleation/coarsening. Many examples can be found in the recent literature regarding Al/ZnO [39] or Ti/TiB [40].

1.4.4.2 Example of Fabrication of Nanostructured ODS Steel Material by LPBF

Oxide Dispersive Strengthened (ODS) ferritic steels are of prime interest for the development of next generation of nuclear reactor because of their very high resistance to swelling at high temperatures. This excellent behaviour is due to the fine distribution of NPs inside the grains of this material that is generally produced by hot extrusion. Using the capabilities of the AM technology could simplify the fabrication of components that implies various welding operations.

The feasibility of such materials has been studied using a LPBF machine (Trumpf TruPrint 1000), starting from a mechanically alloyed powder (Fe-14Cr1W atomized powder with 0.3% wt. Y_2O_3 and 0.3% wt. TiH_2 nanopowders) containing the different elements mixed at a very fine scale.

The influence of the operating parameters (laser power, scanning speed, inter-bead distance, thickness of the powder bed) as well as the characteristics of the powder (particle size, flowability, morphology) on the final properties of the materials produced (density, microstructure, mechanical properties) have been investigated.

One remarkable result concerns the nanoprecipitation of the oxide nanophase. TEM observations coupled with microanalysis allow to conclude that the reaction takes place uniformly inside the grains without specific segregation at grain boundaries as observed in ODS produced by standard process. In addition, from the different analyses performed, it is concluded that the mean size of the NPs was not sensitive to the conditions of fabrication as reported in figure 1.4.11. The pyrochlore phase is the most stable among the ternary system; once formed, there is no indication of further decomposition. Some minor secondary phases have been detected.

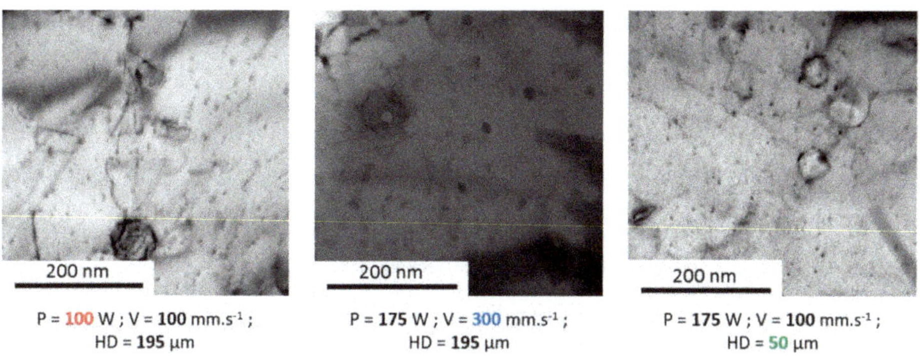

FIG. 1.4.11 – (a) TEM observations of 3 grades of ODS materials fabricated in different conditions (power, scanning speed and hatching distance) [24].

In very harsh conditions, the full densification induces the formation of micronic clusters of oxide phases that are harmful for the mechanical properties [24].

Regarding the effect of the powder in terms of size, PSD, flowability..., a systematic study of different powders concludes that the use of fine powder particles (with a diameter between 10 and 50 μm) and spherical morphology allows dense materials, over a wide range of operating parameters. Beyond 120 J.mm^{-3}, the differences in terms of particle size and morphology of the powder particles are not significant.

1.4.5 Conclusion

This paperwork deals about the fabrication of nanocomposite materials using laser power bed fusion process. The importance of the powder and its characteristics has been studied. It has been shown that no immediate parameter allows to predict the final state in terms of density mainly.

Regarding the effect of NPs inside the grains or on the top surface of the grain, different behaviours are observed. The route to inducing NP formation during the process seems the best promising in terms of microstructural stability and mechanical properties. The interface of NP/matrix is stronger. Nevertheless, this route is still challenging due to the risk of generating uncontrolled reactions.

Acknowledgments

The authors would like to thank both laboratories LTMEx and LISL from CEA with a specific thanks to E. Bouaravong, O. Hercher and J. Varlet.

References

[1] Beilstein J. (2018) *Nanotechnol.* **9**, 1050.

[2] Mishra D.K., Yadav K.S., Prabhakar B., Gaud R.S. (2018) *Woodhead Publishing Series in Biomaterials*, 323.

[3] Nagamine L.C.C.M., Chamberod A., Auric P., Auffret S., Chaffron L. (1997) *J. Magn. Magn. Mater.* **174**, 309.

[4] Chen B., Evans J.R.G., Greenwell H.C., Boulet P., Coveney P.V., Bowden A.A., Whiting A. (2007) *Chem. Soc. Rev.* **37**, 568.

[5] Sanaty-Zadeh A. (2012) *Mat. Sci. Eng. A* **531**, 112.

[6] Pôle Interministériel de Prospective et d'Anticipation des Mutations Economiques, rapport final: futur de la fabrication additive (2017).

[7] Pillot S. (2016) Fusion laser sélective, Techniques de l'ingénieur.

[8] Jadhav S., Dadbakhsh S., Vleugels J., *et al.* (2020) *Eng. Mater.* **22**, 1900946.

[9] Vayre B., Vignat F., Villeneuve F. (2012) *Procedia CIRP* **3**, 632.

[10] Liu B., Wildman R., Tuck C.R., Ashcroft I., Hague R.J.M. (2011) Proceedings, Solid Free Form Fabrication Symposium, Austin, Texas, pp. 227–238.

[11] DebRoy T., Wei H.L., Zuback J.S., Mukherjee T., Elmer J.W., Milewski J.O., Beese A.M., Wilson-Heid A., Zhang A.DeW. (2018) *Prog. Mater. Sci.* **92**, 112.

[12] Baitimerov R., *et al.* (2018) *Materials*, 5.

[13] Ford S., Despeisse M. (2016) *J. Cleaner Product.* **137**, 1573.
[14] Cordova L., Campos M., Tinga T. (2019) *JOM* **71**, 1062.
[15] Boschetto A., Bottini L., Veniali F. (2017) *J. Mater. Process. Technol.* **241**, 154.
[16] DebRoy T., Mukherjee T., Milewski J.O., *et al.* (2019) *Nat. Mater.* **18**, 1026.
[17] Khorasani A., Gibson I., Veetil J.K., *et al.* (2020) *Int. J. Adv. Manuf. Technol.* **108**, 191.
[18] Vock S., Klöden B., Kirchner A., Weissgärber T., Kieback B. (2019) *Prog. Addit. Manuf.* **4**, 383.
[19] Gioia A. (1980), FLODEX Technical Bulletin, Pharmaceutical Technology.
[20] Spierings A.B., Voegtlin M., Bauer T., Wegener K. (2016) *Prog. Addit. Manuf.* **1**, 9.
[21] Karapatis P. (2002) Doctoral thesis, EPFL, Lausanne.
[22] www.tekna.com.
[23] Schulze D. (2008) *Powders and bulk solids: behavior, characterization, storage and flow.* Springer.
[24] Vasquez E. (2018) Thesis INSA Rennes.
[25] Ma Y., Evans T.M., Philipps N., Cunningham N. (2020) *Powder Technol.* **360**.
[26] Tan J.H., Wong W.L.E., Dalgarno K.W. (2017) *Additive Manuf.* **18**, 228.
[27] Khan I., Saeed K., Khan Id. (2019) *Arabian J. Chem.* **12**, 908.
[28] Réau A. (2008) Thèse de l'Université de Paris XI.
[29] www.nanomakers.fr.
[30] Legendre F., Poissonnet S., Bonnaillie P. (2007) *J. Alloys Compd.* **434–435**, 400.
[31] Simondon E., Giroux P.-F., Chaffron L., Fitch A., Castany P., Gloriant T. (2018) *Solid State Sci.* **85**, 54.
[32] de Oliveira P.F.M., Torresi R.M., Emmerling F., Camargo P.H.C. (2020) *J. Mater. Chem. A* **8**, 16114.
[33] Streubel R., Wilms M.B., Doñate-Buendía C., Weisheit A., Barcikowski S., Schleifenbaum J. H., Gökce B. (2018) *Jpn. J. Appl. Phys.* **57**, 040310.
[34] Bai S., Hyeong J.L., Liu J. (2020) *Appl. Sci.* **10**, 3055.
[35] Suryanarayana C. (2001) *Progress Mater. Sci.* **46**, 1.
[36] Simondon E. (2018) Thesis INSA de Rennes.
[37] Dadbakhsh S., Mertens R., Hao L., Van Humbeeck J., Kruth J.P. (2019) *Adv. Eng. Mater.* **21**, 1801244.
[38] Guo J., Fan X., Dolbec R., X S., Jurewicz J., Boulos M. (2010) *Plasma Sci. Technol.* **12**, 188.
[39] Dadbakhsh S., Mertens R., Vanmeensel K., Vleugels J., Humbeeck J.V., Kruth J.-P. (2018) *Procedia CIRP* **74**, 39.
[40] Attar H., Prashanth K.G., Zhang L.-C., Calin M., Okulov I.V., Scudino S., Yang C., Eckert J. (2015) *J. Mater. Sci. Technol.* **31**, 1001.

1.5 Architecture-by-Design: Focus on Ceramic AM

Philippe BELLEVILLE

CEA, DAM, Le Ripault, 37260 Monts, France

1.5.1 Introduction

Since 2010, the French Atomic Energy and Alternative Energy Commission (CEA) has implemented an approach called "The digital materials approach". From numerical simulation experiments, it is therefore possible to predict the mechanical properties (Young's modulus) and thermal properties (thermal conductivity λ) of a material by knowing the intrinsic properties of the material, the solid volume fraction of the object and its architecture. This approach can be applied to an infinity of objects, real or virtual. The design of optimized materials remains a challenge today. Indeed, to guarantee optimized properties of a material, it is important to build a relationship between the architecture and the properties and if possible, to design the material architecture as required by solicitations. Nevertheless, it is not possible to build every optimized structure using conventional building techniques such as molding, machining or even chemical bottom-up approaches. On the other hand, additive manufacturing (AM) techniques and particularly laser stereolithography are perfectly suited to the development of optimized complex objects.

The work described in this chapter relates to the development of oxide-based ceramics using an innovative coupling between additive manufacturing and sol–gel chemistry. The optimization of architecture of 3D open structures used for example as mechanical reinforcements in aerogels for thermal insulation is determined based on the CEA's "Digital Materials" approach. The objectives of this study are to develop new formulations of laser stereolithography photoresins to develop the targeted structures, and to characterize the as-made objects to compare the experimental properties with those predicted numerically. The influence of the resin composition and the fabrication process on the final properties of the material will be detailed.

1.5.2 Development of Ceramics by Additive Manufacturing: State-of-the-Art

By definition, AM is a manufacturing technique by adding material layer-by-layer, unlike conventional techniques which very often require the use of a mold or a machining step and therefore a subtractive manufacturing of the final object.

All AM procedures are based on the same principle. First, the realization of a digital model requires its development by Computer Aided Design (CAD). This 3D model is then cut into two-dimensional parallel slices. Several slicing techniques exist, leading to different file formats (.stl or. slc for example). These files differ in the

DOI: 10.1051/978-2-7598-2446-5.c051

way the objects are described. Finally, these files are transferred to prototyping equipment for formatting the 3D object.

AM methods can be classified into several families according to the initial form in which the material is used: powders, liquids or solids. Five methods are mainly used in AM [1, 2], with sometimes several techniques resulting from these methods. A. Zocca *et al.* reported on the state-of-the-art of these seven main AM techniques in the American Ceramic Society in 2015 [3]. The designation "direct AM" means that each layer of material successively stacks on the previous one to give the final shape of the object. When the term "indirect AM" is used, a first layer of material (liquid or powder) is deposited, then the desired shape is then consolidated using an additional element (laser, binder, etc.…).

Photopolymerization of a liquid polymer, using stereolithography (SLA) and DLP (Digital Light Processing) techniques, has historically been the first technology developed.

The different additive manufacturing techniques by photopolymerization are based on an identical process. The construction of the object takes place by a successive stack of photopolymerized layers by irradiation of UV radiation at a well-defined wavelength. The techniques differ however by the method of irradiation of the resin, thus influencing the size of the objects that can be produced, the resolution of printing and the duration of these prints.

Three main irradiation techniques exist. Irradiation with a laser corresponds to the technique of stereolithography (SLA) and microstereolithography (μ-SLA). Mask projection irradiation is used in the DLP (Digital Light Processing) technique. The third technique is based on focusing in a very small volume of matter from a short pulse laser, and is called the 2PP (Two Photon Polymerization) technique. Stereolithography consists of a laser scan of the resin. The laser beam is moved by a system of galvanometric mirrors controlled by the computer. From the information of the digital file, the printer will automatically manage the movement of the laser to polymerize each layer (figure 1.5.1). The position of the laser cannot be controlled by the user, but is managed by the machine. Various parameters such as the speed of movement of the laser, the pitch along the vertical (which corresponds to the thickness of the printed layers) or even the space between each line of the laser can be controlled when setting up the machine. The size of the laser spot is generally between 10 and 50 μm, so it is not possible to print patterns finer than these dimensions. The μ-SLA process is identical to the "classic" SLA, however the size of the laser spot is smaller, on the order of 1–5 μm.

The techniques with the finest level of detail are not suitable for the production of parts with dimensions of the order of several cm and over. The SLA technique is the one with the greatest flexibility, particularly with the control of print parameters.

When making ceramics by stereolithography, the resin must be loaded with an inorganic precursor, generally ceramic powder. Once the part is printed, it is heat treated, both to sinter the particles, but also to debind the organic part. The printed part undergoes a volume shrinkage during this treatment. This volume shrinkage can be integrated into the model defined by CAD in order to obtain an object with the required dimensions at the end of the process. The debinding and sintering

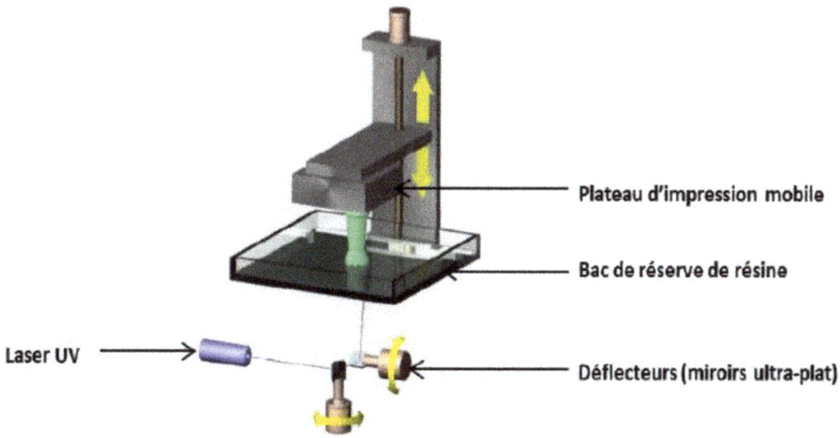

FIG. 1.5.1 – Schematic view of the SLA process using laser irradiation.

parameters (temperature, heating rate, level, duration) influence the final microstructure of the object, as well as the volume shrinkage observed. To obtain a dense part with a minimum of defect at the end of the heat treatment, it is necessary that the resin is heavily loaded with inorganic precursor, up to 50% by volume (or about 80% by mass) [4–6].

The presence of this ceramic charge has several consequences, both in terms of printing and resolution of the printed object. First of all, the viscosity of the suspension increases strongly with the ceramic rate following the Krieger–Dougherty model as presented in [4]. If the viscosity of the resin is too high, the flow of the resin cannot take place completely between printing two layers. The resin surface will no longer be homogeneous, making it impossible to print the object. The incorporation of ceramic particles also influences the printing process. Indeed, the particles absorb part of the UV radiation, but also diffuse part of it. A high concentration of particles prevents radiation from penetrating deeply in the resin and decreases the amount absorbed by the lower layers.

The growing number of scientific papers on ceramic SLA (figure 1.5.2) demonstrates the importance and interest in the subject in the scientific community in recent years.

1.5.3 Formulation of a High-Resolution Printing Stereolithography Resin

In this study, we propose to adjust the composition of the resins according to the geometries of the objects to be printed in order to get more flexibility. The formulation of a "custom" resin for the production of ceramics by SLA will therefore be developed hereafter.

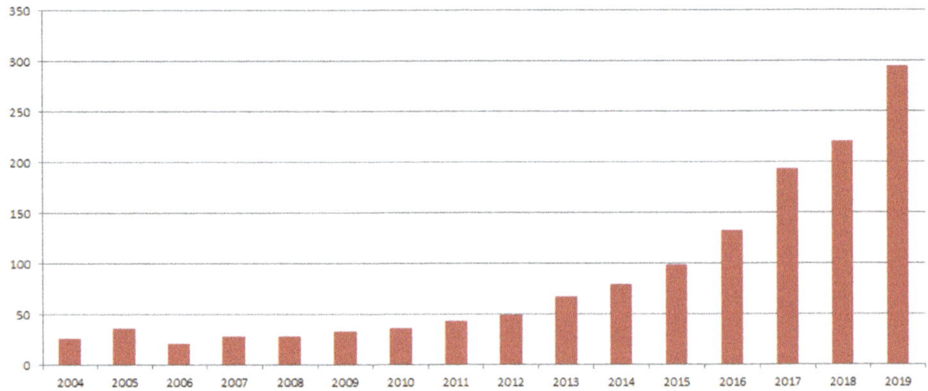

FIG. 1.5.2 – Number of publications per year with the terms "stereolithography" and "ceramic" on Science Direct (as of 08/11/2019).

The formulations of stereolithography resins are relatively standard. A resin contains several components: a photocrosslinkable monomer (mainly of polyacrylate or epoxy type), associated with a specific photoinitiator (respectively of radical or cationic type). A reactive diluent of low molecular weight monoacrylate or polyacrylate type is used to control the viscosity of the resin. The use of monomers having certain specific functions such as urethanes for example brings hardness to the resin. Finally, numerous additives are used to increase the stability of the resin, the achievable resolution or to facilitate the debinding stages, for example.

Figure 1.5.3 summarizes the overall composition of the acrylate-based SLA resins as well as the mass percentage of each component.

Preparing a ceramic part using stereolithography requires the presence of an inorganic precursor in the resin, generally ceramic powders. The characteristics of the powder (chemical nature, particle size) as well as the loading rate and the

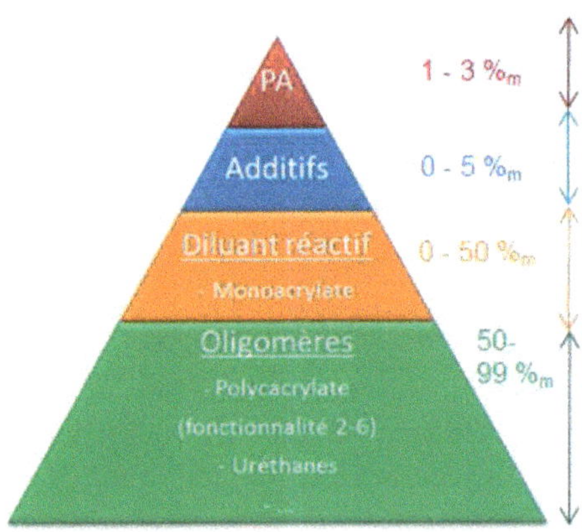

FIG. 1.5.3 – Composition of stereolithography resins based on acrylate polymers.

particle size distribution strongly influence the rheological properties of the resins and the printing process. A new type of material is also increasingly used as a ceramic precursor, and more particularly since the work of Z. Eckel *et al.* in 2016 [7]: these are pre-ceramic polymers.

Most of the work dedicated to the printing of ceramics by SLA concerns the fabrication of dense parts from a resin filled with ceramic powder (SiO_2, ZrO_2 or Al_2O_3). Once polymerized, the resin forms an organic matrix in which the ceramic particles are trapped. A thermal cycle consisting successively of a debinding step and then a sintering step eliminates the organic part and densifies the inorganic part, thus forming the final ceramic part. The nature of the ceramic powder has a great influence on the printing process and on the properties of the final ceramic.

The particle size of the powders also influences the microstructure and properties of the final ceramic. With submicron-sized powders, the proportion of the volume occupied by the grain boundaries and interfaces is very large compared to the volume of the grains themselves [8]. Thus, the finer the particle size of the powder, the more reactive it is. This results in a reduction in the sintering temperature when the particle size of the powder decreases. The initiation of fatigue cracks is also reduced because the intrinsic defects of the material are more homogeneous and smaller. On the other hand, the toughness decreases because the path of propagation of the cracks is less tortuous than with larger grain sizes.

We show that printing of CuG-shape cylinders (figure 1.5.4) was successful using several resins composed of acrylate monomers and a radical photoinitiator (Irgacure 819[3]). It has been possible to identify three main parameters influencing the print resolution: the formulation of the resin, the control of the print parameters and the orientation of the objects that one wants to print. The study on the print resolution as a function of the resin composition showed that the addition of a UV absorber (Sudan I[4]) greatly improves the print resolution by limiting the diffusion of polymerization. The study on the control of the print parameters has shown that the print resolution is also linked to the energy density received by the resin. Different parameters such as hatching or laser movement speed influence this energy density. The study found that the polymerized volume increases with energy density, which results in a decrease in print resolution. On the contrary, a too low energy density does not lead to crosslinking of the polymer with a high conversion rate. The mechanical properties of the part are then greatly altered. Finally, a large difference in print resolution was observed depending on the orientation of the objects to be printed. The larger the surface of the layer to be polymerized, the more energy the resin receives. The polymerization reaction is then likely to proceed in the layers already printed, which increases the polymerized volume and therefore decreases the printing resolution.

[3]Bis(2,4,6—trimethylbenzoyl)phenylphosphineoxyde.

[4]1-(Phenyldiazenyl)naphthalen-2-ol.

FIG. 1.5.4 – Images of different CuG cylinders printed with optimized SLA resin.

1.5.4 Development of Porous Ceramics Using Stereolithography from Ceramic Powders

The incorporation of powders of nanometric sizes in a stereolithography resin is only possible with a low load rate (less than 25% w/w), which is not sufficient to develop resistant ceramics. In order to increase this load rate, the use of a bimodal distribution of particles with micrometric and nanometric sizes is a solution making it possible to achieve a higher load rate. A load rate of 40% w/w. and a particle size distribution of 70% micrometric/30% nanometric corresponds to the optimal composition for printing 3D-shape such as CuG cylinders and maximizing the pressure at rupture (figure 1.5.5). Indeed, the mechanical characterizations carried out on these cylinders have shown that the maximum pressure $P_{\max}$ is the highest when the load rate and the nanometric powder concentration are maximum.

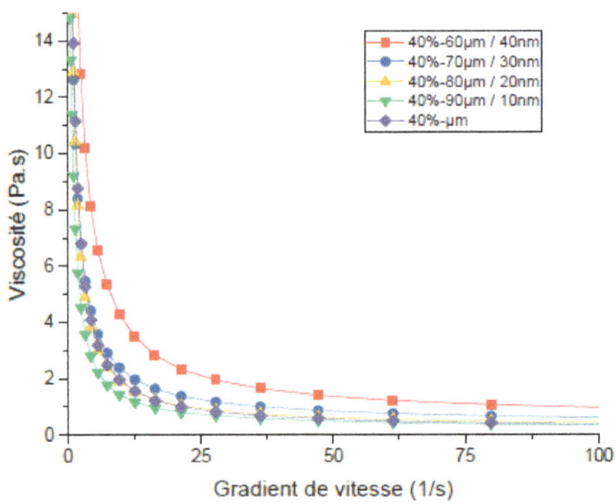

FIG. 1.5.5 – Viscosity of resins with a load rate of 40% w/w. depending on the particle size distribution and applied shear gradient (x-axis).

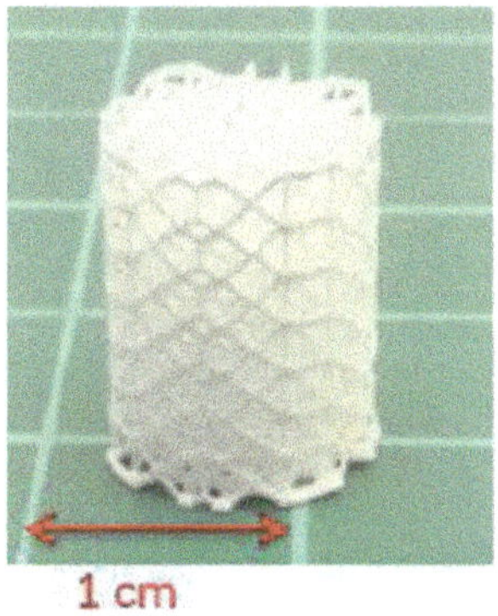

FIG. 1.5.6 – CuG cylinder made from a silica resin with 40% w/w. load rate (70% μm/30% nm).

The formation of crystalline phases during sintering has a great influence on the microstructure and the mechanical properties of the objects prepared. The crystallization increases with the sintering temperature and the hold time. This crystallization could generate cracks which damage the final material. Competition is observed at high temperatures with silica-based part between the densification of the material, which leads to a reinforcement of the material, and the formation of cracks which damages it. A sintering temperature of 1200 °C and a dwell time of 2.5 hr correspond to the optimal conditions identified in this work for silica parts (figure 1.5.6).

However, the density of objects made from these resins is low, around 0.15 g/cm^3. Two solutions can be considered to increase it: increase the diameter of the bars or reduce the size of the structure cell. Tests of printing objects with smaller cell sizes using resins of optimized composition have been carried out. Printing tests of CuG structures with bars of larger diameter were also carried out by adjusting the printing parameters. These tests systematically failed because the resin did not flow and remained trapped in the structure during printing step (the pattern is then no longer respected). In addition, some bars were not printed. One solution for printing these complex patterns is the incorporation of preceramic polymers into the formulation.

1.5.5 Preparation of Oxide and Oxycarbide Ceramics from Preceramic Polymers

The use of preceramic polymers (PdC – Polymer-derived Ceramic) is an innovative solution for the elaboration of ceramics by additive manufacturing. As the literature is showing, silicon precursor PdCs are used in laser stereolithography for the preparation of oxycarbide materials [9–11]. The development of oxide-type ceramics from PdC has so far never been studied to our knowledge. Moreover, these polymers can also be used to control the porosity of the material produced by choosing the

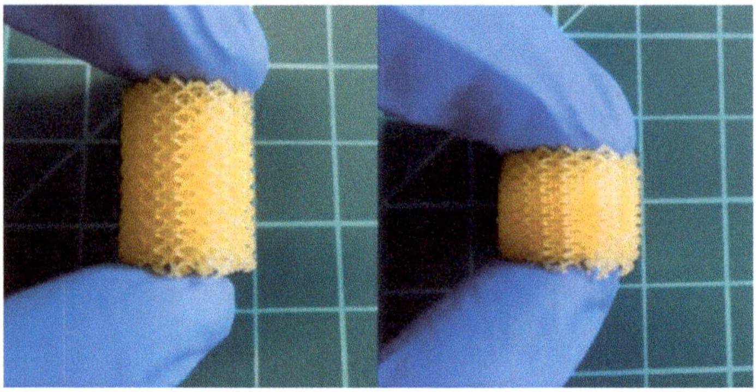

FIG. 1.5.7 – View of a cylinder pattern centered cubic with and without loading made from a resin with a load rate of 70% w/w using PdC H44.

appropriate precursors. The integration of such compounds in our formulation is therefore particularly advantageous in the optics of producing materials with controlled porosity [12–15].

The pre-ceramic polymers are precursors with a high ceramic yield that is to say with a high mass conversion rate into ceramic after heat treatment. Due to its good compatibility with the acrylate monomers, SILRES H44[5] from Wacker® is the PdC chosen for our experiments (figure 1.5.7).

The incorporation of pre-ceramic polymers into the formulations of SLA resins strongly modifies the rheological behavior of these by conferring on them a viscous character very positive for the printing of objects with complex geometries. Due to the ceramic yield of the PdC, a significant shrinkage is observed during the sintering of the objects leading to the production of ceramics with high solid volume fraction and small bar diameter. The charge rate and the PdC ratio allow the microstructure of the elaborate ceramics to be modulated.

When the level of PdC or the diameter of the bars becomes too high, all the gases resulting from the decomposition of PdC cannot escape before the densification of the material, which leads to the incorporation of carbon in the structure and the formation of SiOC type materials (figure 1.5.8). Cracks are also observed in the microstructure of the ceramic. Even if the carbon concentration is less than 0.1% w/w, it directly impacts the mechanical properties of the ceramic part. Indeed, at equivalent density, the failure pressure of these structures is higher than that of oxide type ceramics. This result has never been reported yet.

The incorporation of PdCs into the formulation also allows the formation of silicon oxycarbides with carbon concentrations of up to 30% w/w when the heat treatment is carried out under a non-oxidizing atmosphere (N_2 or Ar for example). A minimum load rate of 40% w/w is necessary for the structures to withstand the heat treatment. The PdC ratio has little influence on the microstructure of the processed ceramic. The main parameter controlling the carbon concentration is the

[5]Poly(methyl phenyl silsesquioxane).

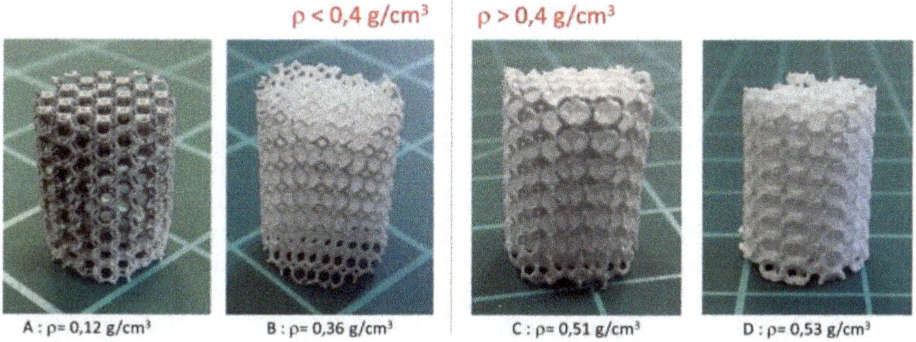

FIG. 1.5.8 – Images of Kelvin hybrid structures, A grayish structure; B, C & D oxide structures.

diameter of the bars. This concentration increases when the diameter and therefore the density of the structures increases. For low densities of structures, the increase in mechanical properties linked to the presence of carbon has not been demonstrated. On the other hand, for structures with densities greater than 0.20 g/cm^3, the $P_{\max}$ of oxycarbide materials is almost twice as high as the $P_{\max}$ of structures with a low carbon concentration (C% $<$ 0.1% w/w) at isodensity. Thus, the presence of carbon is very interesting for optimizing the mechanical properties of ceramics (figure 1.5.9).

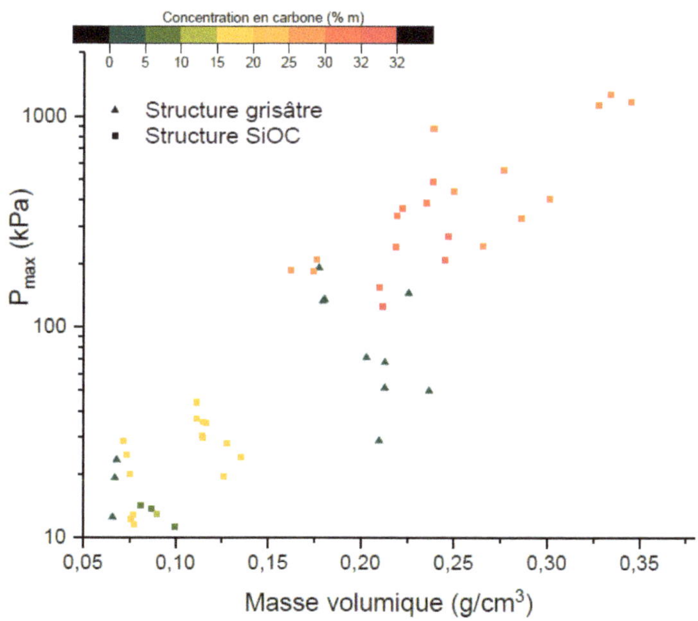

FIG. 1.5.9 – Failure pressure of SiOC and greyish structures as a function of the density of the structure and the carbon concentration.

1.5.6 An Example of Application of the Approach: Optimized Thermomechanical Properties of Thermal Insulators

In order to validate the use of *numerical experiments* to determine the optimal architecture of structures, we have compared the properties predicted numerically for four different patterns (cubic, centered-cubic, Kelvin, Kagome) with the results of free compression experiments performed on same structures. The comparisons between the numerical and experimental values cannot be made directly because on the one hand, the property considered is Young's modulus E while on the other hand, it is the maximum pressure applied before failure (P_{max}). However, the relative values of the properties between the different patterns can be compared.

The results of simulations of the numerical structures as well as the experimental characterizations are presented on figure 1.5.10. With the volume fractions considered, the experiments of numerical simulations make it possible to set up the following ranking: $E_{\text{Cubic}} > E_{\text{Kelvin}} \approx E_{\text{Kagome}} > E_{\text{Cubic-centered}}$.

The experimental results generally confirm the results of the numerical simulations. For example, for a solid volume fraction of 0.07 the order $P_{\text{max-Cubic}} > P_{\text{max-Kagome}} \approx P_{\text{max-Kelvin}}$ is verified.

To compare our resin formulations to others, CuG-shape cylinders were prepared from two commercially available stereolithography resins loaded with silica powder, but also with a resin loaded with alumina powder formulated at CEA. The maximum pressures of CuG structures as a function of the density of the structure are presented in figure 1.5.11. The P_{max} of these structures is compared with that of CuG structures made from resin of powder-based composition and PdC-based resins.

The density of the CuG cylinders produced from commercially available resins is comparable to the density of the cylinders produced from our PdC-based compositions and our resin with a powder load rate of 35% w/w. The maximum pressure

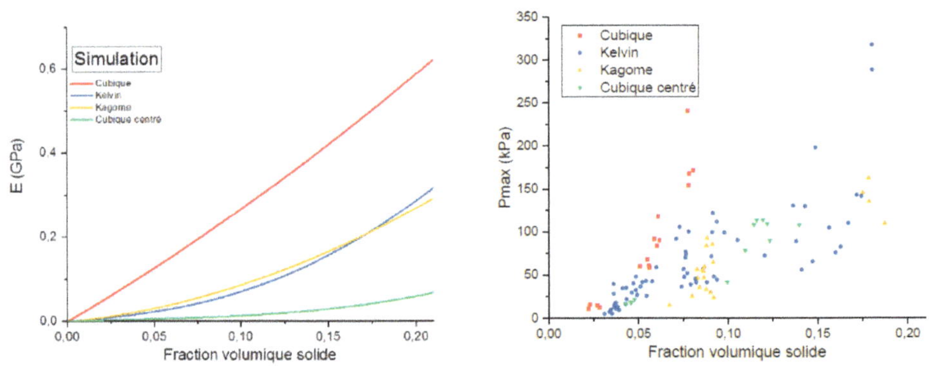

FIG. 1.5.10 – Simulated Young's modulus for the cubic, cubic-centered, Kelvin and Kagome patterns as a function of the solid volume fraction (on left); and results of free compression experiments carried out on these same structures (on right).

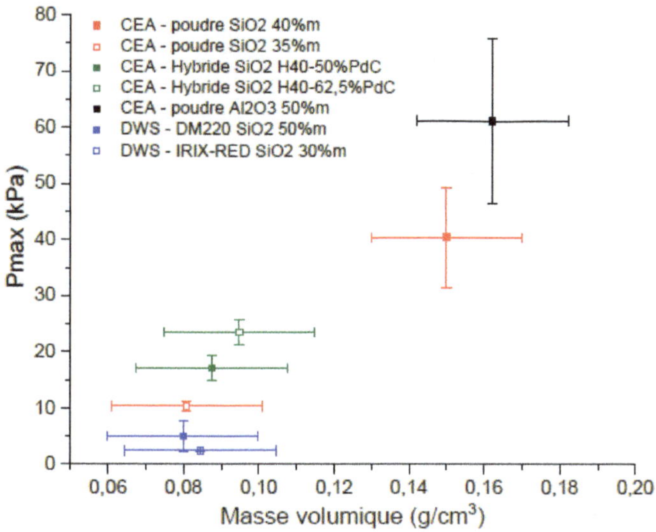

FIG. 1.5.11 – Comparison of maximum pressures of the CuG structure for different materials (CEA resins and commercially available resins so-called DWS).

applicable on these objects is quite low ($P_{max} < 10$ kPa), they are very fragile and are difficult to handle. For equivalent densities, the CuG structures produced from the CEA PdC-based resins are much more resistant, which results in P_{max} values two to three times higher.

The CuG cylinders produced from alumina have a density close to that of the cylinders produced from silica from CEA resins. The maximum pressure of these structures is however higher (60 kPa against 40 kPa for SiO_2 structures), which confirms the interest of alumina formulation to optimize the mechanical properties of 3D open structures.

In this study, we also wanted to investigate the possibility to mechanically reinforce silica aerogels so as to provide better mechanical properties and thus broaden the field of their applications. Aerogels have been produced using sol–gel chemistry. All the syntheses were carried out at CEA Le Ripault [16, 17]. The precursors are metallic alkoxides (Tetraethylorthosilicate-TEOS) in solution in ethanol for the aerogels developed in this study. Reinforcement takes place by carrying out gelation in a mold containing the AM-ceramic structure prepared before the supercritical drying step. Some photographs of aerogel reinforced with ceramic structures are shown in figure 1.5.12.

A non-reinforced aerogel with density d = 0.09 g/cm^3 was characterized using free compression experiments, as well as two additional aerogel reinforced with two Kelvin structures, one of the oxide types with a density equal to 0.11 g/cm^3 and the second of the type "oxycarbide–greyish" with a density equal to 0.22 g/cm^3. The compression module of the reinforced aerogel is 5–10 times higher than the compression module of the aerogel alone. Comparisons of the P_{max} values beside the

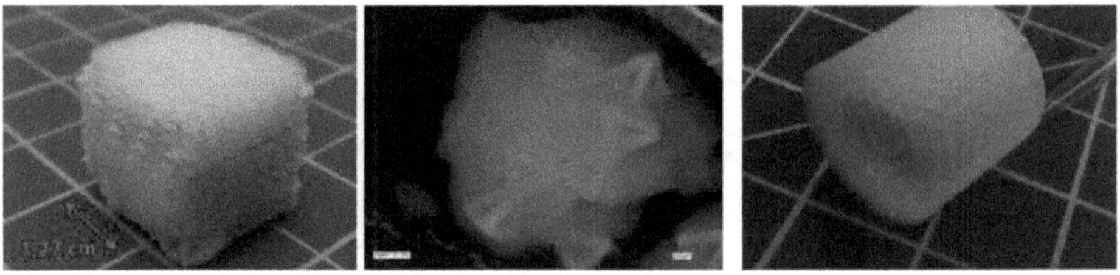

FIG. 1.5.12 – Photograph of silica aerogels reinforced by AM-ceramic structures (Kagome pattern).

aerogel alone confirm the strengthening of the aerogel embedding the ceramic structure. On the other hand, the deformation of the aerogel reinforced by the grayish structure is very low (the $P_{\max}$ is obtained for a deformation of approximately 0.1%), which results in a compression modulus 10–100 times higher than for aerogels reinforced from Kelvin structures of oxide type. At this deformation, $P_{\max} = 195$ kPa, the aerogel is very slightly deformed as long as the pressure applied is less than this value, which reinforces the interest of strengthening aerogels using these 3D AM-structures to facilitate their handling and use.

1.5.7 Conclusion

This study is one of the first to use a coupling between stereolithography additive manufacturing and resin formulation using sol–gel chemistry. For this, a fully organic stereolithography resin with high printing resolution was first formulated. From this organic base, resins allowing the production of ceramic silica structures have been formulated. In order to optimize the mechanical properties of the ceramic formed, a nanometric size powder was used. However, in order to combine high loading rate and rheological properties compatible with the printing process, it was necessary to add a second inorganic precursor to the formulation. Two types of formulations have thus been studied, formulations based on a bimodal distribution of silica powders of nanometric and micrometric sizes and formulations made of nanometric silica powder and a pre-ceramic silica polymer.

The incorporation of preceramic polymer (PdC) into the formulation allowed to show that on the one hand, the rheology behavior of hybrid resins (containing inorganic load coming from a mixture of nanometric powder and PdC) is completely viscous which is very suitable for the printing of complex shapes, and on the other hand, the shrinkage observed during the heat treatment is much greater than what is observed from the formulations consisting solely of ceramic powders. Thus it was possible from these hybrid compositions to develop ceramics with a very complex architecture, a low density (<0.15 g/cm^3), bars of small diameters (less than 175 μm) and good mechanical properties ($P_{\max} = 60$ kPa), but also structures with high density (up to 0.4 g/cm^3) while preserving the original pattern (Kelvin), a bar diameter of the order of 400 μm at maximum and a $P_{\max}$ of 140 kPa. Ceramics made from hybrid resins containing residual carbon have, at isodensity, a higher breaking pressure than equivalent oxide type ceramic. In addition, it is possible to develop ceramics with this microstructure with very high density ($d > 0.5$ g/cm^3) and therefore with very good mechanical properties ($P_{max} \approx 1.2$ MPa). The failure pressures of the structures developed in this study are higher than those of ceramics produced from commercial stereolithography resins with similar load rates. This confirms the importance of the formulation of resins related to the properties of the objects produced and the advantage of using nanometric sized powders in the formulations [18].

Finally, the mechanical characterizations carried out on the silica aerogels reinforced by Kelvin structures of oxide type demonstrate an increase in the compression modulus greater than a factor of 2 compared to a non-reinforced aerogel. When the reinforcement takes place using an oxycarbide ceramic with a low carbon concentration ("grayish" ceramic; C% $<$ 0.1% w/w), the deformation is almost zero before failure which takes place at outstanding pressure of 195 kPa. At equivalent pressure, the deformation of the aerogel alone is 16.5%. These results confirm the effectiveness of mechanical reinforcements to optimize the mechanical resistance of aerogels.

This work illustrates the advantage of controlling the formulation of stereolithography resins in the additive manufacturing of ceramic materials and above all opens the way to numerous adaptations concerning the development of functional materials whether they are used for their thermomechanical properties or in the field of energy.

References

[1] Yeong W.Y., *et al.* (2013) State of the art review on selective laser melting of ceramics, High Value Manufacturing – Bártolo *et al.*, Taylor & Francis Group.

[2] Vaezi M., *et al.* (2013) A review on 3D micro-additive manufacturing technologies, Int. J. Adv. Manuf. Technol. **67**, 1721.

[3] Zocca A., *et al.* (2015) Additive manufacturing of ceramics: issues, potentialities, and opportunities, J. Am. Ceram. Soc. **98**, 1983.

[4] Chartier T., *et al.* (2008) Fabrication of millimeter wave components via ceramic stereo- and microstereolithography processes, J. Am. Ceram. Soc. **91**, 2469.

[5] Montemayor L., *et al.* (2014) Design and fabrication of hollow rigid nanolattices via two-photon lithography, Adv. Eng. Mater. **16**, N°2.

[6] Ovsianikov A., *et al.* (2008) Ultra-low shrinkage hybrid photosensitive material for two-photon polymerization microfabrication, ACS Nano **2**, 2257.

[7] Eckel Z.C., *et al.* (2016) Additive manufacturing of polymer-derived ceramics, Science **351**, 6268.

[8] Greco A., *et al.* (2001) Stereolitography of ceramic suspensions, J. Mater. Sci. **36**, 99.

[9] Mera G., *et al.* (2015) Ceramic nanocomposites from tailor-made preceramic polymers, Nanomaterials **5**, 468.

[10] Bill J., Aldinger F. (1995) Precursor-derived covalent ceramics*, Adv. Mater. **7**, 775.

[11] Pham T.A., *et al.* (2006) Three-Dimensional SiCN ceramic microstructures via nano-stereolithography of inorganic polymer photoresists, Adv. Funct. Mater. **16**, 1235.

[12] Colombo P. (2008) Engineering porosity in polymer-derived ceramics, J. Euro. Ceram. Soc. **28**, 1389.
[13] Lale A., *et al.* (2016) Organosilicon polymer-derived mesoporous 3D silicon carbide, carbonitride and nitride structures as platinum supports for hydrogen generation by hydrolysis of sodium borohydride, Int. J. Hydrogen Energy **41**, 15477.
[14] Colombo P., *et al.* (2010) Polymer-derived ceramics: 40 years of research and innovation in advanced ceramics, J. Am. Ceram. Soc. **93**, 1805.
[15] Mera G., *et al.* (2013) Polymer-derived SiCN and SiOC ceramics – structure and energetics at the nanoscale, J. Mater. Chem. A **1**, 3826.
[16] Phalippou J., Kocon L. (2004) Aérogels Aspects fondamentaux, Techniques de l'ingénieur, États de la matière, base documentaire: TIB109DUO.
[17] Phalippou J., Kocon L. (2004) Élaboration des gels et des aérogels, Techniques de l'ingénieur, base documentaire: TIB489DUO.
[18] French Patent assigned to CEA #1900674 dating from January 25, 2019.

1.6 Some Aspects of Numerical Modelling for Additive Manufacturing

Stéphane GOUNAND, Anaïs BAUMARD, Olivier ASSERIN and Séverine PAILLARD

DES, Service d'Etudes Mécaniques et Thermiques (SEMT), CEA, Université Paris-Saclay, 91191 Gif-sur-Yvette, France

Abstract

This chapter focusses on some aspects of numerical modelling of WAAM, WLAM, SLM and DED processes.

1.6.1 Introduction

Super insulated architectured materials are based on ceramics or ultra-light and resistant hierarchical amorphous structures and shapes that are impossible to manufacture by conventional machining, such as minimal periodic surfaces like gyroids or open-pore organised metallic foams to maximise the solar illumination received for Thermodynamic Solar Power Plants. They are also "cold" fuel pellets to gain margins in case of loss of primary coolant, new Cobalt-free composition gradient hard coatings based on Nickel to replace Stellite, or a no longer manufactured EDF manual valve control, a monobloc grid for holding fuel needles that cannot be made by conventional machining, a rotor lightened by means of lattices, tissues for organ reconstruction or for therapeutic issues.

These are also super materials that can withstand severe environmental constraints thanks to a combination of properties or antagonistic functions such as hardness-ductility that would have been impossible to obtain with a single material. But also components that can no longer be produced or concepts that cannot be realised with conventional processes.

Achieving parts with the expected characteristics and target properties requires high manufacturing quality in a reproducible and cost-efficient manner. This can be achieved by improving productivity, reducing the high cost of raw materials, and increasing the products and manufacturing process performance.

First need is to control the manufacturing process. Additive manufacturing has already demonstrated the ability to produce conform parts comparable to their conventionally processed ones. However, the poor repeatability and reproducibility of the machines are an obstacle. Users would like to make progress in understanding the influence of operating parameters and raw material properties in order to control manufacturing, tailor the properties and functions, and create architectural structures.

Second need is to ensure the quality and performance of the product. Product performance requires a better scientific understanding of additive manufacturing processes, including the influence of microstructure on properties and ultimately on

DOI: 10.1051/978-2-7598-2446-5.c061

the service life of manufactured parts. It is necessary to understand the effect of composition on microstructure and properties under complex thermal cycles.

Process simulation could help in this regard. However, taking all physical phenomena into account requires modelling at different scales; this development will take place over time and requires input data that are not easy to measure and experimental validation.

1.6.2 Overview of Modelling Approaches for Additive Manufacturing

1.6.2.1 Why Model and Simulate?

The reduction in the number of components, functional integration, concatenation of functions, material savings, production as close as possible to the sites, the offer of alternative solutions to certain current sources of supply, and rapid prototyping, all lead to easier manufacturing and control, reduced lead times and costs, improved responsiveness and overall performance and productivity.

However, additive manufacturing ultimately concerns relatively few components. These will be high value-added parts subject to very specific or multiple constraints, or components that can no longer be produced or concepts that are not possible with conventional processes. It is mainly the improvement of performance that is expected, and it is then the cost/performance ratio that establishes the interest of additive manufacturing for the components.

Improving performance concerns the product with antagonistic functional properties challenging the Ashby diagrams and dealing with severe constrains, and also the manufacturing process by simplifying assembly, the achievement of daunting concepts with conventional processes.

Few of the main obstacles to commercial and industrial use of additive manufacturing are the low level of repeatability of the manufacturing process and reproducibility between machines. In order to improve performance, the entire manufacturing process chain must be under control, from the raw material through the process to the finishing treatments.

Thus, it is important to understand the effect of the many parameters that come into play at each stage, for example, the effect of the trajectory and speed of the deposits on the morphology of the beads, on the thermal cycles. We would also like to be able to explain the lack of fusion, the instabilities of the bath, the cracking. Particularly, in SLM, there are also problems of denudation, balling, gas distribution which will have an influence on defects such as porosities, and the control of deformations and stresses, the magnitude of which can lead to the rupture of supports or to important deformations forcing multiple construction stops and scrapping.

1.6.2.2 Modelling Approaches, Multiscale and Multiphysics Aspects

The additive manufacturing process consists of creating objects by a sequence of successive layers of material. The material can be deposited by extrusion (FDM) for polymers, ceramic pastes, or in the form of molten metal drops brought by wire or by a fusible electrode, or sprayed powder (DMD), for thermoplastics or metal alloys. The material may also be already present and then selectively consolidated (coalesced and then solidified or polymerized) by an additional heating source such as a laser or electron beam as in the case of powder bed, or a liquid polymer resin.

For metal fabrication processes with powder or wire such as SLM, DMD, WAAM, WLAM three scales can be distinguished. That of the powder particles which are considered discrete or the so-called microscopic wire, that of the weld-pool or powder bed according to the so-called mesoscopic manufacturing process where the powder is considered continuous and that of the so-called macroscopic part where the material is considered homogeneous. Each of these scales requires special modelling. At the microscopic scale, it is the coating of the powder, the interaction of the material process on the powder grains, the phenomena of denudation, sintering, the shape of the molten zone, the evaluation of the absorption and diffusion of energy that are of interest. At the mesoscopic scale, the material process interaction is considered in the molten pool, but the material is considered continuous. Molten pool dynamics, deposit shape, microstructure, dilution, defects such as porosities, hot cracking, bath instabilities, models of phase transformations, convection, radiation, evaporation, conduction, absorption, reflection, of gravity, magnetohydrodynamics, coalescence, capillarity, wetting, solidification shrinkage are used with various methods such as Lattice Boltzmann (LB), Discrete Element Method (DEM), Smoothed-particle hydrodynamics (SPH), Volume Of Fluid (VOF) and Finite Elements (EF).

In DMD, the powder jet in the nozzle is modelled as well as the laser material interaction at the nozzle exit on the jet and on the workpiece, followed by the material input and the dynamics of the pool.

With wire, the material process interaction is modeled at the wire and part scale, then the material input is sometimes modeled at the drop scale. When the process is an electric arc, electromagnetism is added for the pool dynamics and the arc plasma, the model then becomes magnetohydrodynamics.

At the macroscopic scale, models are based on thermomechanics and thermometallurgy. The treatment is quite similar between material deposition and powder bed processes. In particular, the energy transfer to the workpiece is either the result of microscopic-mesoscopic calculation, or imposed in the form of an equivalent heat source calibrated on previous experiments.

At the mesoscopic and macroscopic scales, solidification, microstructural transformations, residual stresses and deformations are also considered. However, the mesoscopic scale commonly concerns not more than one bead. Thus, changes in these quantities over time and under the effect of other deposits are not taken into account at this scale.

A comprehensive review of all the components and methods of additive manufacturing modelling could be found in Panagiotis Stavropoulos "Modelling of additive manufacturing processes: a review and classification" in *Manufacturing Rev.* 5, 2 (2018) https://doi.org/10.1051/mfreview/2017014.

None of these models has been established, one can find all the DNA of welding simulation with the specificity of the very large number of passes, sometimes very thin beads and stronger thermal gradients (SLM). The developments already made in welding are already a good basis, nevertheless the very large number of passes calls for important numerical challenges to overcome and the very strong thermal gradients require new models and validation experiments.

For powders, morphology (sphericity, size distribution), chemistry will have a decisive impact, as will wires in WLAM and WAAM. Taking these into account requires microscopic and mesoscopic modelling, as the implementation of the discrete element method and thermokinetic modelling as Calphad https://www.calphad.org/ by the tools of thermodynamic calculations (Thermo-Calc, OpenCalphad https://www.opencalphad.org/) and diffusion (DICTRA (module of Thermo-Calcl)).

In addition, the manufacturing machines have many operating parameters whose adjustment is complex by the difficulty to be under repeatability and reproducibility conditions. The user wishes to be able to predict the morphology of the cord and to have an operative control. Experimental designs (screening) combined with simple simulations would make it possible to identify trends.

1.6.3 Focus 1: Multiphysics of Liquid Metal Pool

Main contributor: Stéphane Gounand

In this section, we discuss some aspects of the physical and numerical modelling of liquid metal pools that arise during welding or additive manufacturing processes. Indeed, despite its quite small dimensions compared to the workpiece, the liquid metal pool has an important role in determining the local distribution of temperature. This is due to the fact that intense convective phenomena generally occur in the molten metal.

In section 1.6.3.1, we describe the physical setting of liquid metal pool modelling, first in the context of welding and then in the context of additive manufacturing. This leads us to consider in section 1.6.3.2, the effects of one of the main driving forces which act in the melt pool which is surface tension. From a numerical viewpoint, the small size of the weld pool relative to the workpiece leads to stringent constraints on the mesh, an aspect we discuss in section 1.6.3.3.

Notice that the computation and meshes of this section were obtained from our in-house Finite Element toolbox Cast3M [1].

1.6.3.1 Physical Context

Before going on to Additive Manufacturing (AM), as an example of complex multiphysics model, we describe what physically happens near the heat source in the case of the industrial-grade Tungsten Inert Gas (TIG) welding process.

1.6.3.1.1 Example of Tungsten Insert Gas (TIG) Welding

TIG Welding Process

The TIG metal assembly process uses a Tungsten, refractory electrode, an inert gas (Argon in general) brought through a nozzle which acts both as a plasmagenous medium and as a barrier against oxidation, and a generator to trigger the plasma.

This plasma acts as the heat source for melting both the base metal and a filler wire (figure 1.6.1).

Multiphysical Aspects of TIG Welding

A close-up view on the physical phenomena taking place in the plasma and in the base metal gives the daunting figure 1.6.2 which is not nearly exhaustive: for instance the plasma-weld pool interface deforms and metal evaporation can occur,

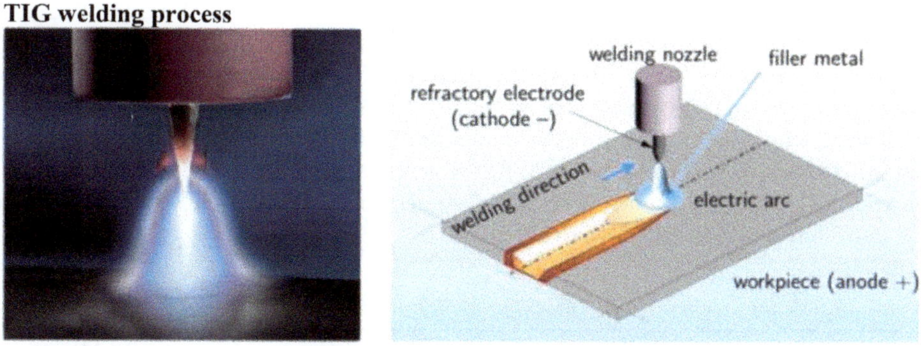

FIG. 1.6.1 – Tungsten Inert Gas (TIG) welding process: arc (left) and schematic diagram (right) [2].

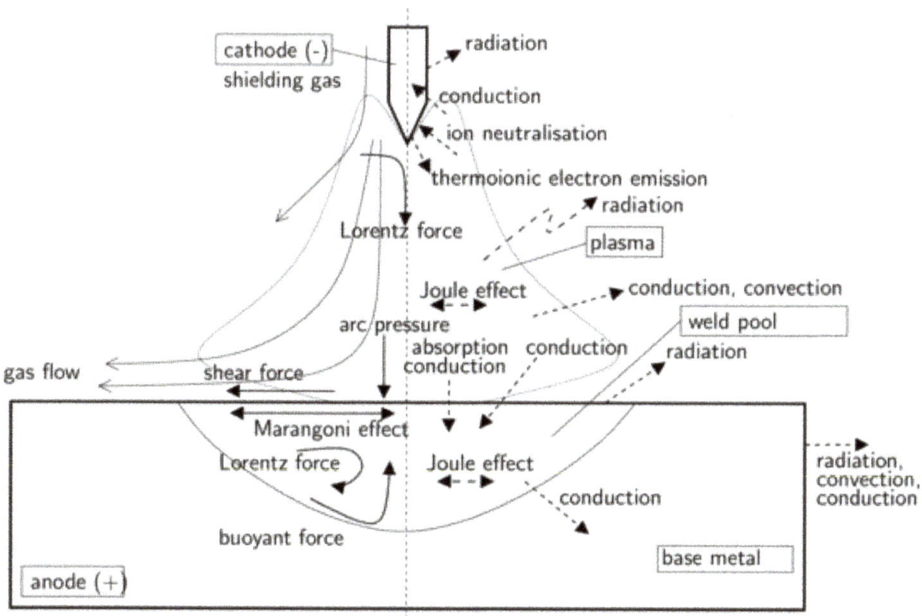

FIG. 1.6.2 – Physical phenomena in TIG process [3].

the weld pool- base metal interface is subject to all the complexity of metal solidification... Thermally important is the fact that both the arc plasma and the weld pool do not remain static and flow quite rapidly: $\approx 300\,\mathrm{m.s^{-1}}$ in the plasma and 0.1 $\mathrm{m.s^{-1}}$ in the melt pool. In general, the main driving force for the flow is the electromagnetic Lorentz force in the arc and the Marangoni force (due to surface tension) in the weld pool.

1.6.3.1.2 From Welding to Additive Manufacturing
Going from welding to additive manufacturing processes, we generally have that very similar physical phenomena occur. For example, the TIG-WAAM (Wire Arc Additive Manufacturing) process and TIG are physically similar. However, going from assembly to manufacturing entails some differences, mainly: metallic powder related aspects and geometry related aspects. In welding processes, the local metal-heat source geometry roughly looks like a plate while for AM (except at the beginning), it will rather look like a fin, which leads to different thermal pumping behavior. Also of importance are the differences in characteristic scales in the process parameters as shown in the table:

Typical characteristic scales.

Parameter	Welding (TIG)	Additive manufacturing (SLM)
$P_{\mathrm{heatsource}}$	1000 W	100 W
$r_{\mathrm{heatsource}}$	5.e–3 m	5.e–5 m
$u_{\mathrm{heatsource}}$	1.e–3 $\mathrm{m.s^{-1}}$	1 $\mathrm{m.s^{-1}}$
Metal addition	Filler wire	Powder

Typically, AM will use faster, more intense heat sources with a smaller length scale compared to welding. This smaller length scale notably implies: on the physical side, higher temperature gradients and surface tension effects and on the numerical side, meshing difficulties. We briefly elaborate on these two aspects in sections 1.6.3.2 and 1.6.3.3.

1.6.3.2 *Surface Tension Phenomena*

We first discuss the physics of surface tension forces before describing its effects on a melt pool.

1.6.3.2.1 Surface Tension Forces in 2D
To get a quick grasp on how surface tension works, it is easier to picture it in 2D in a discrete setting and remember the variational interpretation of surface tension: a force that tries to minimize a γ-weighted surface energy ($\int_S \gamma\,\mathrm{d}S$). γ is the surface tension in $\mathrm{N.m^{-1}}$.

Normal component: First we consider figure 1.6.3 (left) which represents a bent surface of two segments of same γ. The force acting on the end node of a segment is directed inward and tangentially trying to shrink the segment with magnitude

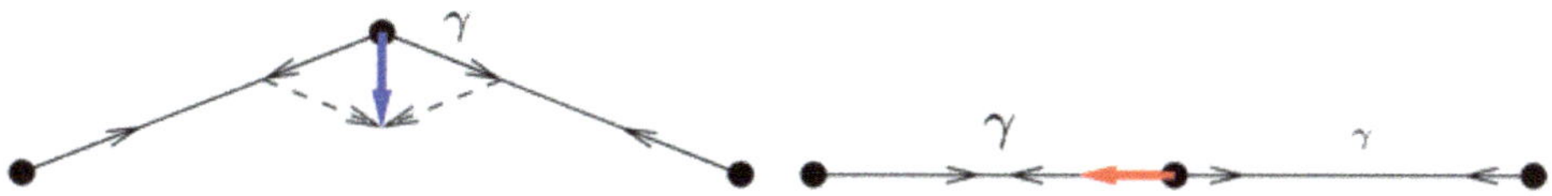

FIG. 1.6.3 – Normal (left) and tangential (right) surface tension forces.

γ (a 2D force has unit $N.m^{-1}$). The net force exerted by the two segments on the middle node is thus obtained by the parallelogram of force shown in the figure. It is directed in the normal direction (in an averaged sense) to the surface with a norm that grows with the variation of slope between the two segments (a discrete curvature). Going back to the continuous setting, it can be shown that following formula for the surface tension force holds: $f_n = \frac{\gamma}{R} n$ where $\frac{1}{R}$ is the curvature.

Tangential component: Second we consider figure 1.6.3 (right) which represents an unbent surface of two segments with different γ (higher for the left segment). Now, the net force exerted by the two segments on the middle node is directed in the tangential direction to the surface towards higher γ with a norm that grows with the variation of γ between the two segments (a discrete gradient). Going back to the continuous setting, it can be shown that following formula for the surface tension force on a flat surface with varying γ holds: $f_t = \nabla_s \gamma$ where ∇_s is the surface gradient.

Surface tension force: Now, adding the normal f_n and tangential f_t components, we get the whole surface tension force $f_{\text{Tension}} = \frac{\gamma}{R} n + \nabla_s \gamma$. The normal component is the *geometric* part, while the tangential component is the *Marangoni* part of the surface tension force and acts as a *shearing* force.

1.6.3.2.2 Liquid Metal Pool, Marangoni Effect

In the case of liquid metal pool, we generally consider that the Marangoni effect is mainly due to the variation of γ with temperature T. Indeed, γ is also sensitive to the concentration of the so-called tensioactive elements (sulfur S concentration is frequently considered for stainless steel materials), but this concentration can be assumed constant in the weld pool due to convective mixing. We then write: $\nabla_s \gamma = \mathrm{d}\gamma/\mathrm{d}T \times \nabla_s T$ with $\mathrm{d}\gamma/\mathrm{d}T$ the temperature derivative of surface tension and $\nabla_s T$ the surface temperature gradient.

Now, $\mathrm{d}\gamma/\mathrm{d}T$ is not constant. Figure 1.6.4 (middle, up) shows the temperature dependency of $\mathrm{d}\gamma/\mathrm{d}T$ for two sulfur concentrations (10 ppm blue and 300 ppm red) for 316L stainless steel. Notice that $\mathrm{d}\gamma/\mathrm{d}T$ even changes sign for the 300 ppm case. This leads to quite different flows in the melt pool together with different melt pool shapes (figure 1.6.5) for the 10 ppm and 300 ppm cases.

The melt pool shapes are obtained for the case of welding in flat position. In AM, it is expected that the melt pool will be smaller but with more intense surface tension effects: higher temperature gradients calls for higher Marangoni flow (velocities of $\approx 10\,\mathrm{m.s^{-1}}$ have been reported) and interface rounding geometric effect. Also, we did not discuss the wetting boundary conditions (triple line between melt pool, base metal and gas) but its importance could be paramount in the case of AM.

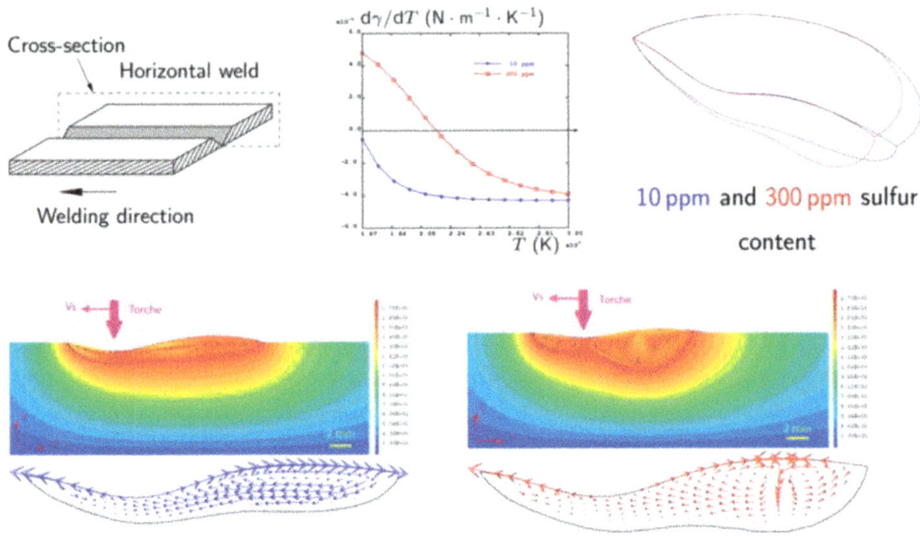

FIG. 1.6.4 – Melt pool shapes [4].

1.6.3.3 Meshing

For a numerical computation of a melt pool to be meaningful, sufficient meshing of the melted part should achieved. This a problem in Additive Manufacturing because the melt pool is quite small relative to the manufactured workpiece.

1.6.3.3.1 Manual Meshing

For a typical melt run on a simple plate geometry with the process parameters of the additive manufacturing case of section 1.6.4, we can come up with the manually tailored mesh of figure 1.6.5. Notice that the melt pool is barely visible at the intersection of the four coloured derefinement parts of the mesh. It is clear that manual meshing of more complex and time-evolving geometries can be very difficult.

1.6.3.3.2 Automatic Meshing

In order to circumvent the difficulty of manual meshing, one could rely on automatic meshing. In figure 1.6.6, we present some preliminary numerical experiments of the use of anisotropic adaptive meshing for a simplified melt run test case (only thermal effects are modelled with no melt pool modelling). Figure 1.6.6 (left) shows the initial non-adapted regular mesh and the corresponding numerical solution for the temperature field. This field shows numerical oscillations (dark blue color) due to insufficient mesh refinement. Figure 1.6.6 (right) shows the anisotropic adapted mesh that was optimised iteratively from the initial mesh and solution. The temperature field is now correctly resolved and free from numerical oscillations with the same number of mesh nodes.

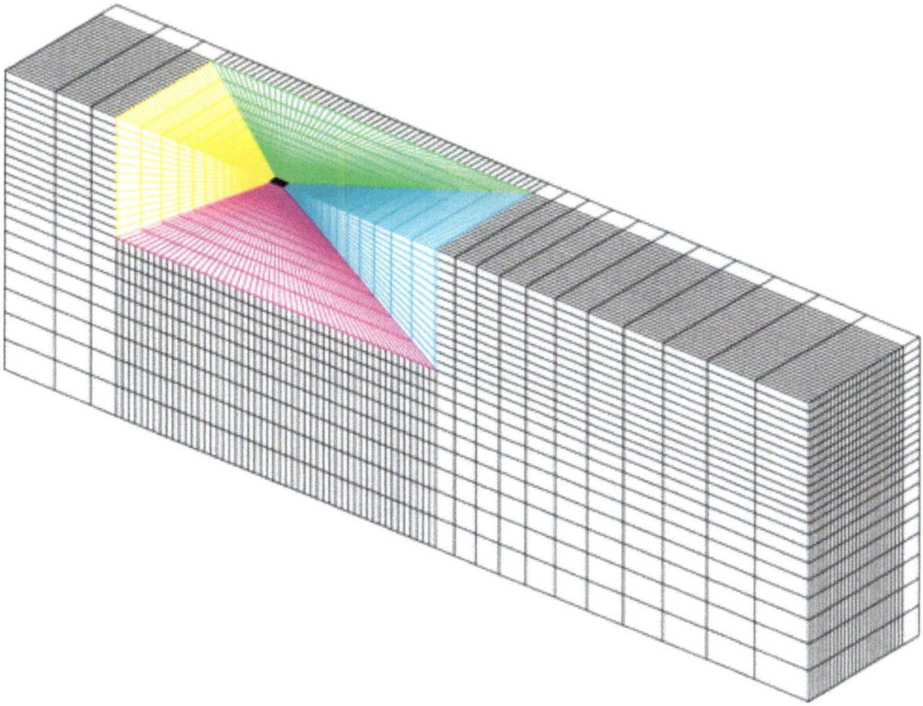

FIG. 1.6.5 – Manually tailored mesh.

Notice that we have focused on conforming meshing strategies (no hanging nodes) but non-conforming meshing is possible. Also, other numerical components are frequently needed to complement the meshing strategy: projection from one mesh to another, error estimators to drive the mesh adaptation process, and modification of the numerical method to account for non-conforming meshes...

1.6.4 Focus 2: Numerical Prediction of Grain Structure Formation in Additively Manufactured 316L Stainless Steel

Main contributor: Anaïs Baumard.

In this section, we discuss a numerical method for predicting grain structure formation during Laser Beam Melting of single track 316L Stainless Steel [5].

1.6.4.1 Context

Additive Manufacturing processes offer the possibility to reduce manufacturing time and material waste, and allows for the creation of structures with complex design. However, an anisotropic mechanical behaviour is frequently observed in components

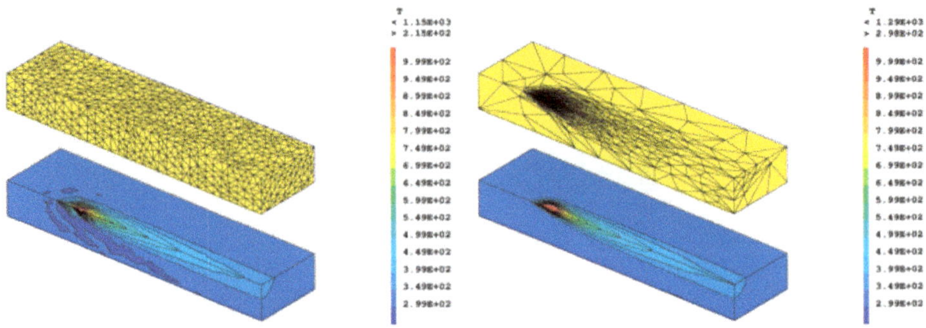

FIG. 1.6.6 – Automatically adapted mesh.

manufactured by AM processes. This anisotropy is directly linked to the component's grain structure characteristics, which are dependent on the process parameters. For this reason, the formation of grain structure during Laser Beam Melting (LBM) of a 316L stainless steel is investigated in this chapter.

For that purpose, an approach combining experiments and numerical simulations is adopted. In section 1.6.4.2, the experimental setup is described. In section 1.6.4.3, the numerical modelling approach follows with a comparison of grain characteristics obtained experimentally and numerically (section 1.6.4.4).

1.6.4.2 Experimental Setup

The experimental part consists in building instrumented one-layer 316L single-tracks by LBM. The study of the single-tracks allows for the characterization of the grains (microstructural analysis is carried out by optical microscopy and Electron Backscatter Diffraction), and the optical instrumentation allows for the observation of the molten pool. The one layer single-track geometry has been chosen in order to limit the building parameters involved in fabrication. Indeed, by choosing a single-track geometry, hatching distance and scanning strategy do not influence the built part. Moreover, as it is a first model validation work, the presented results are limited to melt-runs, which are equivalent to single-tracks but without powder. The process parameters are: laser power (P) of 400 W with a focal spot size of 150 μm, a wavelength of 1030 nm and a velocity (v) of 400 mm.s^{-1}. An airtight box filled with argon is used in order to reduce oxygen content (less than 100 ppm) (figure 1.6.7).

1.6.4.3 Numerical Modelling

The numerical modelling is based on a three-dimensional "CAFE" model, which couples Cellular Automata (CA) and Finite Elements (FE) simulations. It is used to predict grain formation during the LBM process.

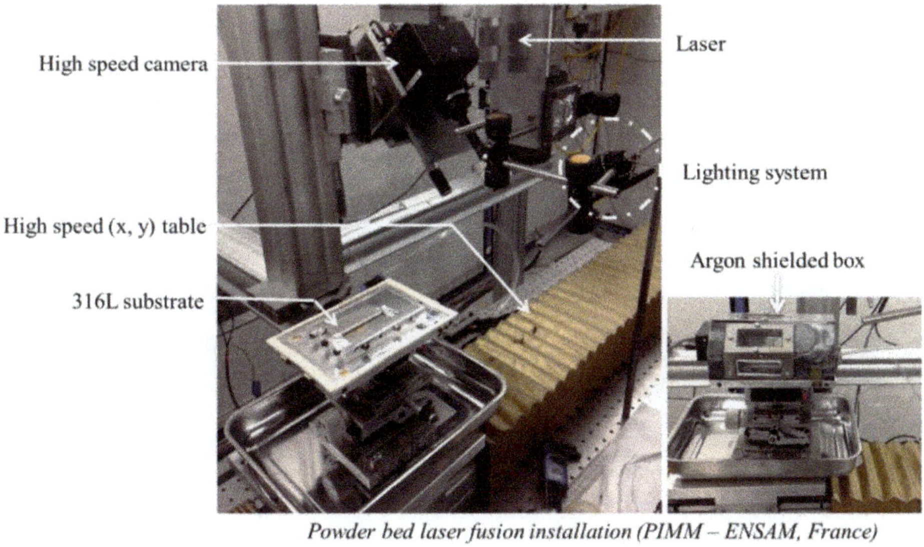

FIG. 1.6.7 – Powder bed laser installation (PIMM-ENSAM, France).

1.6.4.3.1 Thermal Analysis
The first part of the model consists in doing the thermal analysis which consists in solving the heat equation in 3D with a Finite Element method.

$$\rho C_p \frac{\partial T}{\partial t} = \Delta T + q$$

The calculations are carried out with the Cast3M software [1]. The input parameter for this equation (q) is a modelled volumetric heat source that accounts for both the true heat source (laser) and the convective effects in the molten pool. We choose a Goldak (double-ellipsoïd) volumetric heat source that has 4 geometric parameters (half-width, penetration, front and rear half-length of each ellipsoid) and 2 physical parameters (power and velocity). The physical parameters are identical to the laser's. The geometric parameters are chosen so as to approximately fit the experimentally observed molten pool boundary geometry (width, depth and length).

Figure 1.6.8 shows a good agreement between the dimensions of the experimental and numerical molten pool dimensions, validating the use of a Goldak source. Notice that the geometric parameters of the Goldak heat source change whenever a process parameter changes. For example, the geometric parameters are not the same for single tracks and melt runs.

1.6.4.3.2 Solidification Growth Modelling
The grain growth model is based on the cellular automaton (CA) model defined by Gandin and Rappaz [6], which we do not detail here. It focuses on grain structure evolution during solidification accounting for two main physical phenomena, namely

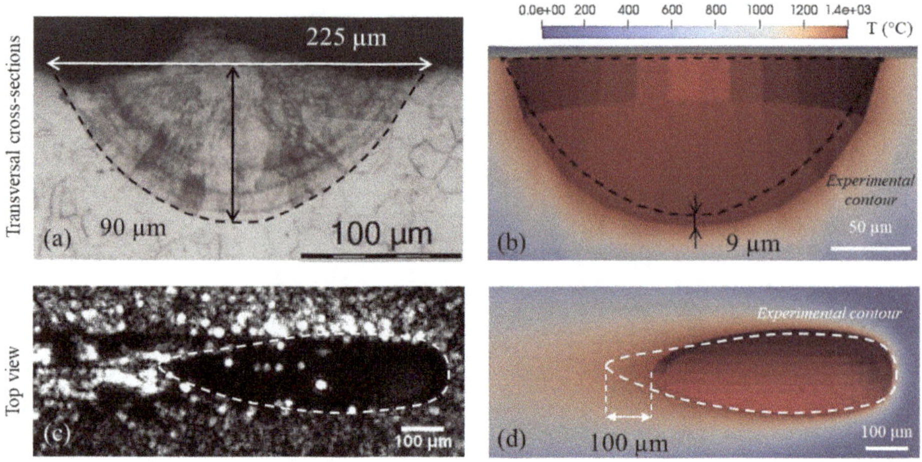

FIG. 1.6.8 – Comparisons of experimental, (a) and (c), and numerical, (b) and (d), dimensions of the molten pool observed in melt-runs.

nucleation and grain growth. The input data for the CA model are the temperatures calculated in the FE model. Notice that we only have a weak coupling between these two models because we assume that the solidification calculations have no influence onto the thermal ones. Also, the cellular automaton mesh is a Cartesian grid which is much finer than the FE mesh: interpolation is used to obtain the CA temperature from the FE computations. The time steps may also be different: they are generally smaller for the CA which is an explicit numerical method.

1.6.4.4 Experimental-Numerical Comparison

Eventually, numerical grain characteristics resulting from the simulations are compared to the experimental ones in figure 1.6.9 on transversal cross-sections. Both experimental and numerical images show visible columnar grains, with perpendicular orientation with respect to the solid interface. The morphology and width of the molten pool are almost identical, but numerical pool's depth is slightly longer. Also, it is clearly visible that the grains are thinner and more numerous in the numerical results. It is assumed that this is because the grain structure of the substrate is not correctly modelled. Indeed, in this current algorithm the substrate is composed of as many grains as cells, and each cell has a specific crystallographic orientation. Regardless, these numerical results are promising.

Future prospects for this study are numerous. Numerically, better modelling of the substrate's grain structure, multi-track configuration with thermal cycling or even multiphysics modeling of the laser and molten pool come to mind. Experimentally, further analysis of grain structure (not only in transversal cross-sections) varying the process parameters would be interesting, keeping in mind the combined numerical experimental approach.

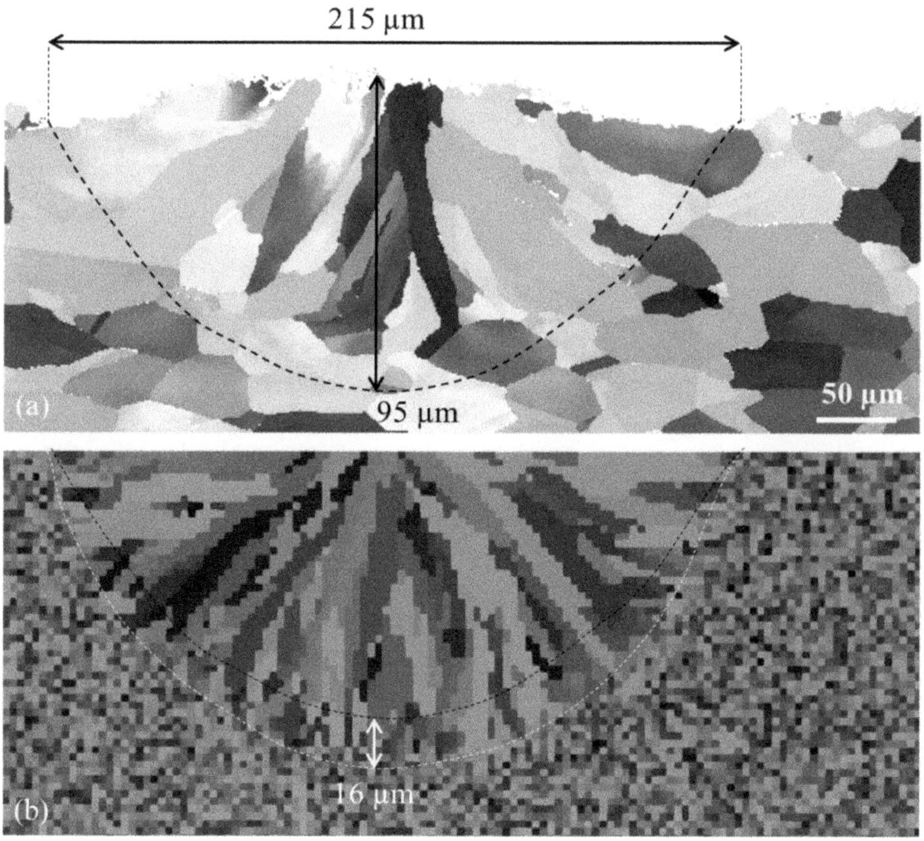

FIG. 1.6.9 – (a) Experimental and (b) numerical transversal cross-sections of grain structure obtained for melt-runs.

References

[1] "Cast3m Web site." https://www-cast3m.cea.fr/, 2020.

[2] Nguyen M.-C. (2015) "Modélisation et simulation multiphysique du bain de fusion en soudage à l'arc TIG," PhD thesis, Université d'Aix-Marseille I.

[3] Brochard M. (2009) "Modèle couplé cathode-plasma-pièce en vue de la simulation du procédé de soudage à l'arc tig," PhD thesis, Université de Provence (Aix-Marseille I).

[4] Nguyen M.C., Medale M., Asserin O., Gounand S., Gilles P. (2017) Sensivity to welding positions in GTA welding with a 3D multiphysics numerical model, *Numer. Heat Transfer Part A: App.* **71**, 233.

[5] Baumard A., Ayrault D., Fandeur O., Bordreuil C., Deschaux-Beaume F., Vetele A.-L. (2020) Numerical prediction of grain structures formation during laser beam melting of single-track 316L stainless steel, *Comput. Mater. Sci.*

[6] Gandin C.-A., Rappaz M. (1994) A coupled finite element-cellular automaton model for the prediction of dendritic grain structures in solidification processes, *Acta Metall. Mater.* **42**, 2233.

1.7 Packaged Electronic Additive Manufacturing

Manuel FENDLER

CEA/DRT/CTREG/DGDE CEA Tech Grand Est 5, rue Marconi, Metz Technopole, 57070 Metz, France

Abstract

The topological optimization made possible by additive manufacturing offers new opportunities for electronic integration and packaging. At a time when the Internet of Things (IoT), for consumers and for industrial applications, is taking up more and more space in design offices, mechanical and electronic co-design is finally taking on its full meaning. This chapter will look at the phenomenon from different angles. The first part will shed light on the first printed circuit boards produced by additive manufacturing. In the second part, we will highlight the interest of the stratification of parts for the integration of electronic functions, in order to embed intelligence in industrial tools, synonymous with migration to the factory of the future. Finally, we will finish with equipment at the leading edge representing a real breakthrough in terms of additive co-manufacturing of mechatronic functions.

1.7.1 Introduction

Before becoming a major center of interest in the sciences of materials and mechanical design, additive manufacturing is the first part of the digital transition of the engineering trades. Within the digital chain, the software capacity to "slice" a 3D design into 2D layers stacked in an optimal direction gave birth to additive manufacturing. On closer inspection, the technology of semiconductor materials, the manufacture of integrated circuits, and on a larger scale, the manufacture of printed circuits, have all already borrowed these principles of 2D + Z design and manufacturing by stacking layers.

Topological optimization, enabled by additive manufacturing, stimulates innovation in the design and manufacture of complex systems, by cross-fertilization between mechanics and electronic integration (prostheses, exoskeletons, drones, droids, etc.). The design of objects by additive manufacturing is therefore gradually becoming obvious in many design offices, whatever their fields of application.

Likewise, it is also revolutionizing the packaging and integration of electronics in objects, most certainly shaping the future of the Internet of Things (IoT), whether they are consumer or industrial.

1.7.2 Additive Manufacturing for Electronics

Current wearable technology encompasses a spectrum of clothing (*e.g.* thermochromic T-shirt, EM-shielding clothes, and capacitive gloves) and low-weight, readily portable accessories or gadgets (*e.g.* watches, glasses, assistive hearing aids, and prosthetics) that (1) enable real-time monitoring of users' biometrics and their

DOI: 10.1051/978-2-7598-2446-5.c071

immediate environment, (2) directly enhance users' physical capabilities, or (3) impart esthetic values. The majority of commercially available wearables are manufactured based on high volume–low-mix economics and their designs are not tailored to specific requirements (*e.g.* body shape, esthetic preference, etc.) of the users. Customization of wearables (*e.g.* hearing aids and prosthetics) using traditional manufacturing techniques often incurs high cost, which is passed on to the end user.

In recent years, the improved availability of reliable, low-cost 3D printers and 3D scanners has made additive manufacturing an increasingly viable, cost-effective option for high-mix–low volume manufacturing of customized wearables. Several groups have leveraged on additive manufacturing tools for prototyping novel wearable electronics and components such as bionic ear [1], antennas [2], light-emitting diode (LED) [3], flex and stretchable sensors [4–6], and electrochemical sensors [7, 8].

The electrically conductive materials inside these wearables were 3D printed using metal-ion or -colloid inks [1, 2], or carbon-based conductive composites [4, 5]. Although metal-ion or -colloid inks such as silver ion or silver nanoparticle (resistivity of silver bulk [9] $= 1.62 \times 10^{-8}$ m) have significantly lower intrinsic electrical resistivity than carbon-based composite [10–13], carbon-based polymer composites are advantageous over metal-based or metal–alloy ink for 3D-printed wearable electronics in three aspects [14]: (1) Carbon-based polymer composites typically do not need additional processing step after printing such as thermal annealing or evaporation of solvent as required by 3D printing of metal-ion and metal-colloid inks. (2) They can be easily manufactured into filament form for direct printing on desktop-sized low-cost FDM (Fused Deposition Modeling) printer. (3) They have significantly longer shelf life (years) than that of metallic ink (*e.g.* silver), which is typically stable for few weeks to months [15, 16]. Carbon nanotubes [17–19] and graphene [20–22] have also been utilized as nanofillers in conductive composites for 3D printing. Commercially available filaments (*e.g.* graphene-based composite sold by Black Magic 3D (figure 1.7.1)) based on graphene–polylactic acid (PLA) composites can achieve volume resistivity of 0.006 Ω.m (or 0.6 Ω.cm) [23], but the budget is quite huge for the moment.

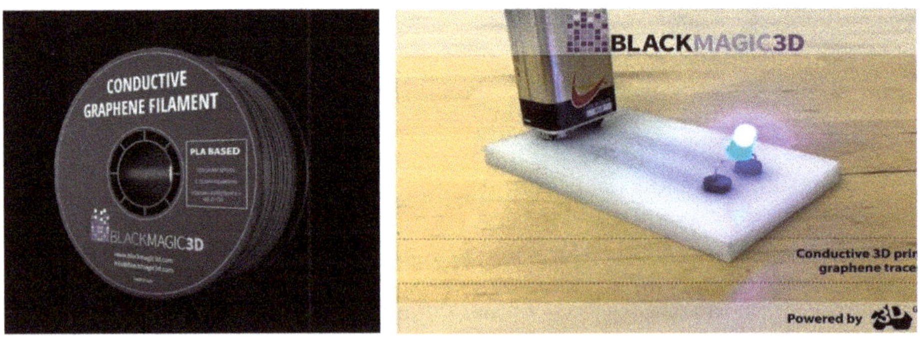

FIG. 1.7.1 – Graphene-based PLA conductive filament for FDM electrical interconnection printing (Black Magic 3D®).

Fused Deposition Modeling (FDM) is also widely used due to the interesting dielectric properties of charged polymers (table 1.7.1, [24]).

A project bringing together CNES and Thales Alenia Space aimed at the production of HF components whose 3D geometry was manufactured by FDM printing, from a monomer loaded with ceramic, and metallized by a double-sided silver coating (figure 1.7.2, [25]). Also identified were microwave filters printed in carbon-charged ABS (IEMN/Thales/CNES/XLIM, figure 1.7.3 [24]), and wide band antennas with dielectric lenses in ABS M30 (EPOD H2020 ST/Orange/Univ Nice, figure 1.7.4 [26]).

As an aside, the Direct Metal Laser Sintering technology (DMLS) making 3D geometries based on powder beds, in particular Aluminum and Titanium, is also typically used in the telecommunication industry to realize RF waveguides (figure 1.7.5) [27].

TAB. 1.7.1 – Dielectric properties of commercially available materials for additive manufacturing (FDM, PLA) [24].

Material reference	Additive manufacturing process	ε' @ 7 GHz	tanδ @ 7 GHz
Ultem 9085	FDM	2.71	3.4.10–3
ABS P430	FDM	2.39	3.6.10–3
ABS M30	FDM	2.46	1.0.10–2
Vero Blue	SLA	2.95	1.9.10–2
PPSF	FDM	2.94	6.3.10–3
PC/ABS	FDM	2.49	4.0.10–3
Polycarbonate	FDM	2.57	3.8.10–3
Accura Xtreme	SLA	3.00	2.9.10–2
Nylon 12	FDM	2.6	–
C/ABS	FDM	10	27.10–2
Cyclo-olefine (COP)	FDM	2.2	6.10–4

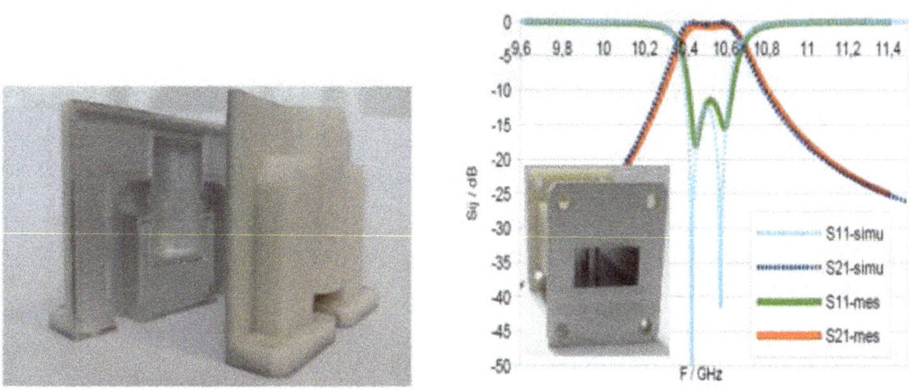

FIG. 1.7.2 – High frequency component (FDM + Metallization) [25].

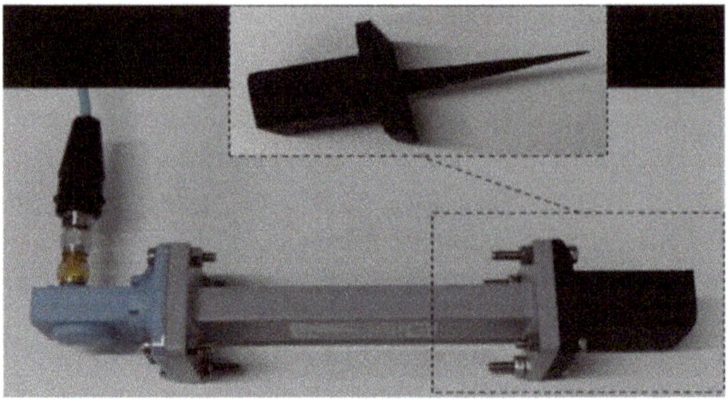

FIG. 1.7.3 – High frequency filter (ABS(C)) [24].

FIG. 1.7.4 – Dielectric Lens Antenna (ABS M30) [26].

FIG. 1.7.5 – High frequency waveguides (DMLS) [27].

Whatever method employed for 3D manufacturing of mechanical parts, the whole procedure is almost the same, and as follows (figure 1.7.6):

- Designing a 3D CAD model.
- Conversion of a CAD model in to a.STL file format.
- Slicing the.STL into thin cross-sectional layers.
- Building 3D parts layer-by-layer.

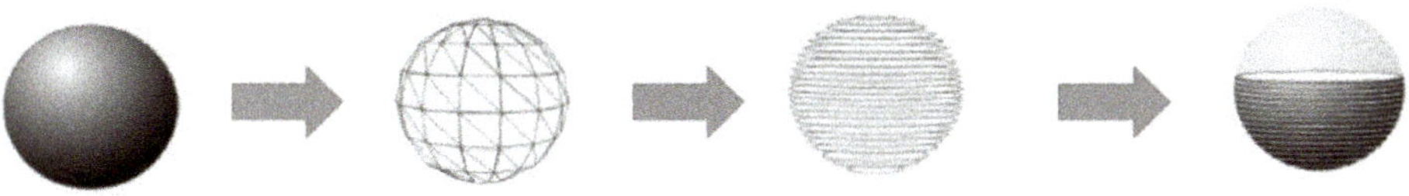

FIG. 1.7.6 – Additive manufacturing design process.

As depicted in Mentor® ECAD design software figure 1.7.7, in electrical engineering, from the design to the manufacturing, the same steps are deployed:

- Designing a 3D Electrical and Mechanical model (ECAD/MCAD).
- Schematic Capture of 2D + Z layouts.
- Design For Manufacturing (DFM) generating each layout Gerber Files (CAM) (figure 1.7.8).
- Construction layout-by-layout (figure 1.7.9).

This is the reason why we can say that electrical and mechanical designs are compatible with additive manufacturing in the same way. Until now, talking about a

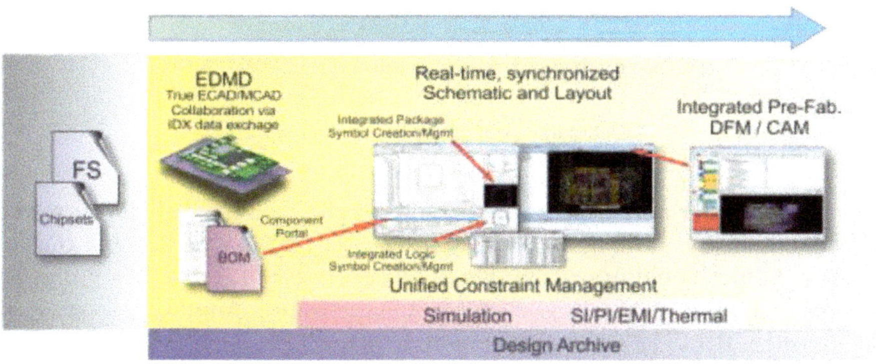

FIG. 1.7.7 – Electronics design process (MENTOR®) [28].

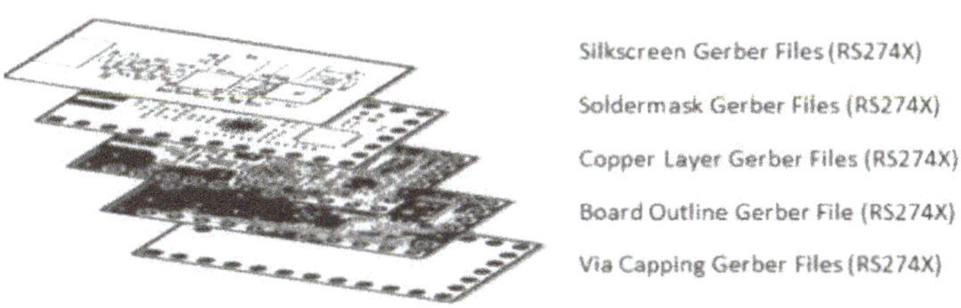

FIG. 1.7.8 – Electronic layers Gerber files [28].

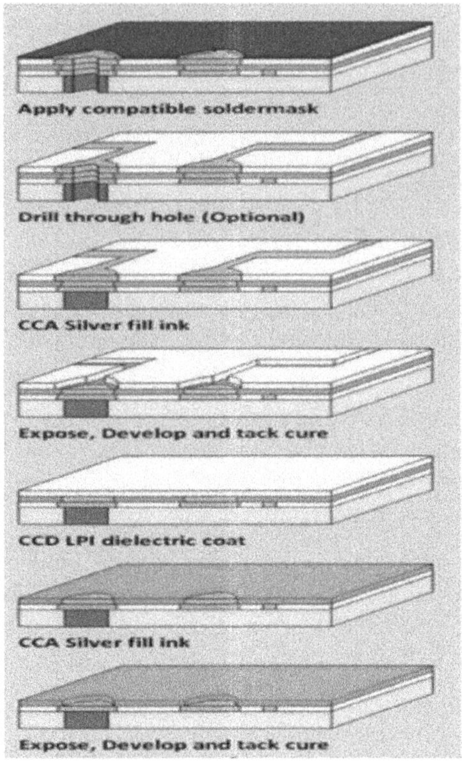

FIG. 1.7.9 – Electronic printed circuit board workflow [29].

printed circuit was an abuse of language because the manufacturing process still required engraving and drilling steps. However, it was not absurd to think that one day a circuit printer could appear in design offices. It was not until 2016 that this concept finally became a reality with a Lights-Out Digital Manufacturing system (LDM, figure 1.7.10). The Nano Dimension DragonFly® 3D printer [30] uses an inkjet deposition system that allows for simultaneous 3D printing of conductive silver nanoparticle ink (metal) and insulating ink (dielectric polymer) in a single print job (figure 1.7.11).

This technology enables product developers to design complex functional components and to print polymers and metals together to create functional parts such as printed circuit boards (PCBs), flexible electronics, antennas/RFIDs, sensors, electromagnets, molded interconnect devices (MIDs) and other experimental circuits right off the printer tray (figure 1.7.12).

Both conductor and substrate are printed in a fully additive process. The object is built up, layer-by-layer (conductive layers, dielectric layers). The full range of PCB features can be 100% 3D printed including interconnections such as vias, through-holes (blind, buried) and complex geometries. Drill and filled vias are performed respectively without additional machining nor electroplating steps (with

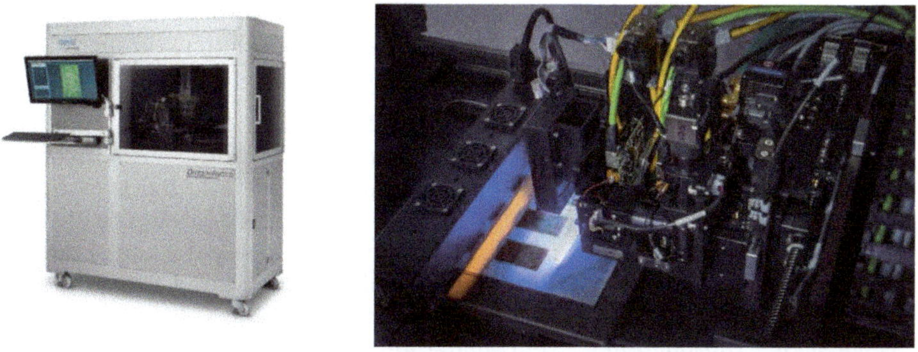

FIG. 1.7.10 – Nano Dimension DragonFly® 3D Lights-Out Digital Manufacturing system [30].

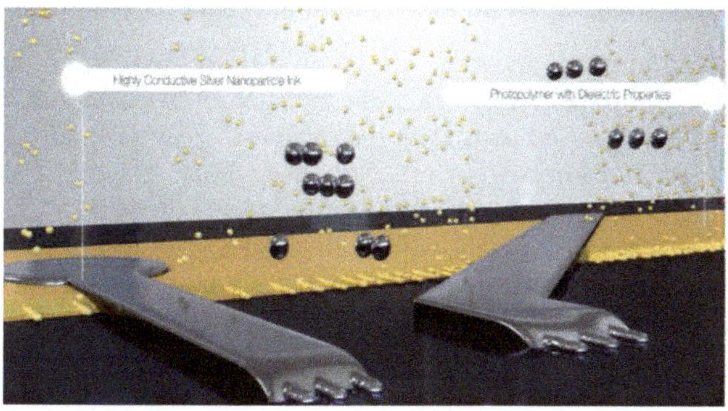

FIG. 1.7.11 – Dual conductive silver nanoparticle ink (metal) and insulating ink (dielectric polymer) printing [30].

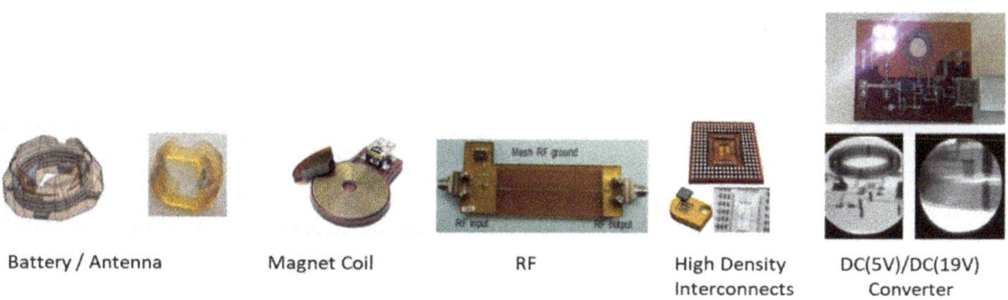

FIG. 1.7.12 – 3D electronics printed parts sampling [30].

hazardous chemical products). It is also a maskless process with direct printed solder mask for further active devices assembly. These technological breakthroughs add-up to the well-known advantages of the additive manufacturing contextualized below:

- *Time*: Reduces development cycles times. Enables on-site prototyping in a matter of hours instead of weeks, even for complex designs.
- *Cost*: Eliminates need for large order minimums. Enables ability to discover design errors in early development stage with agile rapid prototyping.
- *Complex geometries*: Enables increased design capabilities and manufacturability of components. Added agility enables designing, testing, and iterating in real time, on site.
- *Component consolidation*: Multi-material Additive Manufacturing enables functional, compact, denser, non-planar electronics parts.
- *Confidentiality*: Enables retention of sensitive Intellectual Propriety (IP) in-house during development. Eliminates concerns and costs related to IP infringement.
- *Environmental*: Limits environmental impact through optimized design, size, and weight. Reduces waste with additive manufacturing capabilities.

In a survey conducted by the Aberdeen Group for Nano Dimension [30], increased product complexity was the main focus of PCB designers, while improved time-to-market was identified as a primary business objective, ahead of the need to reduce product cost and improve product quality (figure 1.7.13). Time-to-market is a summation of the time required for the printed circuit board (PCB) to be sent for outsourcing and back for concept validation and rapid prototyping. The survey shows that electronics design and development companies need product development cycles that are shorter, more agile and efficient. In addition, decisions need to

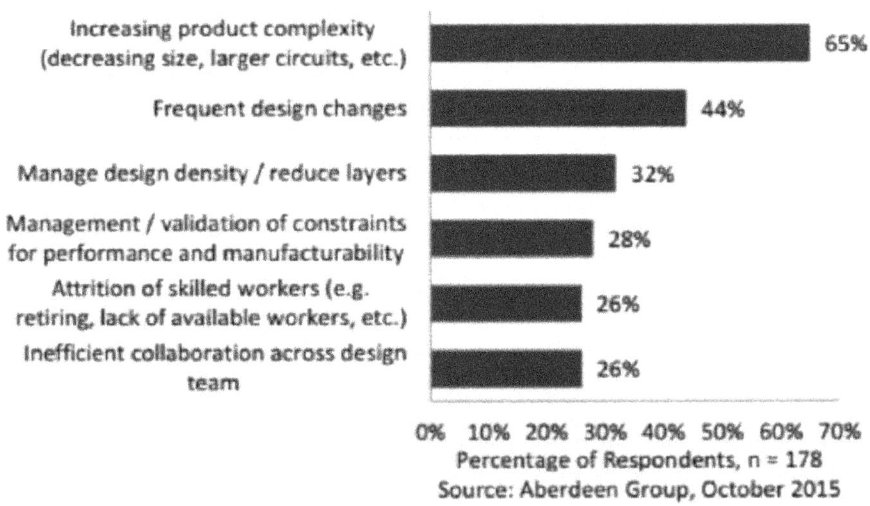

FIG. 1.7.13 – To challenges in PCB design [30].

be made in the early stages of the product development cycle to avoid costly mistakes and rework in the process. This is the reason why additive manufacturing is relevant for rapid prototyping and product launching in electronics.

1.7.3 Additive Manufacturing for Electronic Packaging

The purpose of electronic packaging is to integrate the components within mechatronic devices, and to ensure their robustness and reliability by protecting them from external aggressions. The topological optimization, made possible by additive manufacturing, enables to position the components at the right place in the parts, both in terms of metrological performance and protection. More than anywhere else, the components integrated into the Internet of Industrial Things are subject to strong environmental constraints: vibrations, shocks, temperature, humidity, dust, etc.... Thanks to this technology, we can also perform cavities and corridors allowing energy and signals to be transmitted. Thus, we implement a real mechatronic co-design work, combining mechanical and electronic intelligence. This strategy is particularly interesting in the design of intelligent tools or tooling.

In the family of Laminated Objects Manufacturing techniques (LOM), the additive manufacturing by Stratoconception® is a solid/solid technology based on the assembly of machined plates to obtain the 2D + Z stacking (figure 1.7.14) [31]. The machined plates are glued under vacuum at room temperature, which allows

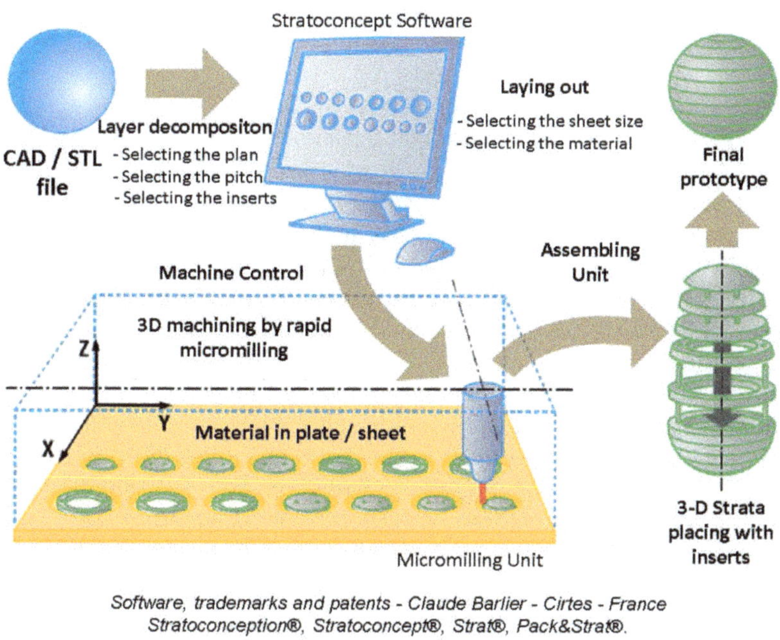

FIG. 1.7.14 – Stratoconception® additive manufacturing technique (CIRTES) [31].

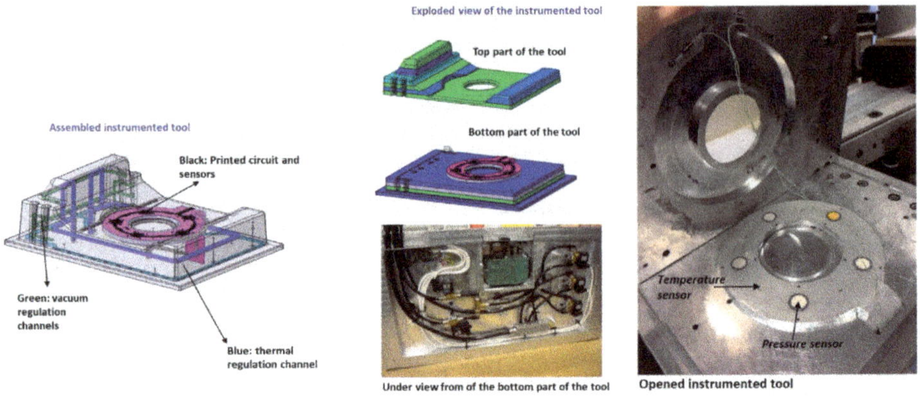

FIG. 1.7.15 – Instrumented thermoforming mold by additive topological optimization (CEA, CIRTES) [31].

the functionalization of the inter-strata by the insertion of a thin plastic insulating sheet with printed sensors (plastronic device).

A patent pending demonstrator has been realized by CEA and CIRTES. It is an instrumented thermoforming mold (figure 1.7.15). Temperature sensors (Negative Temperature Coefficient) and pressure sensors (piezo-resistive gauges) have been located at the last inter-strata, just under the molding surface. Thanks to the flexibility of the plastronic circuit, the interconnection layer escapes at 90° through a corridor allowing to reach the base plate under the mold, interfacing with the readout circuit and the control electronics of the machine, as well as the various supply valves of the fluidic networks.

The spatial distribution of the sensors thus allows to characterize at any point the quality of contact of the polymer sheet formed on the mold, and to spatially map its temperature during cooling. Two corridor networks conforming the 3D geometry of the mold, made possible by topological optimization during its design, allow to optimize the thermoforming process controlled by the sensors. The first vacuum network enables the polymer sheet to be sucked locally, to help it taking the shape of the mold according to the values of the pressure sensors characterizing the contact. The second network of heat transfer fluid enables the mold to be cooled locally according to the temperature readings. This intelligent tool (I-IoT) therefore makes it possible to understand the phenomena involved in thermoforming in order to correct and validate the calculation models, to improve the performance of the process in real time, and to optimise cycle times and material consumption (minimum plate thicknesses to pass the "case corners" at great depths of stamping).

Tests on industrial thermoforming presses have shown that the contact of the polymer sheet with the mold was effectively detected despite the thin aluminum thickness (<0.1 mm) of the last surface layer of the mold (figure 1.7.16). On the other hand, the temperature values recorded were consistent with the measurements made by traditional temperature probes.

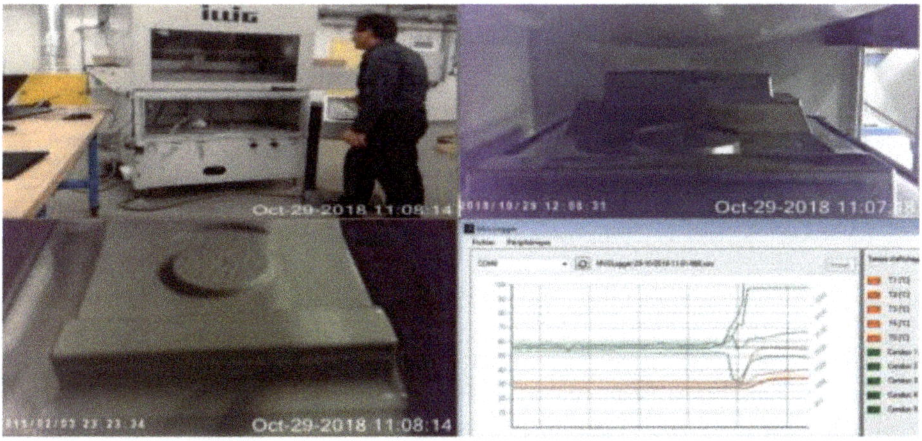

FIG. 1.7.16 – Thermoforming cycle monitoring (CEA, CIRTES); top left: thermoforming machine, top right: instrumented mold and bending hot polymer sheet, down left: polymer sheet in contact with the mold, down right: extracted data from sensors (green = contact points, orange = temperatures) [31].

Finally, these measurements confirm that the plastronic sensors have survived the additive integration process as well as the thermoforming process.

We have opted for Stratofabrication® as a technique for hosting electronics by topological optimization because it is a low-temperature solid/solid additive manufacturing technique (vacuum bonding). Thus, we avoid all the degradation stimuli of other additive manufacturing techniques, *i.e.* dust for powder bed families, and heat and humidity for stereo-lithographic techniques (SLA). Of course, thermoforming can accommodate plastronics and read-out electronics in the vicinity due to the low temperatures and low pressures involved. But, in many applications, it will be necessary to deport the electronics and the use of inorganic printing technologies for interconnections and sensors is mandatory.

1.7.4 Additive Manufacturing for Structural Electronics

As discussed previously, additive manufacturing of smart devices featuring intelligent integrated sensor or electronic systems is a trend that owes its current popularity to the parallel advent of different emerging technologies and application scenarios such as additive manufacturing itself, direct write technologies and the IoT.

Another catchphrase in this context is "structural electronics". The definition of this concept, which is application independent, stresses a twofold role of electronic components, besides providing their functional or electronic capabilities, they are meant to take over a structural role. An often-cited example in this respect is the battery of an electric car, which according to the structural electronics paradigm would not only provide energy to the car, but also contribute significantly to its structural stability. Further examples of structural electronics include smart skins for aircraft, printed OLED lighting devices on the surface of a car's body panel, and also civil engineering structures, such as bridges with concrete-embedded sensors, and electronics for health monitoring. In practice, whether for these industrial applications or for consumer applications, additive manufacturing techniques allow better utilization of a limited design space. The simplest examples in this respect are applications that transform the typically flat PCB that provides the interconnections between electronic components into a three-dimensional object adapted to product shape.

Conventionally, circuit manufacturing technologies, whether on rigid or flexible substrates (plastronics, figure 1.7.17), are also evolving in 2D + Z space. However, they are also referred to as thin-film technologies, as the thickness of material deposits on substrates does not exceed a few tens of micrometers ($<30\ \mu$m). Screen-printing (or stencil printing) accounts for 80% of the activity; inkjet, heliography and iso-engraving technologies make up the remaining 20%. The main advantage of the printing technologies lies in the ability to work on large flat substrates (up to 500×600 mm^2), mainly plastic due to their insulating and flexible characteristics. To achieve structural electronic architecture, thermoforming means are attempting to capitalize on the subsequent deformation of these flat-printed polymer substrates patterned with stretchable inks. This leads to a functionalized 3D object or one that can be shaped into a volume part (figure 1.7.18).

Voltera (figure 1.7.19) and Optomec (figure 1.7.20) companies, offer equipment capable of depositing conductive inks by syringe on, respectively, flat and curved substrates in a half space. Like many laboratory installations, the principle consists in exploiting the architecture of a milling machine for 2D, or a 5-axis CNC machining centre for 2.5D, by replacing the spindle and the tool by a syringe to make micro-extrusion of conductive inks or pastes.

The LDS (Laser Direct Structuring) technology proposed by LPKF results in real functionalized 3D objects (figure 1.7.21). However, this technology requires the use of a laser-compatible resin as substrate, and a constraining chemical revelation (figure 1.7.22). This limits its employability, and thus its real industrial take-off.

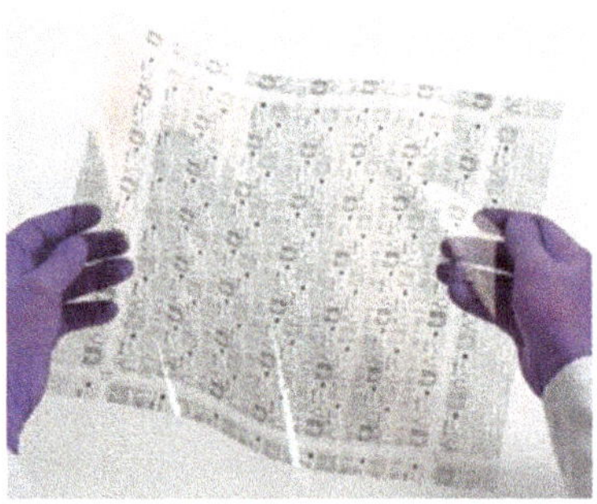

FIG. 1.7.17 – Plastronics (CEA).

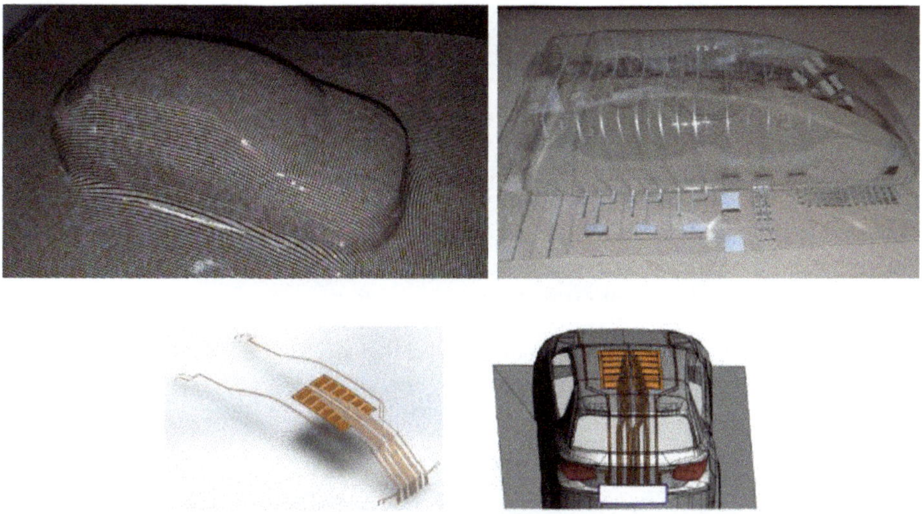

FIG. 1.7.18 – Thermoforming of stretchable electronics (CEA, IPC).

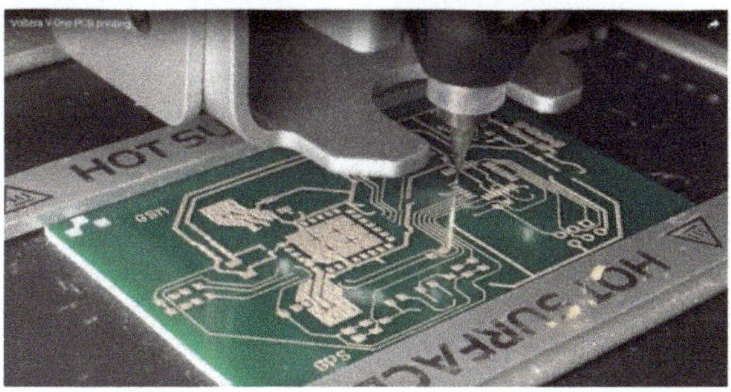

FIG. 1.7.19 – Voltera printing solution [32].

What additive manufacturing offers for structural electronics, besides new geometrical options, is the promise of a simplified approach that realizes functionality based on a single process or at least a single manufacturing system, thus dispensing both with several processing steps and complicated tooling needed in other processes suitable for production of 3D PCB substitutes such as a Molded Interconnect Devices technique (MID).

A printer meant to create objects with integrated circuits, this is what Voxel8 has created. At CES 2016, Voxel8 showcased that their printer could print a plastic body and metal circuitry all at once with the wires actually printed inside the plastic thanks to a dual head extruder. Voxel8's novel conductive inks are deposited by directly writing in the material (figure 1.7.23).

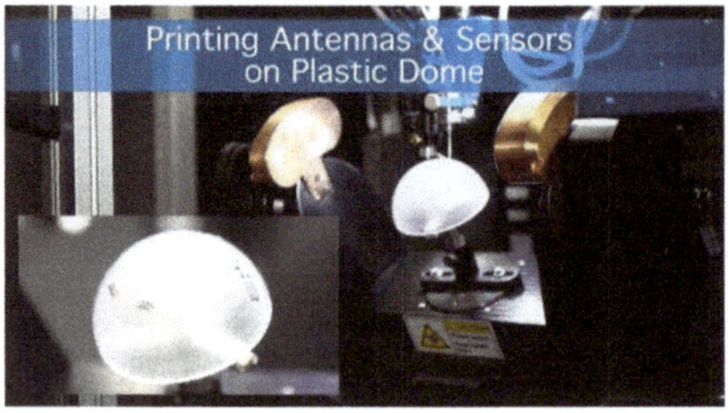

FIG. 1.7.20 – Optomec 3D printed electronics [33].

FIG. 1.7.21 – LDS printed micro-robot (Testo) [34].

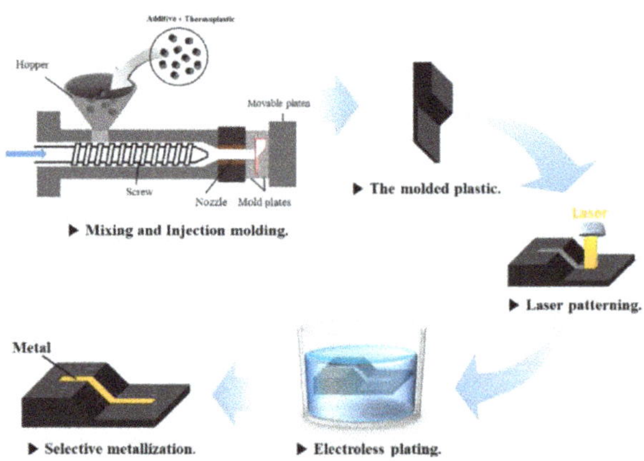

FIG. 1.7.22 – LPKF Laser Direct Structuring (LDS) [35].

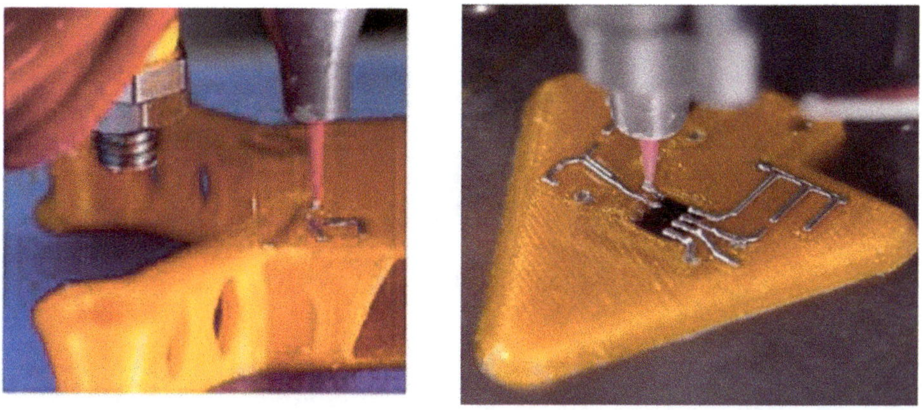

FIG. 1.7.23 – Dual head extruder for mechanical and electrical printing (Voxel8) [36].

Voxel8 have shown that the printer could print out a wireless power disk, an embedded watch, and even a Quadcopter as follows [36].

The wireless power is a 3D printed electronic disk that uses a printable silver coil in order to light up an LED light when placed on a wireless charger (figure 1.7.24). What makes such a small project so significant is the fact that the every component of the disk was printed right on the spot, tying 3D printing technology and electronics together into one simplified process.

In addition to the wireless power disk, Voxel8 also demonstrated an embedded watch at CES 2016 (figure 1.7.25). The 3D printed watch was created with PLA and their conductive ink, and contains an embedded micro-controller and LEDs that are inserted during the printing process. After taking the Embedded Watch off the Voxel8 3D printer all that is needed for the watch to be fully functional is a coin cell battery, which is inserted and removed from the back of the watch.

FIG. 1.7.24 – Printed wireless power disk (Voxel8) [36].

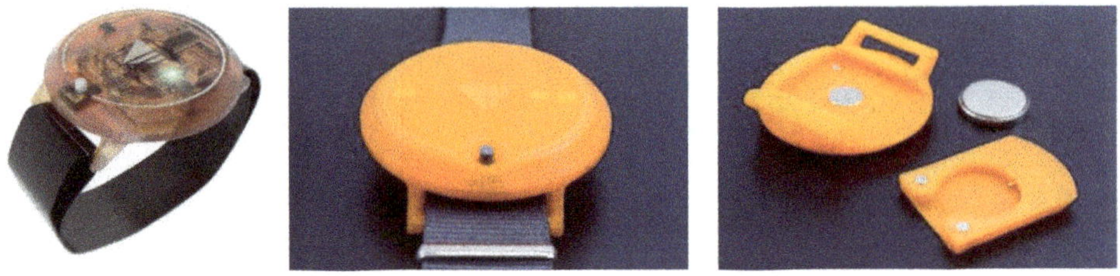

FIG. 1.7.25 – Printed watch (Voxel8) [36].

FIG. 1.7.26 – Printed Quadcopter with embedded electronics control (Voxel8) [36].

The Quadcopter is a fully functional UAV with a 3D printed PLA exterior body, interior wiring using conductive silver ink, and embedded components (figure 1.7.26). The Quadcopter interfaces with an existing PCB embedded inside of a custom 3D model. Magnetic connections ensure contact between the PCB and printed silver interconnects.

As discussed with this previous innovation, for the moment, as usual PCBs are placed into boxes and connected with unwieldy wiring harnesses. A Florida-based manufacturer of micro-dispensing and 3D printing equipment, nScrypt, has developed a tool changer for its multi-head Direct Digital Manufacturing (DDM) machines (figure 1.7.29). nScrypt is working toward a future where the PCB, box, and harness will be depleted so that electronics can be smaller, less expensive, more lightweight, and integrated directly into the structure.

It is the final step to a fully 3D non-planar next generation electronics that either will conform to, or be embedded in an object's structure. The Factory in Tool equipment has multiple tool heads (figure 1.7.27) [37], including nScrypt's nFD for Material Extrusion, the SmartPump for Micro-Dispensing, nMill for micro-milling, and nPnP for pick and place of electronic components, which are placed on a high-precision (1 μm accuracy) linear motion gantry. Multiple cameras allow for automated inspection and computer vision routines, while a point laser height sensor maps surfaces. All of these features add up to allow for successful conformal printing, or micro-dispensing, onto objects. Because everything is combined in one

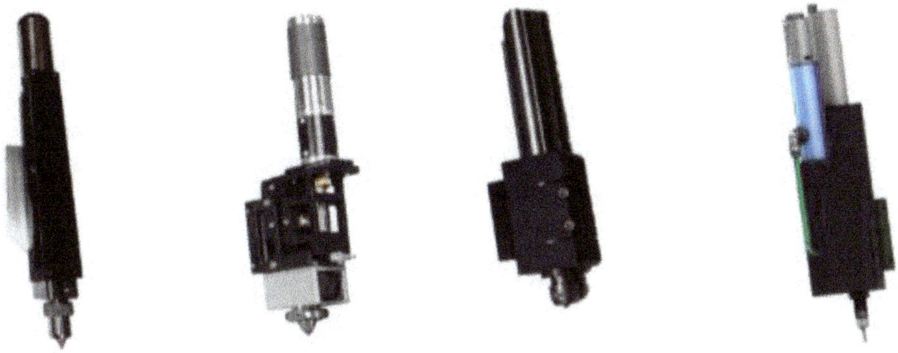

FIG. 1.7.27 – Precision micro dispensing, material extrusion, micro-milling, and pick-and-place tool heads can run on the nScrypt Factory in a Tool ® [37].

FIG. 1.7.28 – Left: 3D printed circuit board with devices assembly in the multi-head system, Mid: chip wire bonding by 3D printing, Right: RF antennas on a military helmet (nScrypt) [37].

FIG. 1.7.29 – nScrypt multi-head Factory in a Tool® [37].

platform, manufacturers of complex structural electronics can create them with the press of a button.

With this equipment high-precision motion and micro-dispensing excels. As an example, each dimension in RF electronics is critical, and if a line is off by even the smallest fraction, the circuit's performance is ruined, and so is that of the device with which it is being used. Nevertheless, the previously mentioned SmartPump offers picolitre volumetric flow control, while the nFD extruder provides precision deposition and the motion platform has 0.5 μm repeatability. This means that nScrypt's platform can produce both conductive and dielectric features to high tolerances ... ensuring successful RF circuits (figure 1.7.28).

1.7.5 Conclusion

As with mechanical design, additive manufacturing is shaking up the design of mechatronic systems, even printed circuit boards and some sensor architectures. The mechanical intelligence offered by topological optimization makes it possible to optimize the reception of electronic components, as well as to imagine original proof bodies for measurement functions. The Internet of Things offers a multitude of development vehicles to test new mechatronic architectures applied to increasingly intelligent objects.

Until now electronics had to do with the remaining space, but today intelligent objects are becoming the synthesis of analytical and intuitive thinking. They are the result of a global approach to collaborative design. Thanks to additive manufacturing, mechanics and electronics finally constitute a set of interlocking spaces.

References

[1] Mannoor M.S., Jiang Z., James T., Kong Y.L., Malatesta K.A., Soboyejo W.O., Verma N., Gracias D.H., McAlpine M.C. (2013) 3D printed bionic ears, *Nano Lett.* **13**, 2634.

[2] Adams J.J., Duoss E.B., Malkowski T.F., Motala M.J., Ahn B.Y., Nuzzo R.G., Bernhard J. T., Lewis J.A. (2011) Conformal printing of electrically small antennas on three-dimensional surfaces, *Adv. Mater.* **23**, 1335.

[3] Kong Y.L., Tamargo I.A., Kim H., Johnson B.N., Gupta M.K., Koh T.-W., Chin H.-A., Steingart D.A., Rand B.P., McAlpine M.C. (2014) 3D printed quantum dot light-emitting diodes, *Nano Lett.* **14**, 7017.

[4] Leigh S.J., Bradley R.J., Purssell C.P., Billson D.R., Hutchins D.A. (2012) A simple, low-cost conductive composite material for 3D printing of electronic sensors, *PLoS One* **7**.

[5] Muth J.T., Vogt D.M., Truby R.L., Mengüc Y.I., Kolesky D.B., Wood R.J., Lewis J.A. (2014) Embedded 3D printing of strain sensors within highly stretchable elastomers, *Adv. Mater.* **24**, 6307.

[6] Guo S.-Z., Qiu K., Meng F., Park S.H., McAlpine M.C. (2017) 3D printed stretchable tactile sensors, *Adv. Mater.* (1701) 218.

[7] Ambrosi A., Moo J.G.S., Pumera M. (2016) Helical 3D-printed metal electrodes as custom-shaped 3D platform for electrochemical devices, *Adv. Funct. Mater.* **26**, 698.

[8] Loo A.H., Chua C.K., Pumera M. (2017) DNA biosensing with 3D printing technology, *Analyst* **142**, 279.

[9] D.R. Lide, Ed. (2005) Electrical resistivity of pure metals, in *CRC handbook of chemistry and physics*, 85th edn. CRC Press, Boca Raton, FL.
[10] Huang J.-C. (2002) Carbon black filled conducting polymers and polymer blends, *Adv. Polym. Technol.* **21**, 299.
[11] Stankovich S., Dikin D.A., Dommett G.H.B., Kohlhaas K.M., Zimney E.J., Stach E.A., Piner R.D., Nguyen S.T., Ruoff R.S. (2006) Graphene-based composite materials, *Nature* **442**, 282.
[12] Tibbetts G.G., Lake M.L., Strong K.L., Rice B.P. (2007) A review of the fabrication and properties of vapor-grown carbon nanofiber/polymer composites, *Compos. Sci. Technol.* **67**, 1709.
[13] Bauhofer W., Kovacs J.Z. (2009) A review and analysis of electrical percolation in carbon nanotube polymer composites, *Compos. Sci. Technol.* **69**, 1486.
[14] Goh K.H., Tan Z.D., Tan S.T., Tjiu W.W., Soh J.Y., Glenn Ng Z.J., Chan Y.Z., Hui H.K., Goh K.E. (2017) Electrically conductive filament for 3D-printed circuits and sensors, *Appl. Mater. Today* **9**, 167.
[15] Ahn B.Y., Duoss E.B., Motala M.J., Guo X., Park S.-I., Xiong Y., Yoon J., Nuzzo R.G., Rogers J.A., Lewis J.A. (2009) Omnidirectional printing of flexible, stretchable, and spanning silver microelectrodes, *Science* **323**, 1590.
[16] Walker S.B., Lewis J.A. (2012) Reactive silver inks for patterning high-conductivity features at mild temperatures, *J. Am. Chem. Soc.* **134**, 1419.
[17] J. Czyżewski, P. Burzyński, K. Gaweł, J. Meisner (2009) Rapid prototyping of electrically conductive components using 3D printing technology, *J. Mater. Process. Technol.* **209**, 5281.
[18] Postiglione G., Natale G., Griffini G., Levi M., Turri S. (2015) Conductive 3D microstructures by direct 3D printing of polymer/carbon nanotube nanocomposites via liquid deposition modeling, *Composites Part A* **76**, 110.
[19] Kim J.H., Lee S., Wajahat M., Jeong H., Chang W.S., Jeong H.J., Yang J.-R., Kim J.T., Seol S.K. (2016) Three-dimensional printing of highly conductive carbon nanotube microarchitectures with fluid ink, *ACS Nano* **10**, 8879.
[20] García-Tuñon E., Barg S., Franco J., Bell R., Eslava S., D'Elia E., Maher R.C., Guitian F., Saiz E. (2015) Printing in three dimensions with graphene, *Adv. Mater.* **27**, 1688.
[21] Kim J.H., Chang W.S., Kim D., Yang J.R., Han J.T., Lee G.-W., Kim J.T., Seol S.K., (2015) 3D printing of reduced graphene oxide nanowires, *Adv. Mater.* **27**, 157.
[22] Zhang D., Chi B., Li B., Gao Z., Du Y., Guo J., Wei J. (2016) Fabrication of highly conductive graphene flexible circuits by 3D printing, *Synth. Met.* **217**, 79.
[23] Black Magic 3D, blackmagic3d.com/Conductive-p/grphn-pla.htm [accessed 02.21.2020].
[24] Arbaoui Y., Laur V., Maalouf A., Quéffélec P., Passerieux D., Delias A. (2016) Full 3-D printed microwave termination: a simple and low-cost solution, *IEEE Trans. Microwave Theory Tech.* **64**, 271.
[25] Jolly N., Tantot O., Delhote N., Verdeyme S., Estagerie L., Carpentier L. (2014) Wide range continuously high electrical performance tunable E-plane filter by mechanical translation, 44th European Microwave Conference, ISBN: 978-2-8748-7035-4.
[26] Luxey C., Bisognin A., Gianesello F. (2015) AG GDR Ondes, Fabrication additive pour les hyperfréquences.
[27] Laur V., Quere Y., Maalouf A., Rius E., Person C., AG GDR Ondes, Fabrication additive pour les hyperfréquences (2015).
[28] blogs.mentor.com/jimmartens/blog/2014/07/01/the-pcb-design-flow-of-the-future/.
[29] Cadalen E. (2016) Conception et fabrication de circuits imprimés rigides, Techniques de l'Ingénieur, E3342 V2.
[30] nano-di.com/, White papers and eBooks [accessed 02.21.2020].
[31] Motta M.P., Pelaingre C., Barlier C., Fendler M. (2019) Instrumented thermoforming tool made by stratoconception process: an additive manufacturing technology enabling the parts functionalization, International Congress of Welding Additive Manufacturing, ICWAM.

[32] voltera.io/product/technology/.
[33] optomec.com/printed-electronics/aerosol-jet-technology/.
[34] festo.com/group/en/cms/10157.htm.
[35] lpkf.com/fileadmin/mediafiles/user_upload/products/pdf/EQ/3D-MID-LDS/brochure_lpkf_laser_direct_structuring_en.pdf.
[36] 3dprintingindustry.com/news/64071-64071/ [accessed 02.21.2020].
[37] nscrypt.com/ [accessed 02.21.2020].

Chapter 2

Nanoobjects: Synthesis, Integration and Application to Energy and Transportation

Abstract

In nanometer scales, materials exhibit new physical or chemical properties. For example, it is well known that the band gap of semiconductors can be tuned by adjusting the structure dimensions. Multifunctional materials presenting non-conventional properties can then be obtained from hierarchical nanostructures. Efforts on the processing of nanomaterials started long time ago, but they dramatically intensified in the last decades. Here is a short overview of the role of nanoobjects in material science, from their synthesis to their integration in device parts, in the frame of their application to energy and transportation.

The first part is devoted to the synthesis of nanoobjects, including a fundamental part on nucleation and growth processes prior to introducing some liquid and gas-phase process examples. In the second part, different integration techniques of nanoobjects in material structures, from the liquid phase, the solid phase and the gas phase, are addressed. The third part is focused on the application of theses nanostructures to energy and transportation. Applications to solar and heat conversion, energy storage and hydrogen technology are developed.

DOI: 10.1051/978-2-7598-2446-5.c002

2.1 Synthesis of Nanoobjects

Magali GAUTHIER, Yann LECONTE, Henri PEREZ and Olivier SUBLEMONTIER

Université Paris-Saclay, CEA, CNRS, NIMBE, 91191 Gif-sur-Yvette, France

Top-down and bottom-up approaches have been developed for the synthesis of nanoobjects [1]. Top-down ones include, for example, milling and attrition for particles in the range from 10 nm to several hundred nanometers in diameter. However, these techniques exhibit large size and shape distributions, and significant amounts of impurities from the treatment medium or defects from the process itself. Bottom-up approaches are much more preferred through many different processes as they generally show more convincing results in terms of uniformity in size, shape and chemical composition distributions. Generally, nanoparticles are synthesized by homogenous nucleation in the liquid phase or in the gas phase, and nanowires or nanotubes are synthesized by heterogeneous nucleation on a substrate. Fundamental aspects of nucleation and growth are presented prior to a presentation of different synthesis processes.

2.1.1 *Nucleation and Growth*

Nucleation is the transition stage between two states of matter organization, and results in the creation of a new phase dispersed in the volume or on the surface of the original phase. Since the historical contribution from Volmer and Weber [3], numerous experimental and theoretical studies have been conducted by chemists, physicists and mathematicians. A standard theory was first proposed by Katz and Widersich [4]. Homogenous nucleation takes place in a supersaturated, continuous and single-phase medium, without any impurities nor interfaces. Nanoparticles can be synthesized through homogenous nucleation in liquid, gas or solid phase. The fundamentals of nucleation and subsequent growth are essentially the same.

2.1.1.1 *Size Control*

Nucleation and growth processes are presented schematically in figure 2.1.1 [5]. When the concentration of grown species increases in time (phase I), either by an increase in the solute quantity or by a temperature decrease below the phase transformation point, nucleation occurs when the supersaturation reaches a critical value $C_{\min}$ (Phase II), above the solubility value C_s (for liquids), which corresponds to an energy barrier for the formation of nuclei. After the initial nucleation, the solute concentration decreases, and when this concentration becomes lower than the critical value $C_{\min}$, no more nuclei are formed. The system comes to phase III, in which growth takes place until the concentration reaches asymptotically the solubility value C_s.

Practically, the concentration of the growth species should increase abruptly, in order to have a rapid nucleation, so that all nuclei are formed in the same conditions. This is essential for the control of the size distribution of the grown nanoparticles.

DOI: 10.1051/978-2-7598-2446-5.c012

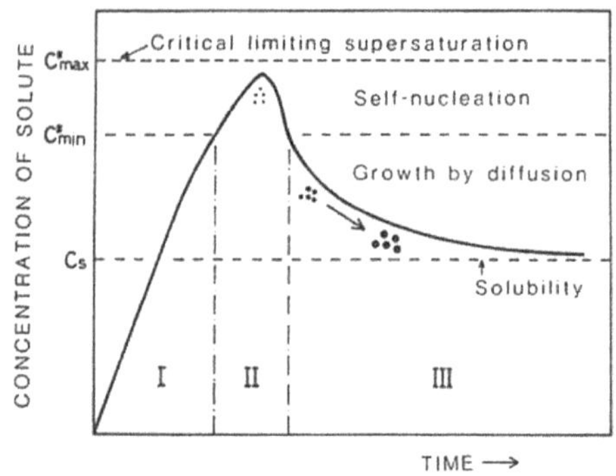

FIG. 2.1.1 – Change in concentration with time before and after self-nucleation [5].

The size distribution width depends also on the growth process, which includes diffusion of growth species in the medium and their adsorption and incorporation on the surface of growing particles. The process can then be diffusion-limited or growth-limited. The diffusion-limited growth is generally preferred to get a monodispersed suspension [6]. In this case, species concentration is kept low in the medium. Increasing the viscosity or controlling the supply of growth species, when a liquid media is used, allows keeping a diffusion-limited growth. When growth species are brought through chemical reactions, it is also possible to control the reaction kinetics by adjusting the concentration of reactants or catalysts [7].

2.1.1.2 Shape Control

Thermodynamics tells us that a solid or a liquid will have a strong tendency to minimize surface energy. Different mechanisms are observed for reducing this surface energy. They can occur at different scales from atomic or surface level, at the individual structure level or at the overall system level. For a given surface area, the surface energy can be reduced through the following mechanisms:

- Surface atoms or ion shift [8].
- Surface restructuring by converting dandling bonds into chemical bonds [9].
- Surface physical or chemical adsorption of chemical species [10].
- Composition segregation by solid-state diffusion [11].

At the individual nanostructure level, the minimization of the surface energy can be brought by the reduction of the overall surface area. If the material is isotropic, like in liquids or amorphous solids, the surface energy is also isotropic, leading to spherical shaped particles. However, in the case of crystalline solids, different crystal facets show different surface energies. The thermodynamically equilibrium shape is then faceted. It is possible to estimate the surface energy of each facet to determine the overall shape of the particle [12], and define the rules for the shape evolution of different types of nanoparticles [13]. Controlling the size and shape, and corresponding uniformities, of nanoparticles is of prime importance as their impact on

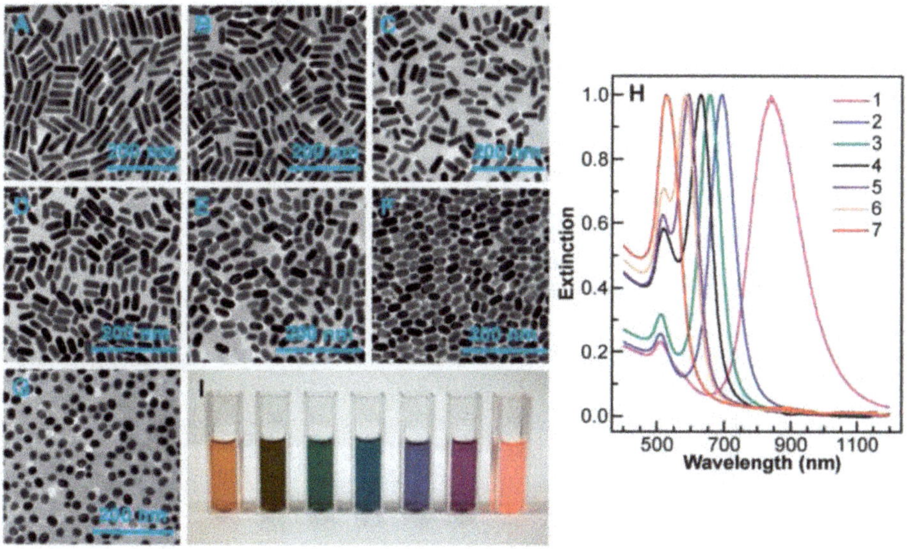

FIG. 2.1.2 – (A–G) TEM Images of Au nanorods, (H) respective extinction spectra of Au nanorods (with curve N°1 corresponding to image A, and so on) and (I) Photograph of the Au nanorods solutions [14].

their properties is decisive. As an example, figure 2.1.2 [14] shows the impact of aspect ratio of Au nanorods on their optical properties.

2.1.1.3 Agglomeration and Aggregation Control

From the overall system point of view, interactions between individual nanostructures can lead to different kinds of processes, including agglomeration, aggregation, sintering and Ostwald ripening. In any case, relatively large structures grow at the expense of smaller ones, obeying to the law of reducing the overall surface energy. In agglomeration, primary particles are sticking together without altering the individual structure and are bonded by weak physical forces. Aggregation happens when primary particles are bonded by strong chemical forces. Agglomerates can eventually be composed of aggregates and/or primary particles. The distinction between agglomerates and aggregates is important because agglomeration is a reversible process, for example, by ultrasonic treatment in the liquid phase, unlike aggregation [15]. In sintering, nanostructures merge together and reshape so that a large part of solid–vapor interface is replaced by solid–solid interface to form a pack-like structure. In Ostwald ripening, two or more particles merge to become a single one. The product of sintering is a polycrystalline material, whereas ripening leads to a uniform one. For solid nanoparticle synthesis, keeping the reaction temperature sufficiently low can significantly prevent the sintering and Ostwald ripening processes.

However, agglomeration and aggregation processes remain an issue when large quantities of nanoparticles are synthesized in a limited volume.

In the liquid phase, molecules are often used to stabilize the colloidal suspension, avoiding sedimentation of particles. The stabilizing agent should be introduced in the solution before the reaction starts. For example, sodium tri-citrate is used in the well-known "Turkevich" synthesis process [16].

In the gas phase, which is widely used in mass production processes of nanomaterials, agglomeration is very difficult to avoid. Distinguishing between agglomerates and aggregates is then critical as it drives the final structure of the subsequent nanomaterial. In high temperature processes like flame or pyrolysis, and also in colloid processes, aggregates with fractal morphologies are often observed. These clusters of primary particles have a self-similar structure over a finite range of length scale, in which the upper limit is the radius of gyration R_g, the root mean square radius that quantifies the overall size of the aggregate. If we make the assumption that all the primary particles are spherical and that they have the same size a, the self-similarity leads to a power law for the number of primary particles N as a function of the reduced size of the aggregate as

$$N = k_0 \left(R_g / a\right)^{D_f},$$

where D_f is the fractal dimension and k_0 is a constant scaling factor. In fractal geometry, the fractal dimension D_f is the translation of the way in which a fractal set fills space, at all scales. It is then related to the compactness of the aggregate. Figure 2.1.3 shows different arrangements of aggregates with corresponding fractal dimensions [17]. D_f is between 1 for a stick-shaped agglomerate and 3 for a compact

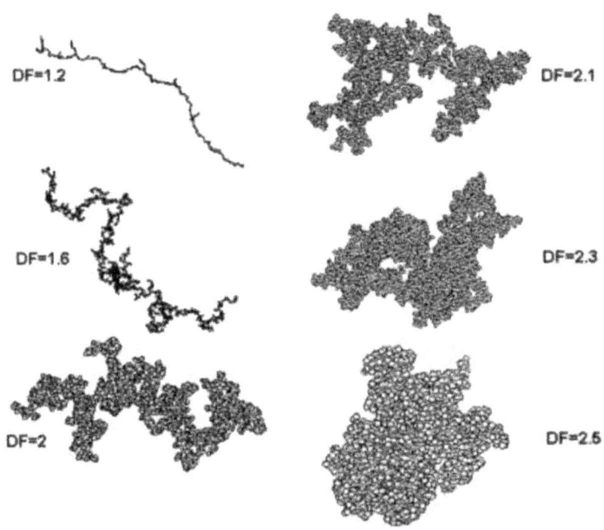

FIG. 2.1.3 – Two-dimensional projections of 4096 primary particles aggregates of different fractal dimensions D_f.

sphere of primary particles. Aggregation processes can be either diffusion-limited or reaction-limited. Diffusion-limited aggregation is observed when repulsive forces between particles are negligible. Reaction-limited aggregation is observed when substantial but not insurmountable forces occur between particles. In the latter case, the aggregation rate is limited by the time needed by particles to overcome this repulsive barrier, for example, by thermal activation. The fractal dimension for diffusion-limited aggregation is generally close to 1.8, while it is close to 2.1 for the reaction-limited case [18].

In conclusion, different synthesis conditions lead to different sizes and shapes of the primary nanoparticles. In addition, the possible agglomeration or aggregation of these particles can have a huge impact on the structure, and then on the properties of the final nanomaterials. Characterizing these building blocks is then of prime importance for the performance level and reproducibility of devices.

2.1.2 *Synthesis Processes*

2.1.2.1 Liquid Phase Processes

Solution chemistry is a bottom-up approach and the simplest method for the synthesis of nano-objects as it generally requires low energy and simple equipment. It is fundamentally based on the insolubility of the product obtained by the chemical reactions involved, therefore resulting in the formation of solids. A high ratio of the atoms forming the nanoparticle are located at its surface. These atoms have an averaged coordination number lower than the atoms inside the volume of the nanoparticle, resulting in an excess of surface energy. The growth of the solid particles beyond nanometer scale will decrease this surface energy excess by strongly decreasing the ratio of surface atoms, and the crystal will grow preferentially through planes with lower surface energies. However, it is worth pointing that for a given material, a "clean surface" is basically unstable and subsequent adsorption or chemical reaction with molecules from the environment will occur. These two aspects of surface energy minimization are even more marked for nanoparticle synthesis in liquid phase. Extremely small nanoparticles (few nanometers in diameter) will have particular shapes, therefore "capping" agent interacting more or less strongly with the surface are introduced on purpose in the reaction media.

There is a very large variety of nanoparticle liquid phase syntheses that in some ways depend on its nature (metal, oxide, sulfide...). Some of the most common methods are detailed below.

Precipitation/colloidal Syntheses

The first paper dealing with nanoparticle synthesis in liquid by precipitation/colloidal method was from Faraday (1857) [19], aiming at the study of light interaction with finely divided gold. About one century later Tukevich *et al.* provided a detailed chemical study of gold nanoparticle synthesis with control of size and size dispersion [7]. This aqueous synthesis involved sodium citrate as both reducing agent and capping agent, provided good stability of the colloidal solution. Several

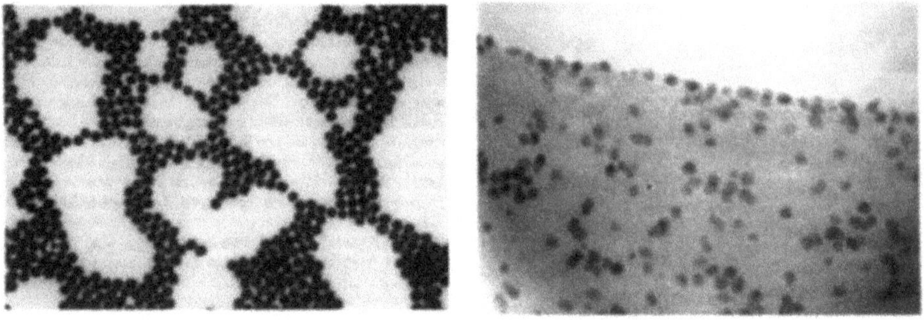

FIG. 2.1.4 – Left: 36 nm ligand-stabilized gold colloids. Right: 3.6 and 3.2 nm Palladium ligand stabilized nanoparticles [20].

authors brought significant contributions later, in particular through both syntheses performed in organic solvents or two-liquid phase medium and second, the use of organic capping agent providing stronger chemical bond with the nanoparticle surface atoms. For example, G. Schmid [20] early described the synthesis of highly monodisperse metal nanoparticles (gold, platinum, palladium) with molecular stabilizers such as phosphine, amine, phenanthroline, allowing reversible isolation of powder nanoparticles at dry state (figure 2.1.4).

Key points in the kind of synthesis described here are the choice of the solvent, precursors and capping agent with respect to the chemical nature of the nanoparticle core. The use of organic thiols for the synthesis of gold nanoparticles published by Brus *et al.* in 1994 [21] leads to the common use of that kind of capping molecule for metal nanoparticle synthesis. In the same period of time, the early use of organometallics precursors and Tri-Octylphosphine as both surface capping agent and solvent provided important report for synthesis of semiconductors nanoparticles such as CdS, CdSe [22].

Microemulsion and Micelles Syntheses

Microemulsion and micelles are actually soft colloidal templates that were exploited to synthesize nanoparticles in the nineties notably with M.P. Pileni's group pioneering contribution [23]. The approach is based on the use of amphililic organics *i.e.* molecules or polymers having a hydrophilic, water-soluble part, and a hydrophobic, water-insoluble part. The self-organization of such compounds in liquids results in the formation of nanosized tanks of various shapes which can be loaded with different kinds of reactants to obtain metal or semiconducting materials (figure 2.1.5). This method offers very good control of syntheses conditions for size and shape. Its main drawback is that nanomaterials are prepared in low amounts.

Sol–Gel Syntheses

The chemistry of the sol–gel process is based on hydroxylation and condensation of molecular precursors and has been initially developed on silica in water. It has been

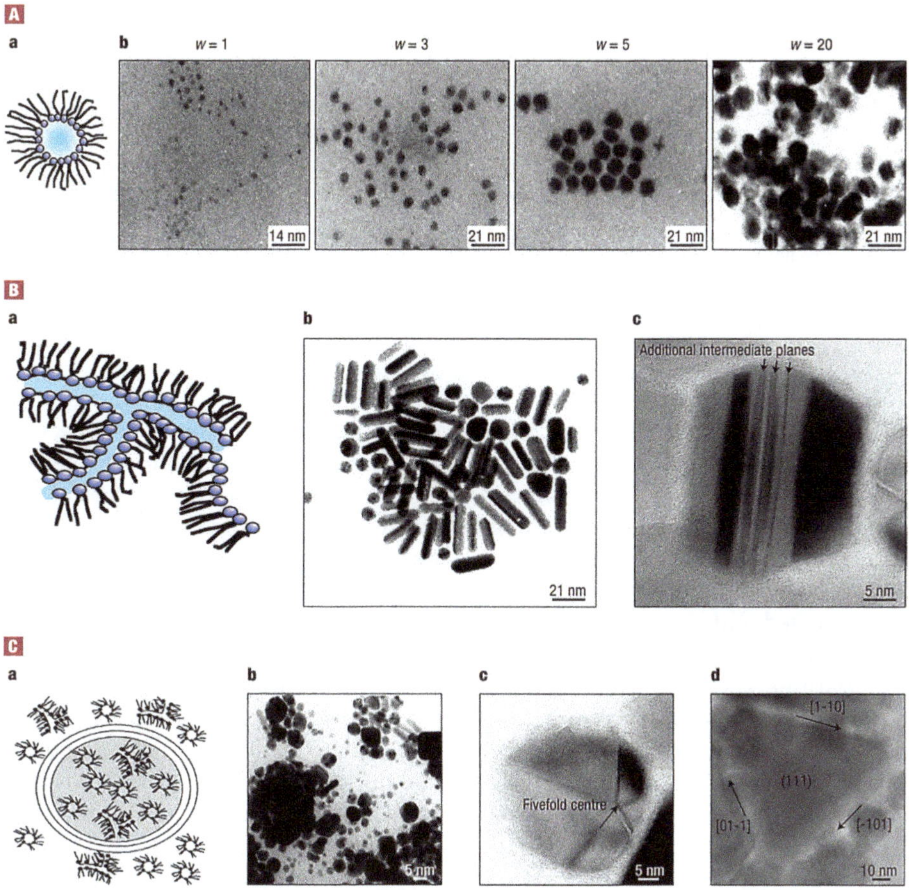

FIG. 2.1.5 – Illustration of the use of surfactant to create various soft templates leading to different nanoparticle sizes and shapes [23].

further extended to transition metal alkoxides [24] thus opening the way to the synthesis of metal oxides (figure 2.1.6). Conducted in aqueous media and driven by pH conditions among other parameters, sol–gel synthesis is a route to randomly graft polymers to nanoparticles.

Although the sol–gel process originally relates to the steps reported in figure 2.1.6, nowadays, it includes a great variety of syntheses involving solution state precursors for reactions conducted also in non-aqueous medium [25].

Solvothermal Synthesis

Solvothermal synthesis (named Hydrothermal when water is involved) refers to a process performed in a closed vessel at a temperature higher than that of the boiling temperature of the solvent in which reactants are introduced [26]. Since the process

$$\equiv M{-}OR + HOH \rightleftharpoons \equiv M{-}OH + ROH \quad (1)$$

$$\equiv M{-}OH + \equiv M{-}OH \rightleftharpoons \equiv M{-}O{-}M\equiv + HOH \quad (2)$$

$$\equiv M{-}OH + \equiv M{-}OR \rightleftharpoons \equiv M{-}O{-}M\equiv + ROH \quad (3)$$

FIG. 2.1.6 – Main steps leading to Metal oxide formation from Sol–Gel process starting from an alkoxide $M(OR)_4$. (1) Hydrolysis of alkoxide, (2) condensation of two hydrolyzed alkoxides, (3) mixed condensation of hydrolyzed and non-hydrolyzed alkoxide [24].

is conducted in a closed vessel, the syntheses proceed at very high pressure (tens to hundreds of bars). The pressure is autogenous since it is defined by the chosen temperature, generally lower than 400 °C, and the filling ratio of the vessel regarding to the solvent. The latter is in supercritical or near supercritical conditions, providing particular and unusual chemical conditions because of strong modification of initial solvent properties such as dielectric constant, viscosity, solvation properties... In solvothermal syntheses, a very high precursor concentration can be reached and thus a high supersaturating condition occurs upon cooling. This favors crystal nucleation and thus the formation of nano-sized particles [27]. Generally, parameters optimization and trial-and-error strategies are applied. Indeed, predictions are difficult and universal approaches are still difficult. Different factors can strongly affect the features of the nanoparticles in terms of size, shape and crystal phase. This includes organic additives, nature of the solvent and precursors, temperature, reaction time, pH. Ji and co-workers [28] reported an example of the pH and concentration effect on the Hydrothermal synthesis of CeO_2 (figure 2.1.7).

Solvothermal processes have the advantage of providing an access to a big variety of nanomaterials, such as numerous kinds of oxides, sulfides, selenides, nitrides, arsenides, phosphides, as well as carbon and silicon based materials.

2.1.2.2 Gas Phase Processes for Nanoparticles

Production of nanoparticles by gas phase processes relies on a common strategy, whatever the process considered: a precursor is decomposed by a heat source into a metastable vapor phase where the nucleation takes place and leads to the growth of particles. The processes differ by the type of precursors they use, and by the source of heat employed to treat the precursors.

When compared to most of the liquid phase processes, gas phase ones present the advantage of operating in continuous production with possible high production rates. They consist in one-step processes, and generally no post-treatment (washing, calcination, crystallization) of the powders is required. They offer the possibility of synthesizing composite powders or core–shell structures in one-step by multiple precursor injection points. Direct deposition on a substrate (coating) is also achievable. Products come usually with high purity as there is no contact with

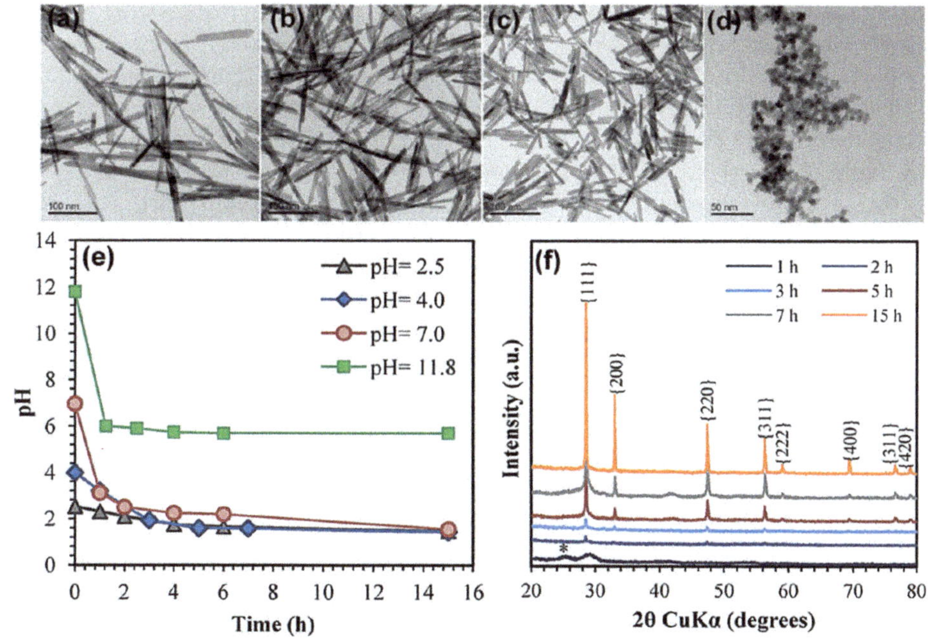

FIG. 2.1.7 – Effect of pH on the morphology and size of CeO_2 nanoparticles produced by hydrothermal synthesis [28].

reactors. The powders are collected in dry condition, without any organics on their surface which means that they are agglomerated (see previous section dealing with growth mechanisms). Particle characteristics (size, crystalline phase, agglomeration) are mainly driven by synthesis parameters such as pressure, temperature or residence time. Drawbacks of gas-phase processes *versus* liquid-phase ones are: (i) they are more expensive to exploit, (ii) agglomerated powders are difficult to disperse into stable suspensions of individual objects, (iii) dry condition leads to more severe safety issues.

Aerosol flame synthesis (AFS) is widely used for the synthesis of oxide nanopowders and appears as the process of choice for this kind of materials. It is indeed the cheapest gas phase process and it is already employed at the industrial scale for the production of silica, alumina or zirconia nanopowders. It is able to produce homogeneous particles in terms of size distribution or chemical composition, even for complex crystalline structures. In this process, the heat source is the combustion: the liquid precursor is decomposed by a pilot flame. AFS processes are classified into three main categories, depending on the state of the precursor being fed to the flame [29]. When it is vaporized, the process is called vapor-fed AFS (VFAFS). When it is fed as an aerosol, two cases are distinguished: if the precursor is not flammable, it is called flame assisted spray pyrolysis (FASP); if it is flammable, it is called flame spray pyrolysis (FSP). With FASP, depending on synthesis conditions, nanoparticles can be obtained (when the droplet is vaporized and reaction takes place in the gas phase), as well as hollow or large particles when reaction occurs within the liquid droplet. On the contrary, FSP appears more efficient for nanoparticle production because of the high operating temperature (3000 K) obtained by the combustion of the precursors in addition to the pilot flame that leads generally to gas phase reactions. As oxygen is present in the reactor for the pilot flame, AFS processes are almost only used for oxide nanoparticle synthesis (or soot or carbon black production directly from a fuel-rich combustion).

When it comes to non-oxide nanoparticle synthesis, several processes appear more efficient than AFS techniques. These processes do not need oxygen for heat production, and allow working under neutral atmosphere. They are also able to produce oxide particles, but less efficiently and in a much more expensive manner than AFS. They are thus mainly employed for non-oxide ceramics or metallic nanoparticle synthesis.

The most versatile process is the plasma: almost all materials, metals or ceramics, can be obtained with this process with high throughput. Because of the very high temperatures achievable by a plasma torch (>9000 K), even solid raw powders are admissible as precursors. These latter, either gaseous, liquid or solid, are introduced in the center of a plasma torch where they are completely vaporized. The plasma is generated by discharge (DC plasma) or induction (ICP or RF plasma). The absence of electrode in the latter case provides higher purity in the obtained products. Despite the high velocity of gas in the reactors, the relatively long residence time in the hot region may lead to the formation of large particles with strong agglomeration. To avoid this, a quenching system is often used to inject radially and perpendicularly to the plasma axis a cold gas able to cool down very rapidly the gas phase [30]. Drawbacks of this process are related to the high temperature gradient between the center and the side of the plasma that sometimes leads to the presence of secondary phases (especially in the case of complex crystalline phases), and the use of reactive species such as H_2 that tend to modify the material composition by removing O or C atoms. Obtaining amorphous materials or tailored hollow large particles from liquid droplets remains hard to address with high temperature plasmas. Nevertheless, ICP is already developed at the industrial scale for "basic" materials production.

When a more accurate control of complex crystalline phase (or amorphous materials) is required, or when one wants to take advantage of liquid droplets reactions as in AFS technique, laser pyrolysis technique appears attractive. Operating at lower temperature than plasma, this process uses a laser beam (usually CO_2 at 10.6 μm) to decompose the gaseous or liquid precursors. In addition to other gas-phase synthesis parameters, it is possible to control accurately the way energy is delivered to the system by tuning the laser parameters (pulse duration, power density) and thus control precisely the characteristics of the nanoparticles [31]. As the reaction zone is roughly delimited by the laser beam overlap with the injection nozzle, temperature and velocity gradients are not as important as in plasma torch, and the obtained products are more homogeneous even for complex crystalline phases. Limitation of the process comes from the mandatory resonant absorption of at least one of the precursors at the laser wavelength. When this condition is not fulfilled, a sensitizing gas must be employed (C_2H_4, NH_3, SF_6), decreasing production yield and potentially contaminating the products. On the contrary, when the precursor absorbs by itself as in the case of SiH_4 for Si-based materials, production rates and yield are high enough for industrial applications. Because of the lack of suitable precursors, laser pyrolysis is barely employed for the synthesis of metallic nanoparticles.

The last gas-phase process presented here is maybe the simplest one: the spray pyrolysis technique. The liquid precursor is sprayed into a furnace, which enables the synthesis of a wide range of materials as there is no condition for energy coupling in the precursor. This process can be easily used for large-scale production as a cost effective technique in contrast to plasma or laser based processes. It operates at lower temperatures than these latter processes, which can limit some precursors complete decomposition or some materials' crystallization. Yet, this lower temperature is partially compensated by a much longer residence time, as this latter parameter is related to the length of the hot zone in the furnace (in addition to lower gas flow rates) instead of the pilot flame, the plasma torch or the laser spot. Heating rate of the precursor droplet is also slower. The consequence is that the morphology of the obtained powders is generally different: instead of the common chain-like type primary particles agglomerates encountered with FSP, plasma or laser, larger spherical particles (sometimes several hundred nanometers in size) are obtained, usually constituted of smaller crystallites. Indeed, depending on solution characteristics, different growth mechanisms are observed [32], mostly being related to solvent evaporation, diffusion of species and precipitation in the drying droplets. These general mechanisms are similar to the ones encountered in FASP and lead generally to the formation of one particle per droplet after drying and thermal reaction of the precursors inside the dried particle. Nevertheless, the rather dense particles obtained are poorly suitable for dispersion when primary particle suspensions are required. On the contrary, this structure may be of particular interest for the fast preparation of dense green bodies for ceramics elaboration by sintering.

2.1.2.3 CVD Processes for Nanowires and Nanotubes

Chemical Vapor Deposition (CVD) is a widely used process in industry to produce high quality and high performance solid materials, including nanowires and nanotubes. Here is a non-exhaustive overview of the synthesis of nanowires (NWs) and carbon nanotubes (CNTs) by CVD for energy and transportation applications.

CVD for NWs

NWs are one-dimensional (1D) nanostructures with lengths of microns and diameters in the order of tens to hundred nanometers. Unlike classical thin film technologies, lattice mismatch strain in NWs can be relaxed elastically at their free surface side without dislocations. This capability is often used to grow III-V-like semiconductor NW heterostructures on Si substrates, for example. Properties like quantum-confinement or light-trapping lead to performance enhancement and new functionalities for optoelectronic devices [33], which are expected in applications including energy conversion [34].

The most used synthesis process from CVD is the Vapor–Liquid-Solid method (VLS) [35]. A catalytic liquid nano-droplet heated above the eutectic temperature for the metal–semiconductor system adsorbs a vapor of gaseous precursors to supersaturation levels and the crystal NW grows from nucleated seeds, after diffusion in the liquid nano-droplet, at the liquid–solid interface. The principle of the VLS is schematically illustrated in figure 2.1.8 [36].

Au droplets were first used as catalytic agent, but other metals or alloys can be used. They can be prepared by depositing a metal thin film on the substrate by sputtering or evaporation, and annealing the film to form droplets. The size of the droplets, which is critical for the subsequent NW diameter, is controlled by the thickness of the film. Freestanding crystalline NWs of semiconductors [37] or metal oxide materials [38] can be produced with fully controlled nucleation sites and diameters by VLS. However, other methods including laser-assisted CVD, oxide-assisted CVD, thermal CVD, metal-catalyzed molecular beam epitaxy (MBE) and Chemical beam epitaxy (CBE) can also be used [39]. Growth of NWs is still a very active research field [40].

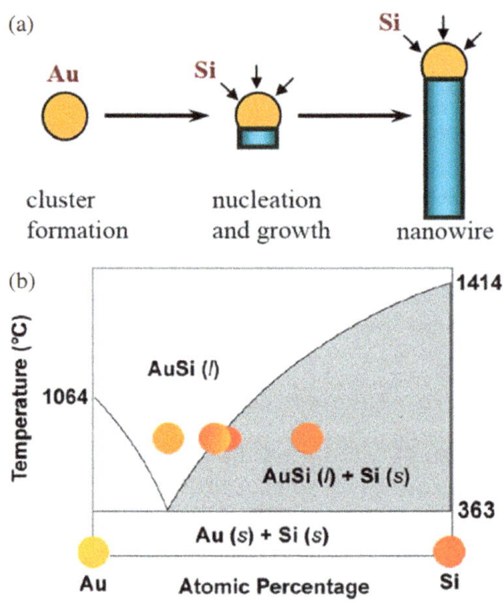

FIG. 2.1.8 – Schematic of VLS growth of Si nanowires. (a) A liquid alloy droplet AuSi is first formed above the eutectic temperature (363 °C) of Au and Si. The continued feeding of Si in the vapor phase into the liquid alloy leads to oversaturation of the liquid alloy, resulting in nucleation and directional nanowire growth. (b) Binary phase diagram for Au and Si of VLS growth [36].

CVD for Carbon Nanotubes

Carbon nanotubes (CNTs) are allotropes of carbon composed of cylindrical graphite sheets that are self-rolled up in a tube-like structure [41]. CNTs composed of a single graphene sheet are single-walled carbon nanotubes (SWCNTs), and those composed of multilayers of graphene sheets are Multi-walled carbon Nanotubes (MWCNTs). MWCNTs and SWCNTs are depicted in figure 2.1.9 [42].

The method of choice for the synthesis of CNTs is Catalytic Chemical Vapor Deposition (CCVD) [43]. This method allows control of the morphology and structure of the CNTs and the growth of aligned CNTs in a desired direction [44]. CNTs are the subject of many applications in various fields including sensors [45], energy storage [46], microelectronics [47], coatings [48] and biotechnology [49]. Among the CCVD techniques, two variants stand out: the two-stage syntheses and the one-step syntheses. For two-stage syntheses, the metal catalyst is pre-deposited on the substrate in a first step, then the carbon precursor is injected. In the one-step syntheses, the carbon precursor and the catalytic precursor are injected simultaneously. The two-stage CVD is the most widely used method in the literature [50]. Iron is the most commonly used catalyst, but other transition metals or mixtures of transition metals such as Co, Ni, Cu or even Mo can be used. At low temperatures, the commonly used substrate is aluminum [51]. The carbon precursor is generally acetylene, but it can be replaced by ethanol or ethylene for example. The one-step CCVD method is based on the simultaneous and continuous injection of a catalytic precursor, generally in the form of a metallocene, and a carbon precursor, generally a hydrocarbon in liquid or gaseous phase [52]. Ferrocene is generally used as catalytic precursor and toluene as a carbon precursor but it can be replaced by xylene, ethylene or ethanol. CCVD is currently

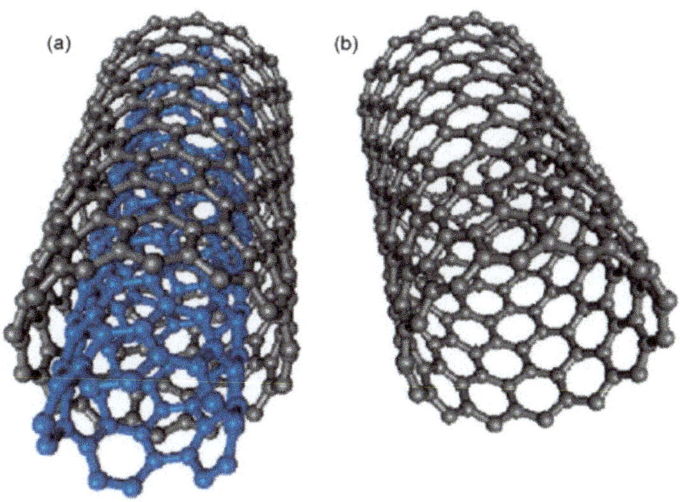

FIG. 2.1.9 – Structure representation of (a) multi-walled carbon nanotubes (MWCNTs); (b) single-walled carbon nanotubes (SWCNTs).

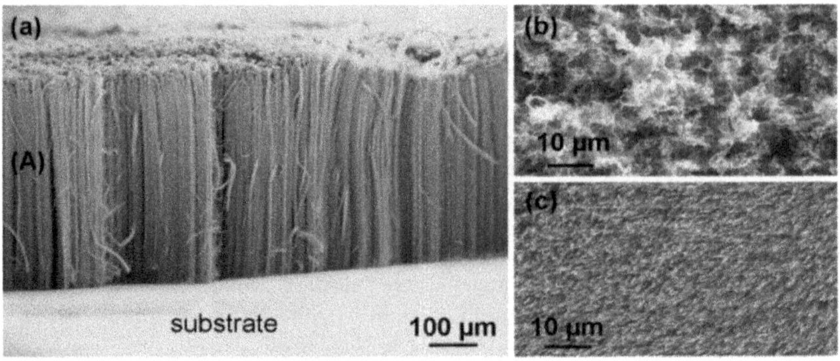

FIG. 2.1.10 – SEM Images of (a) the cross section of a typical carpet of VACNTs, (b) the top of the carpet, (c) the base of the carpet after scratching from the substrate [52].

used for industrial production (Nanocyl, NawaTechnologies...) of Vertically Aligned Carbon Nanotubes (VACNT). SEM images of VACNTs are presented in figure 2.1.10.

2.2 Integration of Nanoobjects

Magali GAUTHIER, Yann LECONTE, Henri PEREZ and Olivier SUBLEMONTIER

Université Paris-Saclay, CEA, CNRS, NIMBE, 91191 Gif-sur-Yvette, France

In this part, we are interested in the different ways of handling nanomaterials aiming ultimately at their incorporation into solid-state structures, in general as more or less thick deposits. These different routes are classified according to the processes used, depending on whether they consider handling liquids, solids or gaseous phases. None of these methods appears specific to the implementation of nanoobjects into solid-state structures. The use of one method rather than another seems essentially guided by the "functionality" of the structure, for example, the fact that the deposit constitutes one of the elements of a final structure and/or necessitates a separate characterization, or the fact that the deposit brings together several elements whose properties are expected to be exploited in synergy.

2.2.1 The Liquid Phase

Many methods (figure 2.2.1) exist, from a liquid phase containing nanoobjects, to produce a solid structure in the form of a nanostructured film. The manipulation of nanoobjects introduced into a liquid phase presents several advantages, among these the fact it is not affecting their feature, as well as the possibility to add in this media several types of objects to combine several properties. The more or less marked drawback of this approach is the elimination (drying) of the liquid phase serving as a vehicle for the nanoobjects, which can induce unwanted features of the deposit such as crack formation or adherence problems.

While some examples involving many different methods for the integration of nanoobjects can be found in the literature, some are more preferred in the field of energies. Indeed, for photovoltaics, batteries, or fuel cells, the process must be as simple as possible while providing the highest performances. Energy-related applications imply mass production ability and reasonable cost for high dissemination. Therein, methods such as ink jet printing, screen printing, or spray coating are of particular interest. However, an important point in some crucial energy production devices is that they are complex systems in which charge (electron, hole, ions) and mass exchange and transport occur at different scales. This makes the way the integration of nanoobjects are incorporated a very important point. This can be illustrated using two examples in the field of fuel cells. In the Proton Exchange Membrane Fuel Cell (PEMFC) technology the formation of the active layer by spray coating is reported by A.M. Chapparo *et al.* [53]. The liquid media are a slurry containing platinum nanoparticles deposited on carbon black and Nafion® electrolyte at various contents (figure 2.2.2). Performances of the elementary fuel cells prepared by electrospray are better than those prepared by airbrush, and the latter are not sensitive to the Nafion® content. The authors explain this result by the way the electrolyte component is located in the final structure resulting from drying

DOI: 10.1051/978-2-7598-2446-5.c022

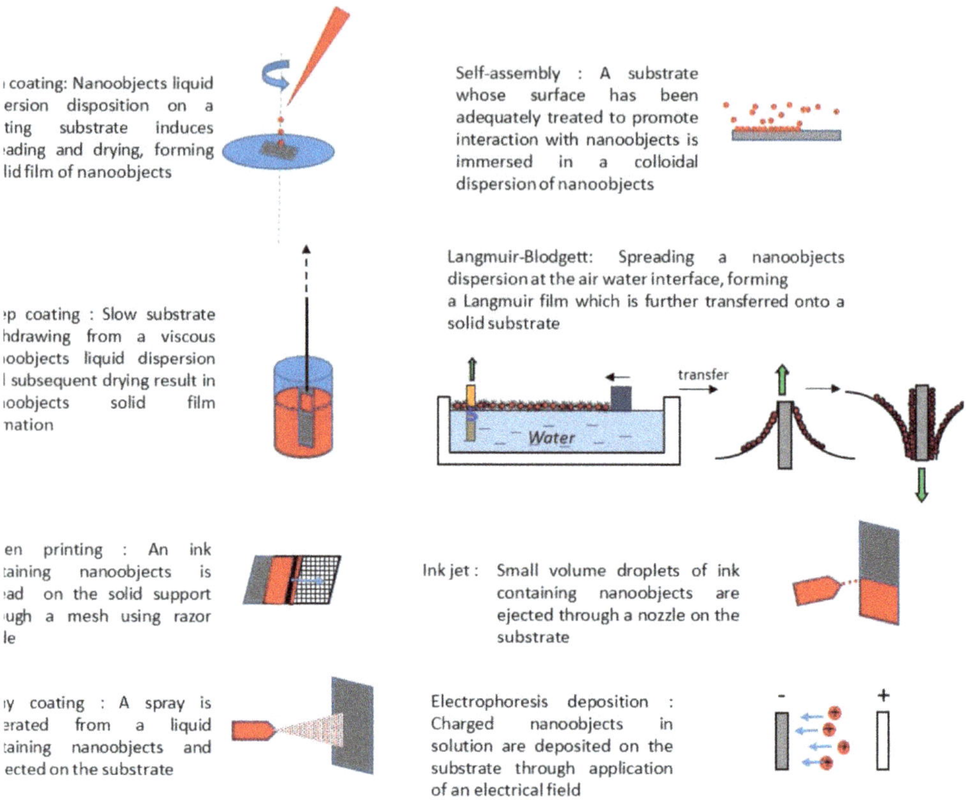

FIG. 2.2.1 – Examples of deposition methods using liquid media applicable for the integration of nanoobjects into solid-state structures.

conditions. This shows how using a given declination for a given deposition method, for a particular nanoobject liquid dispersion, might be of paramount importance.

The ink-jet deposition method has been reported for application of a thin intermediate cathode layer in Solid Oxide Fuel Cells (SOFC) [54] using nanoobjects prepared by ball milling. This work is an illustration on how the combination of different deposition methods may strongly affect the final performances of the device. Three different cells were prepared in which the anode and electrolyte are identical and cathode are different (figure 2.2.3).

The polarization curves for the different cells show dramatic differences. The best performances are those in which the ink-jet intermediate cathode is combined with a painted cathode of the same material.

2.2.2 The Solid Phase

Sintering

Consolidation processes starting from a powder constitute the most common route used to produce ceramic parts, usually named powder metallurgy processes.

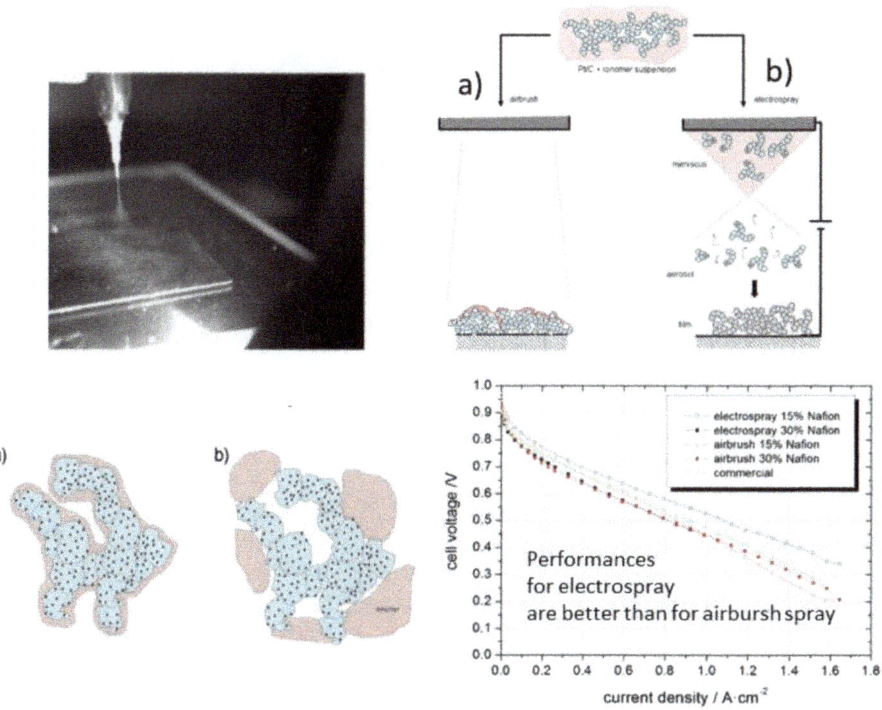

FIG. 2.2.2 – Influence of different spray coating methods (Airbrush and Electrospray) on the structure of the active layer of PMFC and its performances [53].

These processes are also used for metallic parts, as the obtained microstructures (and thus properties) differ from the ones obtained with conventional metallurgy processes implying melting or plastic deformation. The consolidation process, named sintering, aims at creating irreversible cohesion between the grains of the starting powder together with eliminating intergranular porosity in order to obtain a dense and solid final part.

Sintering process starts with the preparation of the initial part made of the powder, called the green body. This step includes powder preparation (dispersion, granulation, mixing with additives or binders when necessary) and green body formation (die pressing or casting, injection molding, machining). The green body is then subjected to a high temperature heat treatment, the sintering step by itself, where solid or liquid state diffusion phenomena lead to the densification of the part. Typical sintering temperature is in the order of magnitude of half of the melting temperature of the considered material, and can exceed 2000 °C for refractory carbides, for example. At the end of the process, a shrinkage of the part is observed resulting from the elimination of porosity.

The driving force for sintering is related to the surface energy of the grains: some materials being difficult to sinter (low specific surface, covalent bonding, refractory),

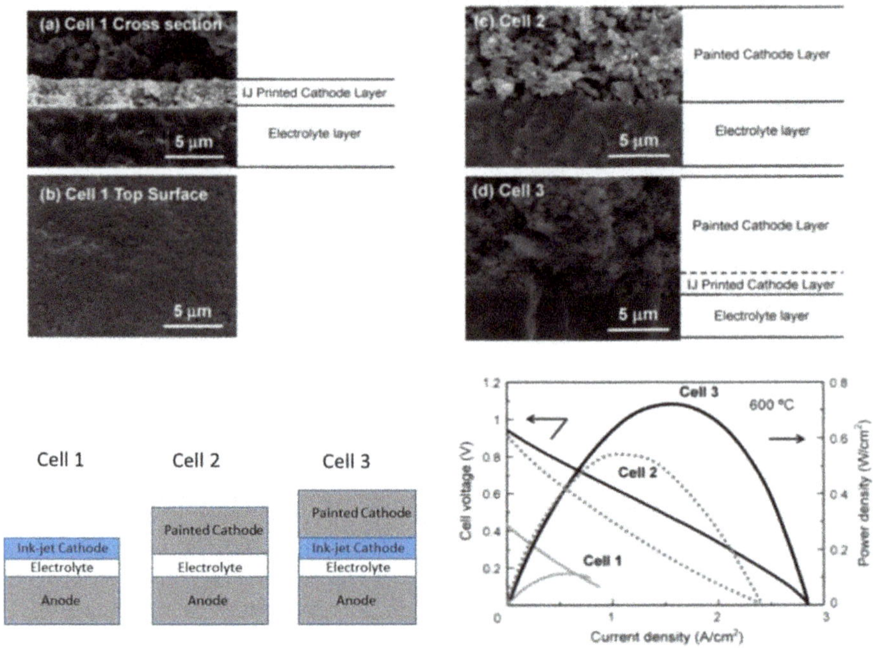

FIG. 2.2.3 – Use of ink-jet nanoparticle deposition for thin cathode layer in SOFC and impact of the configuration on its efficiency [54].

the application of a mechanical pressure is sometimes mandatory to reach satisfying densification. In this context, the high specific surface area developed by nanopowders helps in decreasing the required sintering temperature, because nanoparticles are more reactive than coarse particles. In a simplified scheme, densification can be described through two processes: porosity elimination and grain growth. The latter phenomenon has to be well controlled in order to keep a fine microstructure able to provide good mechanical properties. In the specific case of nanostructured materials, the final grain size has to remain as small as possible. Regardless of the initial powder size distribution, sintered materials with mean grain sizes lower than few hundreds of nanometers are difficult to achieve by conventional sintering processes (natural or pressureless sintering, uniaxial hot pressing), because the porosity elimination step cannot be separated from the grain coarsening one. New sintering processes were thus developed, aiming generally at decreasing the sintering time or temperature to limit grain growth. Three main processes are presented here. The first one uses very high pressures (hot isostatic pressing, HIP), the second one uses electric current (field assisted sintering techniques, FAST), and the third uses microwave energy (microwave sintering, MWS).

HIP consists in using pressure as high as 300 MPa applied through a gaseous or liquid media. In this case, pressure is distributed in an isostatic way, unlike conventional hot pressing where it is applied along the axis of the press punches. The obtained microstructure is then more homogeneous, and the elevated pressure allows to shorten the sintering time down to less than 1 h. To avoid destruction of the green body by the infiltration of the pressure fluid in the open porosity, a sheath has to be used. For high sintering temperatures, this sheath has to withstand the heat treatment and eventually expensive materials have to be employed (Ti or Ta). In this case, HIP can be performed in two steps, a first one at low temperature with cheap sheath to close the open porosity, followed by a second one at a higher temperature (without sheath as only closed porosity remains) for final densification. It can also be used to finish the densification after a first sintering step by another process (post-HIP). Cold isostatic pressing (CIP) is also achieved to prepare the green body before another sintering process when initial green body density has to be high.

FAST denomination covers several current assisted processes differing in the way current is delivered to the green body. The most famous one is the spark plasma sintering technique (SPS): it is basically a hot press system with uniaxial pressing between two punches, but heating is provided by running a pulsed current through the graphite punches and the conductive mold where the green body is placed. Depending on the conductive or isolating nature of the material considered, the current will rather flow in the sample or in the mold, leading to different heating kinetics and electrical phenomena in the porous material [55]. High heating rates are obtained, and sintering times can be as short as few minutes. A very fine microstructure is achievable with very limited grain growth while obtaining full densification (see figure 2.2.4) [56]. Advanced structures are addressed, as multi-layers with sharp interfaces or functionally graded materials. Despite the uniaxial mold configuration, complex shapes are obtained without machining. The limitation comes from the nature of the mold that will allow either working at high pressure or at high temperature but never both.

MWS is generally compared to natural sintering, even if some furnaces are now equipped with pressing features. The green body is placed directly in the MW cavity without mold or sheath, which simplifies the elaboration of complex shapes. When compared to conventional furnace heating, MW heating is faster and comes from the

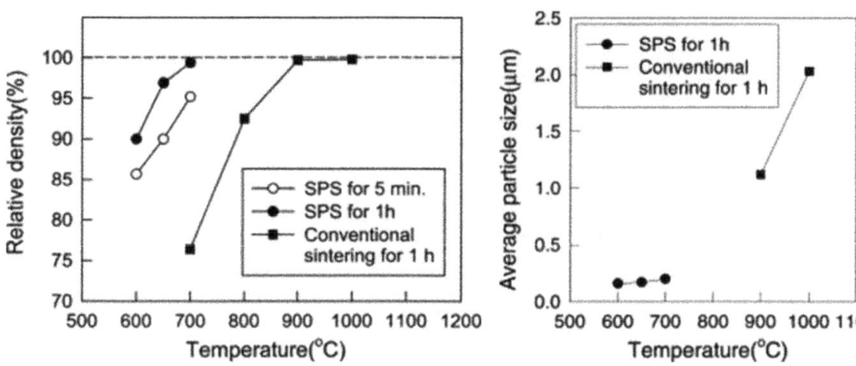

FIG. 2.2.4 – Effect of SPS process on the densification and grain growth of TiO_2 compared to conventional sintering [56].

inside of the green body, diffusing towards the borders (inverse of the conventional one). Difficulties come from the MW absorption by the material: when a too low susceptor has to be employed around the green body to transfer the heat to the sample, losing part of the MWS volume heating advantage.

Phase Segregation

In the previous section, the elaboration of composite films from a suspension of nanoobjects *via* wet routes was presented. Instead of integrating nanoobjects independently synthesized into a matrix, it can be interesting to grow these nanoobjects directly inside the matrix during its synthesis. In this case, there are less process steps and the hazardous handling of nanoobjects is avoided. Moreover, protected in the matrix, the nanoparticles are never exposed to air or pollutants.

The basic idea is to grow a matrix with an incomplete stoichiometry in an element or with the presence of an additional element that can lead to a secondary phase. In both cases, an element in excess is able to precipitate into nano-inclusions to recover the matrix stoichiometry upon annealing. A famous example is the case of silica layers containing Si quantum dots obtained after annealing of SiO_x ($x < 2$) films deposited by magnetron sputtering. The lack in oxygen in the film obtained *via* deposition conditions results in an excess in Si *versus* SiO_2 stoichiometry. After annealing, this Si excess precipitates into Si nanoparticles dispersed in the matrix (see figure 2.2.5), this latter's stoichiometry being recovered to O/Si = 2. The size of the inclusions and their distribution can be controlled by synthesizing multilayer system such as SiO_x/SiO_2. The stoichiometric SiO_2 layers are used to limit the growth of the Si clusters to the thickness of the non-stoichiometric layer. The periodicity of the multilayer film enables the elaboration of quantum dots arrays with specific optical properties. A similar successful approach was reported for other composites where the inclusion is not made of the same element as the matrix: Sn

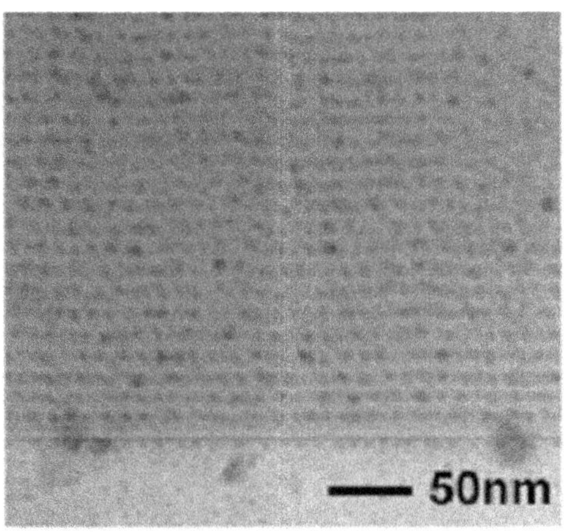

FIG. 2.2.5 – TEM cross section image of Si quantum dots embedded in a SiO_2 matrix [57].

nanoparticles embedded in a silica matrix [57]. Such composite layers are achieved by co-sputtering of two different targets to enrich the silica growing layer with Sn atoms. In that case, attention has to be paid to the phase diagram and stoichiometry to avoid the formation of a ternary phase upon annealing.

2.2.3 The Gas Phase

Processes based on the gas phase present decisive advantages for nanostructured coatings. These methods, involving aerosols, require controlled deposition of nanoparticles for homogenous film elaboration on large surface areas. Supersonic expansion of gas under vacuum has been used for a long time as a support for nanoparticle jets and further deposition on a substrate. Physical Vapor Deposition (PVD) is widely used for classical and nanostructured coatings. After presenting briefly nanoparticle jets and PVD, we will focus on hybridization of these technologies for manufacturing nanocomposite coatings in the form of nanoobjects embedded in a matrix of a different material.

Jets of Nanoobjects

Deposition of nanoparticles on quite large surface areas is feasible by dragging nanoaerosols in a supersonic gas expansion under vacuum [58]. The supersonic beam is obtained by a huge pressure drop between a gas container and vacuum through a nozzle of a few tenths of a millimeter in diameter. In optimized processes, the nanoparticle beam is directly connected to the synthesis process in the gas phase. An example is given in figure 2.2.6 [31], where a laser-pyrolysis reactor is used for the synthesis of nanoparticles. The reactor is placed in a source chamber and has a hole, which serves as a nozzle for the vacuum expansion. The beam of nanoaerosols is formed in the supersonic expansion and the deposition can be achieved, after passing through a skimmer, in the deposition chamber, on a substrate placed on the beam trajectory. An optional version of the system includes a third chamber for *in situ* measurement of the size of primary particles by Time-of-Flight mass spectrometry.

The advantages of supersonic beams are that particles reach a high velocity, depending on the gas nature and temperature before expansion. This leads to higher film densities and the possibility to incorporate nanoparticles in soft material matrixes [59].

Electrospray is also applied to powder production and surface coating [60].

Thermal spray [61] and supersonic cold spray [62] are also used for coating deposition or additive manufacturing. In thermal spray, materials are deposited in a molten or soften condition on substrates. Supersonic cold spray can deposit coatings of micrometer and nanometer metal particles, ceramic particles, clays, polymer, hybrid materials, graphene oxide and metal–organic frameworks, using various types of non-molten materials from solid powders or suspensions.

For small metallic nanoparticle (10 nm or less) fabrication and deposition, the combination of gas aggregation source with magnetron sputtering is a method of choice [63]. In this process, shown in figure 2.2.7, sputtering vaporization is combined with gas condensation in controlled volume filled with inert gas. The size of

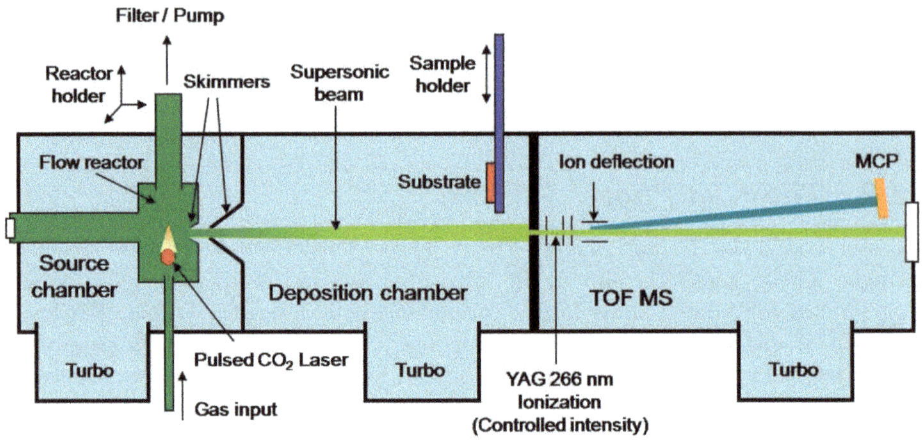

FIG. 2.2.6 – Experimental setup equipped with (i) a source chamber (left) including a laser pyrolysis reactor as a source of nanoaerosols, (ii) a deposition chamber (middle) with a removable substrate holder and (iii) an analysis chamber (right) including a Time-of-Flight Mass Spectrometer for *in situ* primary particle diameter measurement [31].

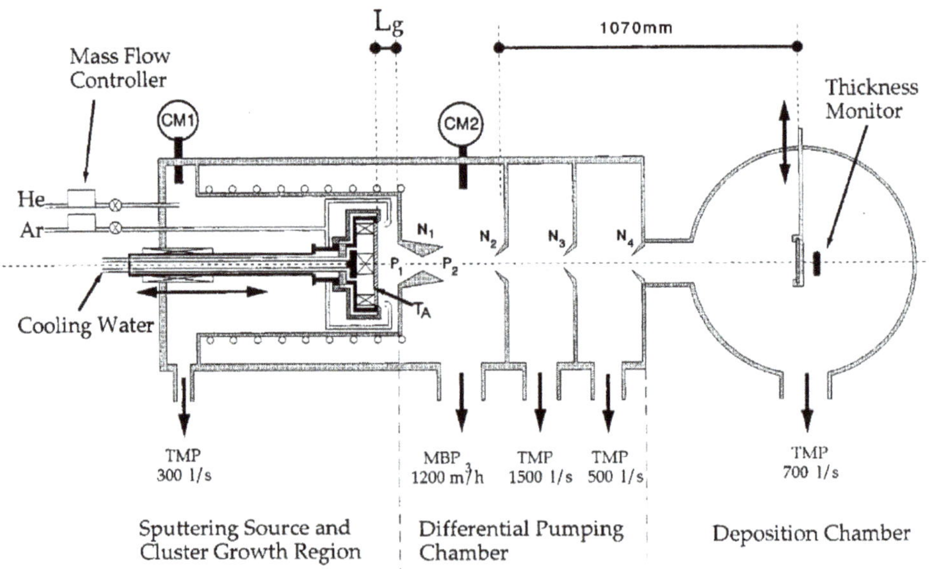

FIG. 2.2.7 – Schematic arrangement of sputtering-gas-condensation nano-cluster source and deposition [63].

the nano-clusters can be varied by adjusting process parameters like the volume of the condensation chamber, the pressure and the power applied to the sputtering cathode. The condensation volume is directly connected to an extraction nozzle to form a diffusive jet of nano-clusters, which can be further deposited on a substrate placed in the beam.

This technique is now used in industrial product (Mantis Deposition Ltd) and can be used for core–shell type nanoclusters.

The famous aerodynamic lens system is a very promising tool for producing nanostructured particulate films by gas-phase synthesis [64]. An aerodynamic lens system allows the transfer of small aerosols from atmospheric pressure to vacuum while producing a dense and thin particle beam. It is composed of three parts: a flow control orifice, an aerodynamic focusing stage, and an acceleration nozzle. The flow control orifice adjusts the gas volume flow through the whole system. Its diameter is generally set around one to several hundreds of microns, depending on the internal geometry of the lens system. The aerodynamic focusing stage is composed of at least one elementary lens, which is made of a buffer chamber and a diaphragm. Several elementary lenses are generally used. Aerodynamic particle focusing is achieved by successive compression and expansion of the carrier gas through the series of lenses. Particle trajectories in this part rely on aerodynamic drag of the particles by the carrier gas. The inertia of particles is responsible for the difference between their trajectories and those of the gas flow. Although theses system were originally made for producing collimated beam under vacuum, it was recently demonstrated that controlled divergent beams of nanoparticles were also possible, opening the way to homogenous deposition on large surface areas [65], as shown in figure 2.2.8.

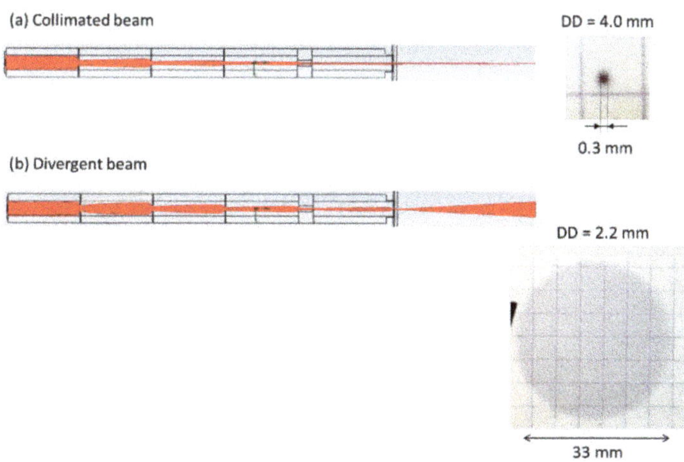

FIG. 2.2.8 – Trajectories of nanoparticles in (a) a collimated-beam and in (b) a controlled-divergent aerodynamic lens system. Figures on the left are computational fluid dynamics calculations with 40 nm gold nanoparticles. Figures on the right are photographs of deposits of 40 nm gold nanoparticles were DD is the last diaphragm diameter [65].

Hybridization of Nanoobject Jets with PVD for Nanocomposite Coatings

The hybridization of nanoparticle jet techniques with PVD, especially magnetron sputtering, is of great interest for producing nanostructured coatings in the form of nanoparticles embedded in a matrix of a different material, as show in figure 2.2.9.

The coupling of a gas aggregation source with magnetron sputtering, shown in figure 2.2.10 has been realized in order to produce nanocomposite films of Pt nanoparticles in a Pd matrix for an application in fuel cells [66].

Hybridization of aerodynamic lens systems with magnetron sputtering is also possible [67, 68]. This technique presented in figure 2.2.11a is promising, because it allows virtual infinite combinations in the nature of nanoparticles and matrixes, and both can be deposited in the same time, leading to homogenous stochastic dispersions of particles in the matrix, as shown in figure 2.2.11b.

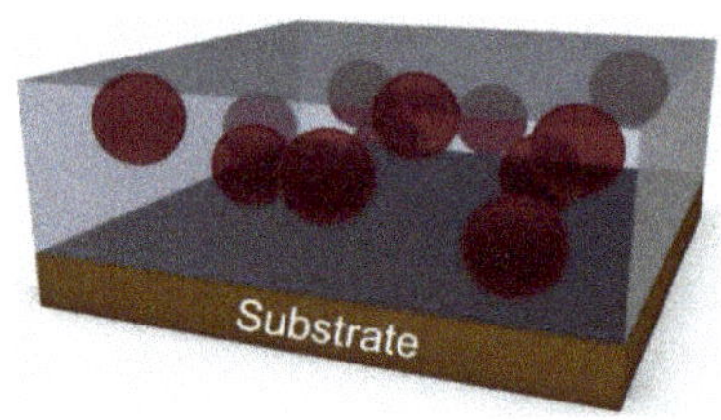

FIG. 2.2.9 – Schematic of a nanostructured coating in the form of nanoparticles embedded in a matrix.

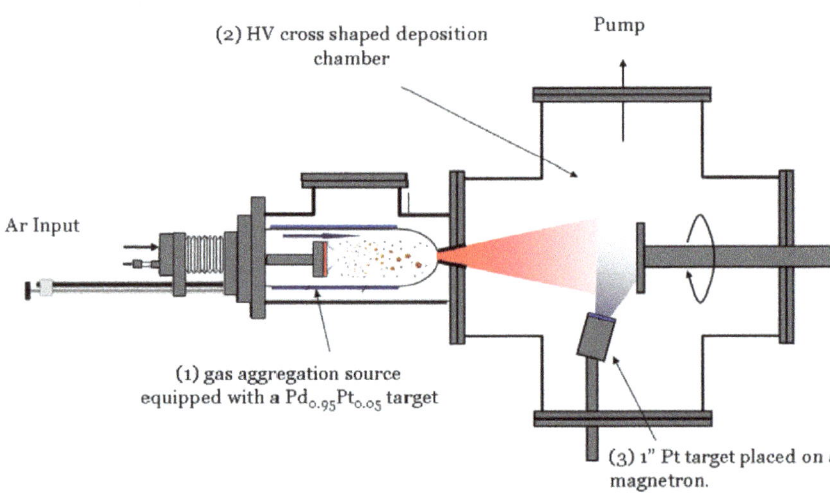

FIG. 2.2.10 – Schematic of the combination of gas aggregation source with magnetron sputtering for nanocomposite fabrication [66].

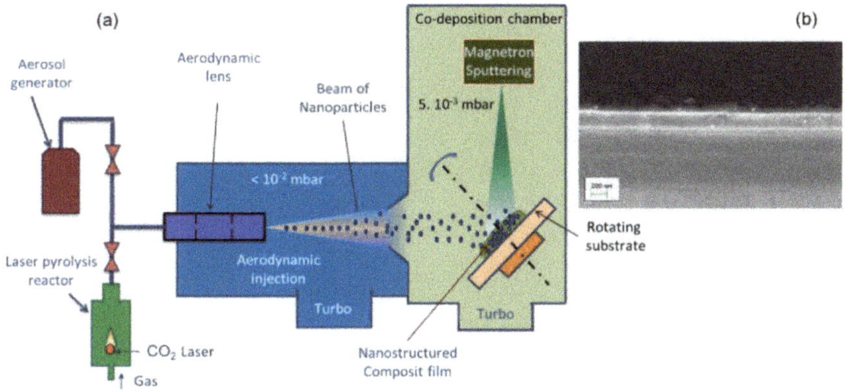

FIG. 2.2.11 – (a) Principle of the hybridization of a nanoparticle jet produced by an aerodynamic lens system with magnetron sputtering. (b) SEM image of a nanocomposite film of Au nanoparticles embedded in a SiO_2 matrix [67].

Different kinds of sources for the nanoparticles can be used in this process. It is possible to start with a suspension of colloids (for metallic particles) that is further atomized and dried before entering the aerodynamic lens system. It is also possible to connect directly a gas-phase synthesis apparatus like a laser pyrolysis reactor (for ceramic particles). A large range of nanoparticle types is then possible.

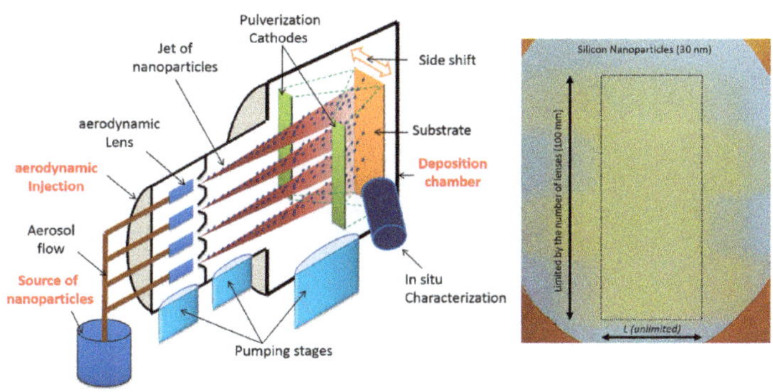

FIG. 2.2.12 – (a) Principle of nanocomposite coatings on large surface areas by combining a multiple aerodynamic lens system with magnetron sputtering and side shifting of the substrate. (b) Demonstration of the deposition of 30 nm Si nanoparticles on a large surface area with four aerodynamic lens systems [70].

Homogenous deposition of nanostructured films was demonstrated with a multiple divergent aerodynamic lenses system [69], combined with smart masking of the jets and side shifting of the substrate [70], shown in figure 2.2.12a. An example of homogenous deposition with 4 aerodynamic lens systems is presented in figure 2.2.12b. Inhomogeneity of less than 8% in thickness was achieved with a deposition of 30 nm Si nanoparticles with 4 aerodynamic lens systems in parallel.

2.3 Application of Nanoobjects to Energy and Transportation

Magali GAUTHIER, Yann LECONTE, Henri PEREZ and Olivier SUBLEMONTIER

Université Paris-Saclay, CEA, CNRS, NIMBE, 91191 Gif-sur-Yvette, France

2.3.1 Solar and Heat Conversion

2.3.1.1 Photovoltaics

Semiconductors nanocrystals (or quantum dots, QD) show the peculiar property of enabling the control of their electronic band structure through the design of their size, thanks to the so-called quantum confinement phenomenon. It is thus actually possible to modify their optical properties in terms of absorption and emission through the modification of their bandgap. Nanoparticles can also change the interactions between a luminescent ion and its host matrix in order to favor interesting relaxation pathways (enhanced luminescence lifetime, lower non-radiative recombination…). Finally, metallic nanoparticles are also known to enhance optical phenomena for given wavelengths through plasmonic effects. In the field of photovoltaics (PV) where the optimal conversion of light over the whole solar spectrum is targeted in order to increase the cells efficiency, such optical and electronic properties tunability appears as a striking advantage.

PV history relies on semiconductor single or multiple junctions, and yet today this technology still ranks among the top performances (for single junction configuration). Several sources of loss contribute to limit these cells' efficiency. Among these sources, two take their origin in the absorption properties of the active material: (i) photons with energy lower than the bandgap are not absorbed, and (ii) excess energy of high-energy photons is lost through thermalization process (these latter loss being responsible for roughly 30% of the total loss in Si). Different strategies were thus developed to overcome these limitations, some of them being achieved owing to nanoparticles peculiar properties.

First idea was to convert the incident solar spectrum into the most efficient conversion wavelength range of the active cell material. A conversion layer made of a transparent dielectric matrix embedding QDs and luminescent atoms can therefore be added to the cell structure. Concerning low energy photons, up-conversion mechanisms are addressed aiming at the emission of one higher energy photon after absorption of two or more lower energy ones (see figure 2.3.1). To obtain this feature, lanthanides ions are usually employed (Er^{3+}, Yb^{3+}, Tm^{3+}) inside a nanostructured matrix used to optimize the up-converted emission [71, 72]. Inverse approach was followed for high-energy photons aiming at getting two lower energy photons from a higher energy one (down-conversion, or quantum cutting). In this case, the QD-containing layer is used to absorb the solar light and subsequently transfer excitation to surrounding luminescent atoms. When the incident energy is

DOI: 10.1051/978-2-7598-2446-5.c032

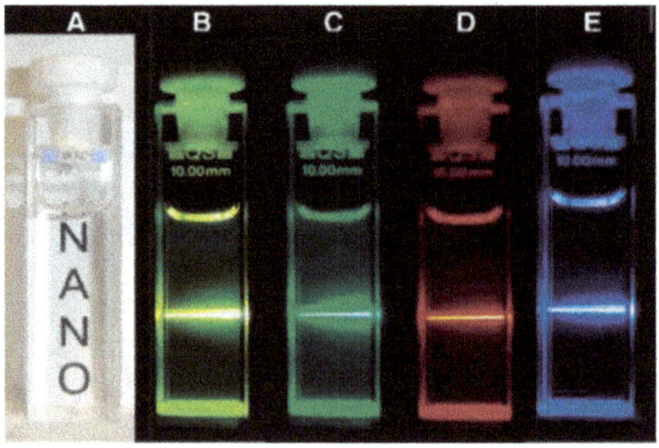

FIG. 2.3.1 – Illustration of the up-conversion process. (A) NaYF4 doped with 2% Er^{3+}, 20% Yb^{3+} solution under visible light. (B) same suspension under 977 nm excitation. (C and D) same viewed through green and red filters, respectively. (E) NaYF4: 2%Tm^{3+}, 20% Yb^{3+} suspension under 977 nm excitation [72].

high enough, the generated exciton in the QD will recombine through two successive Auger processes, each of them exciting a neighboring luminescent rare-earth ion [73].

Instead of converting incident light through luminescent processes with limited quantum efficiencies, it is possible to take direct benefit of a wider range of photons in the IR or UV domains. For low energy photons, absorption below the bandgap can be achieved by adding a QD layer to the cell structure. This layer, when properly chosen, generates an intermediate energy band in the active material bandgap, enabling the absorption at lower energy. Upon a second low energy photon absorption, the electron generated in the intermediate band after the first absorption will pass to the active material conduction band [74]. For high energy photons, an idea is to generate, from one single incident photon, several excitons with lower energies. This process, called multi excitons generation (MEG), relies on reverse Auger effect (or impact ionization). It can be observed in bulk materials but with energy threshold much higher than in QD [75]. Another possibility is to collect high energy excitons before they thermalize (hot carriers solar cell), which is a key issue as this latter process is very fast. To be successful, absorbing material must be able to slow down this thermalization, and contacts with selectivity in energy have to be employed to avoid the collection of high energy carriers with lower energy ones [76].

The last approach presented here, the preparation of multigap or tandem cells can be addressed with QDs. Basic idea is to stack several layers of different materials (*i.e.* different bandgaps), each of them being dedicated to the efficient conversion of one part of the solar spectrum. This concept allows reaching top performances for the cells (close to 40% [77]) but also leads to expensive elaboration processes and sometimes faces issues related to the use of different materials (lattice mismatch) which forbids the use of Si substrates ("cheap" technology) and requires expensive Ge or III–V substrates. Using QDs with tunable bandgaps can reduce production costs and also facilitate layers integration by keeping the same material [78]. On the other hand, lattice mismatch issues can be overcome by the use of nanowires that can be grown directly on Si substrates. In this case, lattice constraints can be relaxed at the free surface of the wires, unlike conventional continuous films. Despite the low active material volume, it has been demonstrated that the absorption coefficient of nanowires array can reach the one from thin films or even exceed it as each single wire acts as an optical waveguide. Resonant absorption is observed for different wavelengths depending on wire diameters. For PV devices, radial or axial junctions are directly grown within the wire [79].

In parallel to the development of PV systems based on semiconductor junctions, alternative routes were also addressed, inspired by nature photosynthesis. These cells are constituted by two electrodes separated by an electrolyte. Photoanode is made of a nanostructured porous oxide electrode (usually TiO_2) covered by a dye working as light absorber. Upon absorption, the excited dye transfers an electron to the oxide that can be extracted to the working circuit. The oxidized dye is regenerated by the means of a redox couple in the electrolyte that closes the circuit with the counter electrode (cathode). These electrochemical cells, originally known as Grätzel cells [80] and then dye sensitized solar cells (DSSC), take advantage of the large surface area developed by the nanoparticles in the porous electrode that ensures a larger available surface for light harvesting by the dye, and also a larger contact surface between the dye and the electrolyte. Best performances for these cheap cells reach efficiencies close to 12% [81]. For DSSC, and moreover for organic solar cells where carriers mobility is very low, one dimensional nanoobjects such as nanowires, nanotubes or nanorods are expected to provide direct pathways for electrons collection, hindering trapping and recombination when compared to nanoparticles films [82].

Metallic nanoparticles can also help in increasing solar cell efficiency. For all the types of cells mentioned before, additional improvements can be obtained by the enhancement of photon-to-electron processes (light absorption, excitation transfer, hot carrier injection...) provided by the coupling with an electromagnetic field at the vicinity of a metallic nanoparticle surface. This intense field is generated by surface plasmon (collective oscillation of free electrons near the surface of metallic particles in a dielectric environment). Plasmonic solar cells are being developed, trying to take advantage of different mechanisms enabled by the surface plasmon well described in a 2016 review [83]. An example is given in figure 2.3.2 dealing with the improvement of external quantum efficiency (*i.e.* number of collected electrons over number of incident photons at a given wavelength) for a Si PV cell upon deposition of a mixture of Au and Ag nanoparticles on its surface.

As described above, numerous strategies are employed to enhance the performances of PV devices thanks to nanomaterials' specific properties. Nevertheless, important research and development work remains to be done on the synthesis and integration of the nanoobjects, as well as on the device elaboration by itself to overcome limiting drawbacks faced by nanomaterials (surface recombination issues for charge transport and low quantum yield for optical conversion for example).

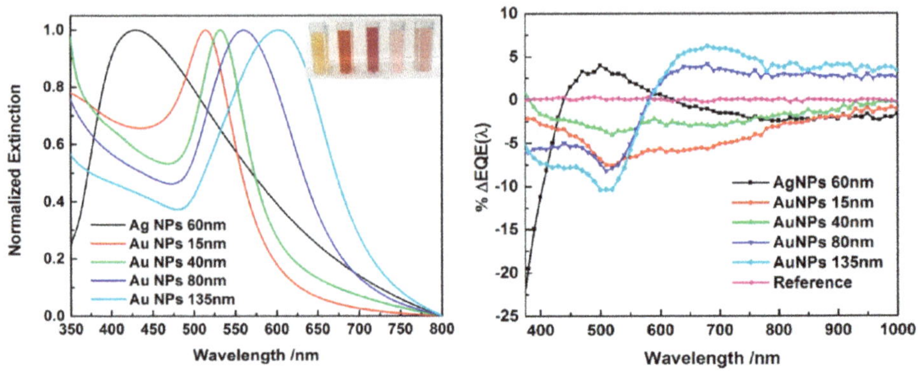

FIG. 2.3.2 – Left: Extinction spectra of aqueous dispersions of Ag and Au nanoparticles (inset: suspensions in glass cuvettes: 60 nm Ag NPs, 15, 40, 80, and 135 nm Au NPs from left to right, respectively). Right: Evolution of %ΔEQE (external quantum efficiency) plots for Si PV device covered by Ag or/and Au nanoparticles [84].

2.3.1.2 Solar Thermal Energy

Another way to benefit from solar energy is to use it as a thermal source for electricity production. In this case, solar light is "simply" used to heat a liquid and produce steam that can be used to drive a turbine as in coal, gas, or nuclear power plants. When compared to simple collector plates placed on the roof for house hot water generation, solar heat power generators operate at higher temperature and thus require sunlight to be concentrated by means of mirrors. These plants, called concentrated solar power (CSP), generally use parabolic trough collectors (PTC): curved reflectors focus sunlight on a tube located in the focal line of the mirrors where the heat is collected upon absorption by a coating and transferred to a steam generator (or to a storage system) through a transfer fluid. Nanomaterials can be employed in the absorbing coating as they can be easily deposited by cost-effective wet routes (sol–gel with nanoparticles dispersion for example) over large surfaces. More interestingly, they can also show peculiar advantage for the working fluid: suspension of nanoparticles can actually modify drastically the heat transport properties of a liquid. Such suspensions, called nanofluids, can enhance the heat transport from the absorbing layer or can be used as absorbers themselves.

This enhancement seems to depend not only on the nanoparticle nature itself, but also on their size, shape or concentration in the fluid. Physical phenomena that drive thermal properties modification in nanofluids are multiple (heat transfer, radiation, friction, Brownian motion...), and their respective influences depend on the nanofluid characteristics (particles morphology, concentration): lot of models were proposed, being still a matter of debate [85]. Even experimental data reports show discrepancies, making the choice between models difficult. Beyond these basic science difficulties, practical issues also exist concerning the stability in time of the suspension, the deposition of particles in pipes and the wear of tubing in contact with the hard particles. Nevertheless, the observed enhancement in heat transport

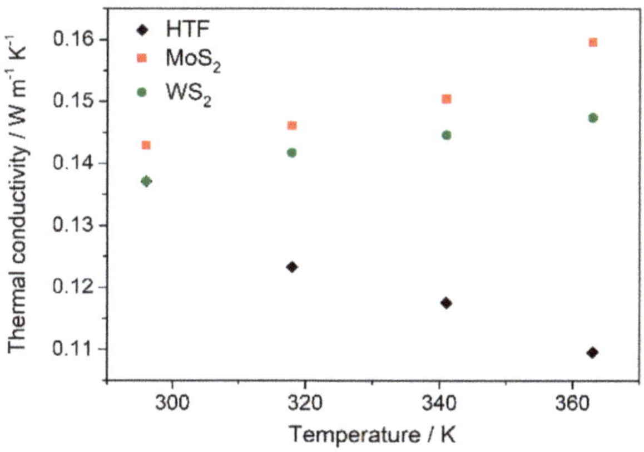

FIG. 2.3.3 – Evolution of the thermal conductivity of nanofluids *versus* temperature for 3 samples: reference heat transfer fluid (HTF), MoS_2 based nanofluid, and WS_2 based nanofluid [87].

performances (see figure 2.3.3) when compared to conventional heat transfer fluids still motivates research works [86].

2.3.1.3 Thermoelectrics

Heat, whatever its source, can also be directly converted to electricity without steam intermediate. Thermoelectric devices rely on the conversion into an electric current of a temperature gradient (Seebeck effect) applied to the extremities of p-type and n-type semiconductors. In both types, carriers (holes and electrons, respectively) diffuse from the hot zone to the cold one, creating a difference of potential (thermovoltage) proportional to the temperature difference through the Seebeck coefficient. This coefficient is governed by the diffusion and the interaction with the phonon of charge carriers in the materials. The efficiency of a thermoelectric device can be described through a dimensionless figure of merit (ZT) which depends on three characteristics of the materials: electrical conductivity, thermal conductivity, and Seebeck coefficient. ZT value should then be higher than 1 in order to reach an energy conversion efficiency of 10%. An ideal thermoelectric material should then exhibit high Seebeck coefficient (increase in thermovoltage through Seebeck effect) and electrical conductivity (limitation of heating by Joule effect), but low thermal conductivity (conservation of temperature gradient between heat source and sink). Unfortunately, these properties are interconnected and it is difficult to modify them independently.

Thermal conductivity depends on two contributions, one coming from the electrons, the other from the lattice vibration (phonons). An efficient way to decrease the lattice contribution for a given material is to favor the phonon diffusion by multiplying the diffusing elements such as grain boundaries or nanoinclusions [88]. Downsizing of the grains results in increasing the grain boundary density, and

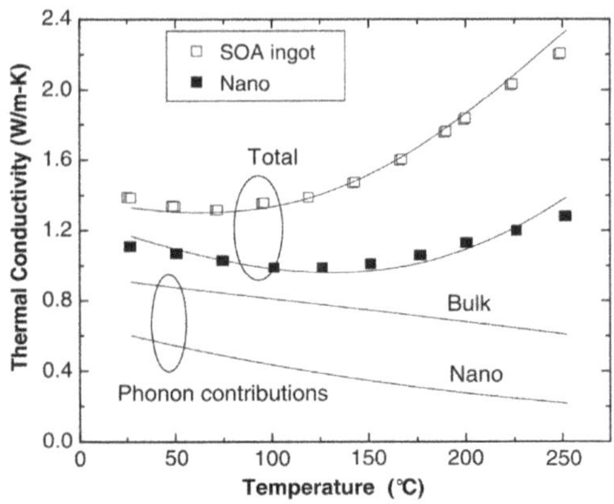

FIG. 2.3.4 – Thermal conductivity of bulk (SOA ingot) and nanostructured $Bi_xSb_{2-x}Te_3$ alloy. Squares represent experimental results while solid lines represent corresponding calculations of the total and lattice contributions to the thermal conductivity [88].

thus nanostructured materials exhibit much lower thermal conductivity than their micron-scaled counterpart (see figure 2.3.4). The drawback is the potential degradation of the electrical conductivity as grain boundaries act as efficient traps. Nanocomposites (dispersion of nanoparticles in a matrix or nanowires-based materials) are also considered, expecting a lower carrier transport property degradation. For example in Si, phonon mean free path is in the order of magnitude of 100 nm, while electron one is 10 nm. A Si nanowire with diameter between 10 and 100 nm sees actually its thermal conductivity decreasing while keeping its charge transport ability almost unchanged [79].

Increase in Seebeck coefficient and electrical conductivity is another issue in thermoelectrics. Most of the time, it is recommended to follow strategies aiming at increasing the density of state and at narrowing the electronic band structure. As previously mentioned in the PV application part, nanostructured materials experience strong modification of their electronic band structure when quantum confinement phenomenon occurs. Such modification observed in quantum dots or nanowires is thus expected to play a key role in enhancing the performances of thermoelectric devices through the improvement of Seebeck coefficient and carriers transport [89]. Carbon nanotubes are also strongly investigated as their electronic properties should give them interesting values of Seebeck coefficient and electrical conductivity. They could constitute a viable alternative to the expensive or toxic elements used in common thermoelectric devices on the market. They actually show promising performances, whether they are used alone or in a composite material [90]. Other composites can also improve Seebeck coefficient values, by means of a dispersion of metallic nanoparticles. In this approach, the metallic inclusions create potential barriers that enable an energy selection among the carriers (favoring high-energy ones) that is beneficial for Seebeck coefficient [91].

2.3.2 Electrochemical Storage

2.3.2.1 Nanomaterials for Batteries

The lithium ion battery involves the reversible exchange of Li ions and electrons between two electrodes separated by a non-aqueous electrolyte. The battery converts a chemical potential into electrical energy through redox reactions. These redox reactions imply the diffusion of electrons and ions in the solid and in some cases variation of volume inside the material. These properties strongly influence the overall electrochemical behaviour of the materials and are critical for the battery performance [92].

In this regard, nanomaterials have attracted a strong interest in the battery community over the past two decades as their reduced size can favour high charge/discharge rate and high power [2–6]. The benefits of nanomaterials in batteries are multiple [92, 93]:

- They increase the specific surface area and enable large contact surfaces with the electrolyte and a high ion flux at the electrode/electrolyte interface.
- They trigger some reactions and phase transitions impossible in micrometric particles and may modify the thermodynamics (change in electrode potential).
- They alleviate the strain caused by insertion of ions.
- They increase the electron transport within the particles.
- They finally enhance the kinetics and diffusion of ions and consequently the rate of ions insertion/removal by reducing the ion diffusion length. The diffusion time constant is in fact proportional to the square of the particle size:

$t = L^2/D$, where L is the diffusion length and D the diffusion constant.

Yet, it is important to keep in mind that nanomaterials are not the magic remedy for batteries as they suffer from important limitations. Their main drawbacks are (i) their cost and difficult processability, (ii) their high surface area causing significant reactivity with the electrolyte, especially a high first cycle irreversibility and (iii) their lower density compared to the micrometric particles. This creates a penalty in terms of electrode volume for the same mass of material and decreases the volumetric energy density [93, 94]. One innovative approach that we will address in this chapter is the hierarchical assembly of nanostructures into aggregates or clusters at the microscale to overcome the limitations of the nanomaterials [94].

Herein, far from proposing an exhaustive list of literature dealing with nanomaterials for batteries, we aim at describing few examples illustrating the influence of nanostructuration on some Li-ion and post Li-ion materials.

The case of the positive electrode material $LiFePO_4$ (LFP) is a perfect example of how nanostructuration can strongly affect the electrochemical behaviour. When downsized to the nanoscale, $LiFePO_4$ nanoparticles show fast intercalation kinetics, and more interestingly present a complex change in their electrochemical response compared to their micrometric counterparts. Masquelier *et al.* [95] synthetized 40 nm LFP nanoparticles using a low temperature precipitation method (figure 2.3.5a, insert). They showed that the two-phase reaction in LFP can be driven to a single-phase (solid solution) reaction. The solid solution is clearly evidenced on the evolution of the potential as function of capacity (figure 2.1.1a), where a sloppy profile is observed in contrast to the well-defined plateau observed for micrometric LFP [95]. Reducing the particle size of layered $LiCoO_2$ also affects the electrochemical behaviour but is in this case detrimental to the cathode performance. Through a hydrothermal reaction, Okubo *et al.* [96] varied the size of $LiCoO_2$ particles from micrometric to nanoparticles from 17 to 6 nm. The decrease in particle size leads to a sharp decrease of the capacity and to almost linear potential curves, denoting a surface charge storage behaviour like supercapacitors [94, 96]. Other oxide materials take advantage of a decrease in the particle size. For example, nanostructured $Li_{4+x}Ti_5O_{12}$ exhibits increased cycling performance compared to its micrometric equivalent [92]. Here, the enhanced performance results solely from improved kinetics and diffusion as $Li_{4+x}Ti_5O_{12}$ does not suffer from any strain or volume change upon lithiation [92].

Nanomaterials greatly contributed to the development of alloy materials such as silicon, germanium or tin to replace graphite at the negative electrode of Li-ion batteries. Silicon has a theoretical capacity of 3579 mAh g^{-1}, ten times higher than the capacity of graphite. Yet, lithiation of Si is accompanied by a large volume expansion ($\sim$300%) while the particles shrink upon subsequent removal of lithium, leading to cracking and pulverization of the material [94]. By tailoring the silicon electrode with the use of Si nanowires, nanotubes and core–shell and yolk-shell particles, the volume expansion and shrinkage of the electrode can be alleviated thanks to the mitigation of mechanical constraints and fracturing [92, 94]. Si nanowires directly grown on a stainless steel by a vapour–liquid–solid (VLS) method are able to accommodate the volume expansion without cracking thanks to the their narrow diameter, good electronic contact with the current collector and 1D conduction of electrons [92]. Core–shell nanoparticles such as Si@Carbon are also efficient to improve the capacity retention of Si electrodes. Sourice *et al.* produced 30 nm Si@C nanoparticles through a one-step synthesis using laser pyrolysis [100]. Si@C electrodes demonstrate a high and stable capacity of 2400 mAh g^{-1} at C/10 (charge in 10 h) and up to 500 mAh g^{-1} over 500 cycles at 2C. The excellent capacity retention is attributed to the carbon shell, protecting Si from direct contact with the electrolyte. Yet Si nanoparticles and nanowires suffer from too low coulombic efficiencies due to the large surface area developed, resulting in large reactivity with the electrolyte. One strategy to overcome the limitations of nanoparticles in terms of reactivity and low density is to cleverly aggregate nanomaterials to produce larger structures at the micrometre scale. This has been applied by Magasinski *et al.* [97] who hierarchically assembled Si–C micrometric granules (figure 2.3.5b). They used chemical vapour deposition (CVD) to deposit Si nanoparticles on the surface of carbon nanoparticles, later hooked together into chains by high-temperature annealing. The small clusters formed are then branched into large porous spheres by deposition of carbon by CVD [97]. The intrinsic porosity accommodates the volume swelling of Si and allows for remarkable capacities of more than 1500 mAh g^{-1} at 1 C [97]. A pomegranate-like yolk-shell structure has also been proposed by Cui *et al.* [98] (figure 2.3.5c). The composites, from 500 nm to 10 μm, are formed by encapsulating Si nanoparticles in a carbon conductive shell containing some room to accommodate volumetric changes.

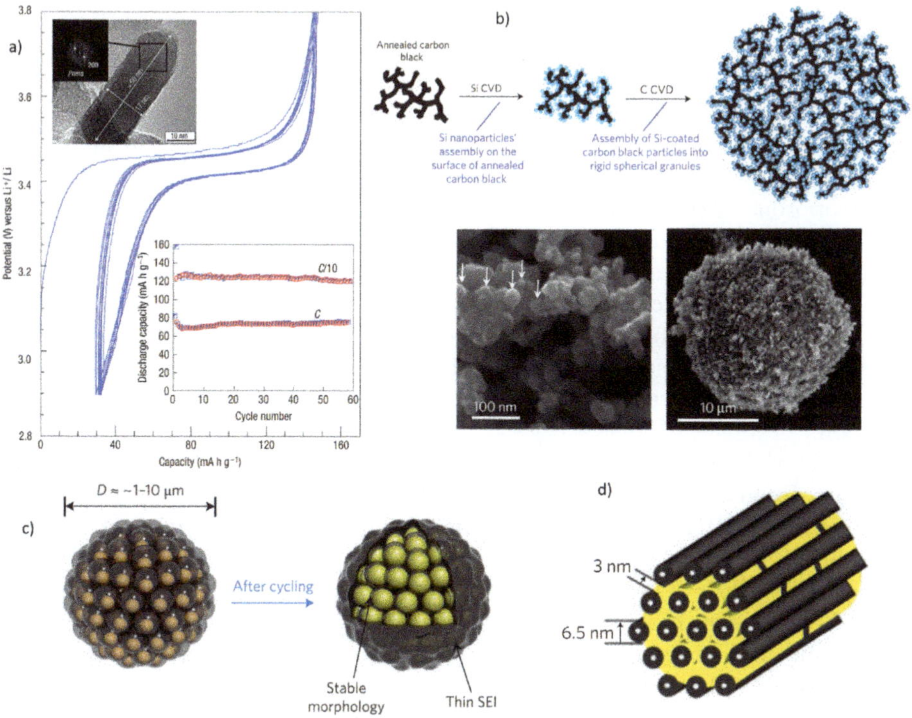

FIG. 2.3.5 – Examples of applications of nanomaterials in batteries. (a) Cycling performance of nanosized $LiFePO_4$ [95]. (b) Bottom-up assembly of Si–C composite granules using CVD and corresponding SEM images [97]. (c) Schematic of pomegranate-like microparticles constituted of Si nanoparticles encapsulated in carbon with well-defined void spaces to accommodate volume expansion [98]. (d) Schematic of sulphur hosted in mesoporous carbon, formed from carbon nanotubes and carbon nanofibers [99].

These individual particles are then aggregated in a thick carbon layer to form micrometric beads [98]. These beads present a 97% capacity retention after 5000 cycles at 0.1 C. While the decrease in size of Si particles prevents from cracking during lithiation, carbon maintains the global mechanical integrity and electronic conduction of the particles and serves as a barrier to electrolyte [98].

Post Li-ion batteries such as Li-air and Li–sulphur (Li–S) batteries are no exception when it comes to the use of nanomaterials. The use of nanostructured materials to favour oxygen reduction and oxygen evolution reaction in Li-air is well documented [96, 100]. Challenges and strategies are usually similar to the one developed for proton exchange membrane (PEM) fuel cells for the synthesis of catalysts; this will be detailed in section 2.3.3.1. Li–S batteries offer important theoretical capacities of 1673 mAh g^{-1} and 2500 Wh kg^{-1} [94]. However, sulphur cathodes are subjected also to a large expansion (80%) upon lithiation. More importantly they suffer from dissolution of polysulfides formed upon lithiation in the electrolyte, and finally from the low ionic and electronic conductivities of lithium sulphide [94, 96]. Graphene and carbon-based materials are the most common positive electrodes for Li–S batteries, as carbon provides a large surface area and high electronic conductivity [96]. Using nanomaterials in the case of Li–S batteries is a powerful strategy to improve the overall performance by improving the Li ions diffusion and by trapping polysulfides inside a porous structure [96]. Nazar *et al.* [99] proposed a nanostructured carbon–sulphur composite where sulphur is immobilized in a mesoporous carbon framework, formed from carbon nanotubes and carbon nanofibers (figure 2.3.5d). The channels of the carbon structure constrain the growth of sulphur products inside the electrode and provide at the same time sufficient electronic conduction. In a different way, Campidelli *et al.* [101] used a grafting chemistry to functionalize carbon nanotubes with thiolated molecules containing electrochemically active disulphide groups. The covalent bonding of the molecules contains the reactions with lithium to the electrode to avoid sulphur dissolution.

Nanomaterials participate also to the fervent hype for solid-state batteries seen in the past decade. Using solid-state electrolytes (SSE) in place of the liquid electrolyte in Li-ion batteries, the use of metallic lithium at the negative electrode could be possible, offering high energy densities and safe batteries [102]. Three main classes of SSE exist: polymers, ceramics/glasses, and the hybridization of both. Ceramic SSEs present high ionic conductivity at room temperature, high mechanical strength, and low flammability. Yet they are also fragile and their interface with electrodes is difficult to master. Polymers in contrast are flexible and easily processed, however they exhibit a low ionic conductivity at ambient temperature. Combining ceramics and polymers in hybrid materials is a way to obtain decent conductivity with a good processability [102]. Adding nanoparticles or nanowires of ceramic SSE such as $Li_{6.75}La_3Zr_{1.75}Ta_{0.25}O_{12}$ or $Li_{3x}La_{2/3-x}\square_{1/3-2x}TiO_3$ in a poly (ethylene oxide) (PEO) network is one of the efficient strategies [102]. The use of nanoparticles is not limited to hybrid systems but can also help the processability of the ceramic. The use of nanopowders of $Li_{6.25}Al_{0.25}La_3Zr_2O_{12}$ produced by liquid-feed flame spray pyrolysis (LF-FSP) has been shown for example to improve the densification process of the final ceramic [103].

Looking to an application point of view, the cost of nanomaterials appears still as a barrier to their large development. Some other challenges remain, such as the toxicity of the nanomaterials. In this regard, carbon and silicon appear not harmful to the environment [94], but the toxicity and environmental impact of new compounds need to be assessed carefully. The scalability of the complex syntheses is another concern. However, the emergence of start-ups and companies such as Sila Nanotechnologies in the US [94] (producing the particles shown in figure 2.1.1b) and Nanomakers in France (Si@C by laser pyrolysis) demonstrates that processing nanostructures on a large scale is possible for Li-ion batteries.

2.3.2.2 Nanomaterials for Supercapacitors

Supercapacitors, also called electrochemical capacitors, are power devices that present lower energy density compared to batteries, but that can however be charged and discharged in seconds over more than thousands of cycles. They are as a result the technology of choice for high power applications and can also complement batteries. Three types of supercapacitors exist: electrical double layer capacitors (EDLC), pseudo-capacitors and hybrid capacitors. EDLC store energy by adsorbing ions from an electrolyte onto a high surface area electrode. In the case of pseudo-capacitors, the charge storage is based on voltage dependent fast and reversible redox reactions [92, 104]. Redox reactions are usually confined at the electrode–electrolyte interface. The third class of supercapacitors, hybrid capacitors, combines the latter two to take advantage of capacitance from both pseudo-capacitive and double layer storages. Nanomaterials play a critical role in the field of supercapacitors, providing high surface area for the double-layer formation and pseudocapacitive reactions, while reducing the diffusion distances for ions [92]. Consequently, nanomaterials exhibit higher capacitance and overall performance compared to their bulk equivalents.

For EDLC, high charging of the double layer is achieved with a high surface area and highly conductive electrodes. Typical materials are thus activated carbon, nanotubes, carbon aerogels, carbon onions and graphene (figure 2.3.6a) [92, 104]. Carbon nanotubes show high capacitance, yet the absence of micropores in their structures limits their surface area. Presence of defects on the CNT walls and surface treatments are ways to improve their performance [92]. For graphene, the theoretical

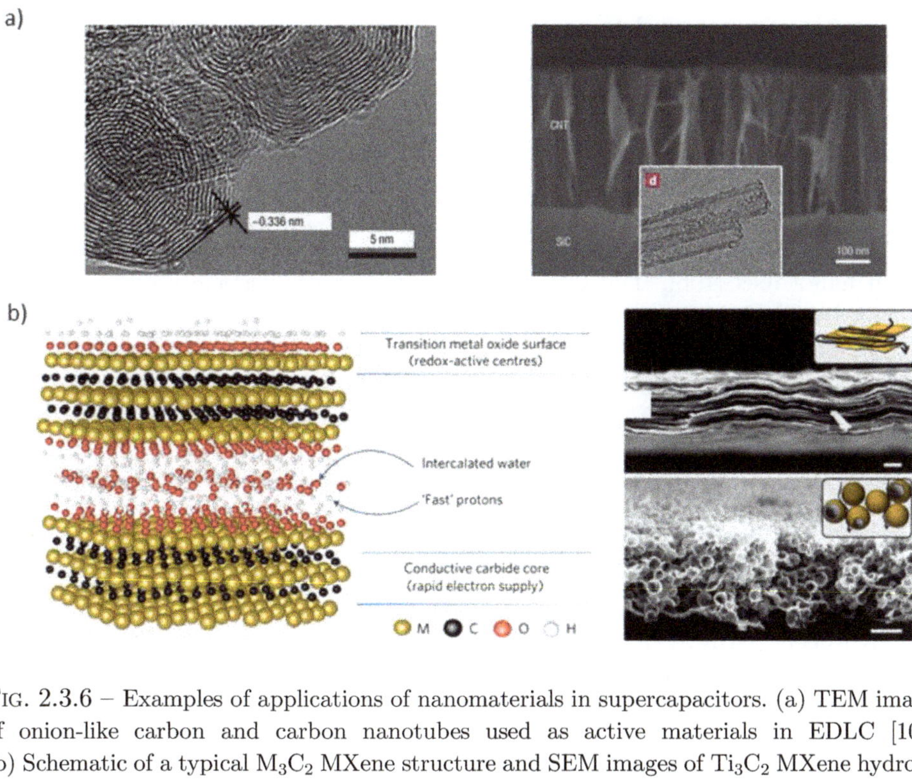

FIG. 2.3.6 – Examples of applications of nanomaterials in supercapacitors. (a) TEM images of onion-like carbon and carbon nanotubes used as active materials in EDLC [104]. (b) Schematic of a typical M_3C_2 MXene structure and SEM images of Ti_3C_2 MXene hydrogel and macroporous Ti_3C_2 electrode cross-sections. Scale bars 5 μm [105].

capacitance is quite high (550 F g^{-1}), however graphene sheets have a tendency to agglomerate, thus reducing the surface area and limiting the capacitance to ~100 F g^{-1} [92]. Introduction of CNTs in between graphene sheets has proven effective to increase the capacitance to more than 300 F g^{-1} [92].

For pseudocapacitors, transition metal oxides, nitrides and sulphides as well as electronically conductive polymers are materials of choice. Typical oxides include RuO_2, MnO_2, Fe_3O_4 or Co_3O_4, which offer high charge storage compared to carbon materials, but they unfortunately present lower conductivity compared to carbon-based materials [92, 104]. Nanostructuring these oxides in the form of nanowires, nanoporous materials or nanotubes help increase the performance [92]. Similarly, conductive polymers (such as polypyrrole, polyaniline and polythiophene) are produced in the form of porous thin films to improve their capacitance [92]. Recently, MXenes materials have also emerged as an efficient material for supercapacitors as they possess both high electronic conductivity and high ionic conductivity [94]. MXenes are 2D metal carbides, nitrides or carbonitrides obtained through exfoliation of layered ternary transition metal carbides or nitrides. Figure 2.3.6b presents the structure of a typical 2D MXene. Already more than 30 MXene have been discovered in the literature [94] and the best example is the material Ti_3C_2 [106]. Lukatskaya *et al.* [105] have formed either hydrogel or macroporous Ti_3C_2 nanomaterials (figure 2.3.6a, right) and demonstrated impressive capacitance and performance in aqueous electrolyte. The Ti_3C_2 macroporous MXene is able to outperformed the best carbon supercapacitors with up to 210 F g^{-1} at high rates, while the hydrogel exhibits a capacitance of ~1500 F cm^{-3} competing with RuO_2, which previously had the highest volumetric performance [105].

Combining a pseudocapacitor compound with a nanostructured EDL material provides a synergy between the two materials and improves the global capacitance of the electrode. Associating metal oxides such as MnO_2 with CNT or polymers can lead to capacitance values ranging from 210 to 415 F g^{-1} [92]. One example is the deposition of a thin layer of a polythiophene polymer on long thick vertically aligned carbon nanotubes [106]. The CNTs are directly grown vertically on the current collector by CVD, improving the contact resistance.

Finally, all these examples show that the overall range of nanostructures from nanoparticles, nanotubes to 2D sheets and porous structures has clearly provided new properties, not achievable with conventional bulk materials. Further research on synergistic effects between different nanomaterials should help reduce the gap in energy density between supercapacitors and batteries while maintaining high power densities.

2.3.3 Hydrogen

The hydrogen pathway is presently offering unprecedented interest [107], as it potentially provides a significant contribution to solve the severe problems due to global climate warming, triggered by the massive and still increasing consumption of fossil fuels. This concerns especially energy consumption sectors that are difficult to decarbonize such as transportation. The chemical energy stored in the hydrogen molecule can be converted into electrical power with good yields, while side products resulting from its combination with oxygen only consists in water. Regarding transportation, fuel cells have just entered into the "real world". Recently, first commercial mass-produced hydrogen passenger cars appeared, heavy vehicles such as buses are in use, while trucks powered by fuel cells are being under development. However, it is worth noting that the stringent urgency due to visible and strong impacts of climate warming provides to the hydrogen pathways a real interest beyond application to transportation.

Starting from water, hydrogen can be produced in good yields through the reverse reaction that stands in fuel cells, *i.e.* water splitting, thus providing a fuel, from a so far highly available resource. When produced from electrical power, hydrogen is thought to be able to facilitate renewable energy deployment. Most of the latter are intermittent and the time at which electricity can be produced does not necessarily fit the demand period. Therefore, hydrogen production is a way to store renewable energies into chemical energy, which can be converted back into electrical power on demand, using fuel cells (figure 2.3.7). This is the basis of the power to gas concept.

Another point related to hydrogen pathway development is related to the issue of storage. The current ways favor the high-pressure storage in tanks. However, other ways are considered through the study of various kinds of materials and approaches. An example of the latter, not developed here, is the transformation of hydrogen into methane (Sabatier reaction) which consumes CO_2 and produces water as side

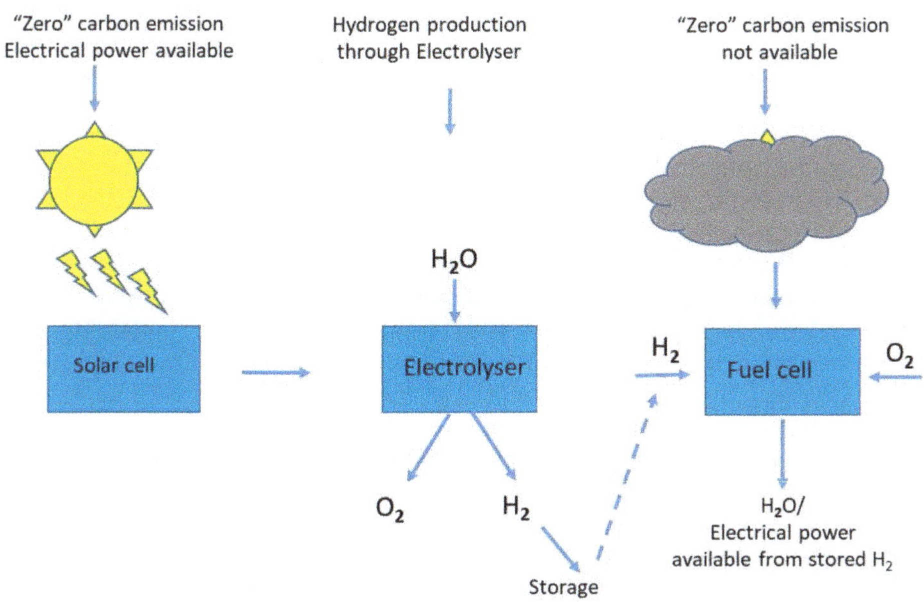

FIG. 2.3.7 – Hydrogen pathway and solar "zero" carbon emission energy source combination illustrating the power to gas concept.

product. Hydrogen is thus stored as methane; the latter can be used in transportation, home heating, and can be stored in an easier way than hydrogen.

Finally, they are multiple aspects for which nanomaterials and nanoobjects are studied and developed in the frame of the hydrogen pathway. Key-points of this particular interest are the high surface to volume ratio developed by nanoobjects, and the specific properties that can be tailored to reach the highest performances for a given targeted property. In the different aspects of hydrogen pathway, high specific surface area is most often desirable; this is true especially in some aspects of hydrogen utilization in fuel cells, as well as for production and storage as evocated below. Nanomaterials in fuel cell devices are particularly addressed in this part, hydrogen production and storage are considered to a lesser extent.

2.3.3.1 Nanomaterials and Nanoobjects for Fuel Cells

There are historically different kinds of fuel cells among those three main types: PEMFCs, Alkaline Fuel Cells (AFCs) and SOFCs. In fuel cells, hydrogen is oxidized at the anode, and oxygen is reduced at the cathode. An electron insulating but ion conducting material *i.e.* an electrolyte, separates both electrodes from each other. These kinds of fuel cells use different electrolytes. The role of the latter is to prevent mixing of the gases that feed the electrodes to prevent internal electronic shortcuts, and to conduct the ionic species generated by the redox reactions from one electrode to the other one (figure 2.3.8), thus counter-balancing the electron flow in the external load. Research on fuel cell is an extremely active field of research and the very last years provided spectacular progress and hopes for highly performing devices, based on new nanomaterial development. A way to describe fuel cell types can be as low temperature fuel cells having a polymer or liquid electrolyte, and high temperature fuel cells, having an inorganic solid electrolyte. With the most recent development, each kind of fuel cell can be based on a proton-conducting electrolyte or an anion-conducting electrolyte (figure 2.3.8).

Low temperature fuel cells work at about 100 °C while high temperature ones work at 400–1000 °C. Each technology has some advantages and drawbacks, which are related to the constraints imposed by the electrolyte involved. The most obvious, but not exclusive, illustration of nanoobjects and nanomaterials interest in fuel cells development is related to the incorporation of catalysts into devices. Here, catalyst are solids at the surface of which catalysis occurs. Therefore, and potentially, for a given catalyst material, the smaller the solid the higher the optimization of catalyst use and efficiency in the fuel cell device. The role of the catalyst is to improve the yield of conversion of the chemical energy of hydrogen into electrical power. This effect is characterized by the decrease of the so-called activation overpotential, which is the difference between the standard potential of the redox reaction (1.23 V for O_2/H_2O redox couple) and the potential at which a redox current is recorded. Actually, overpotentials develop at both the cathode for the Oxygen Reduction Reaction (ORR) and the anode for the Hydrogen Oxidation Reaction (HOR). First, it is worth pointing to the fact that the tension cell drop due to overpotential is the sum of the overpotentials at the cathode and the anode, and second, that the electrode,

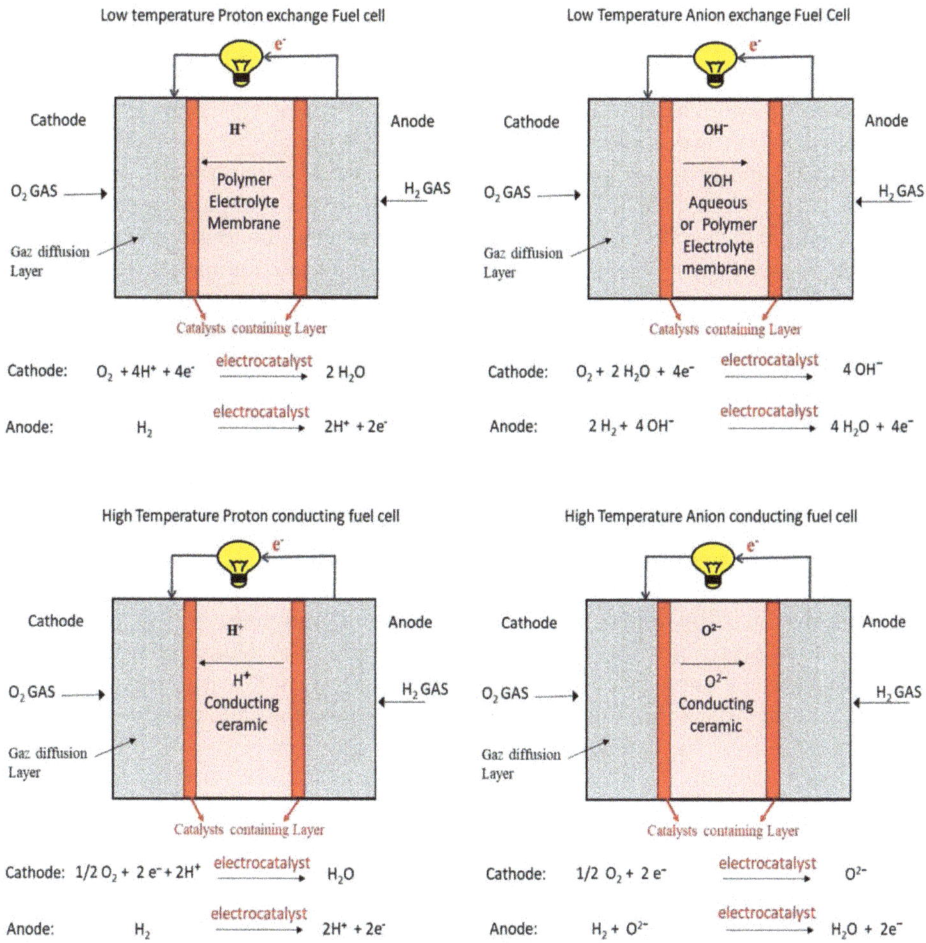

FIG. 2.3.8 – Schemes of low and high temperatures with proton or anion conducting electrolyte and electrochemical reactions involved. For proton conducting electrolyte, water is produced at the cathode and for anion conducting electrolyte, water is produced at the anode.

anode or cathode, at which the overpotential is critical (because it is high) is pH dependent.

Low temperature fuel cells are PEMFCs, Alkaline liquid electrolyte fuel cells and Anions Exchange membrane fuel cells. Alkaline electrolyte fuel cells were the first kind of practical fuel cell devices in service about seventy years ago. The interest for this kind of fuel cell decreases as a result of emerging simpler PEMFC technology in the sixties. This could be amplified by recent progress in anion conducting exchange membranes. The catalysts involved in PEMFCs are based on platinum, which is generally supported by carbon black particles. PEMFCs work in highly acidic conditions and with platinum the HOR is very fast, (overpotential is low) therefore

PEMFCs require only very low platinum loading at the anode (few tens of μg of Pt per electrode cm^2) to operate efficiently. In contrast, the ORR is particularly sluggish in highly acidic medium, therefore cathode have to be loaded with much higher Pt contents (in the range of few hundreds μg Pt per electrode cm^2) to keep acceptable overpotentials and significant currents at high-tension cell. Pt is a scarce and expensive material, therefore there is a tremendous literature on strategies aiming at the decrease in Pt use at the cathode or its complete replacement by non-noble metal catalysts. An excellent recent review can be found on these aspects by Shao *et al.* [108]. One way to decrease Pt loading is to combine it with transition metals, which can be done by alloying or by core–shell structure formation in which the core is a non-noble metal and the shell is Pt. Pt alloying not only decreases Pt use but also increases the electrochemical activity of the catalysts. This improvement originates from structural effect (changes in Pt–Pt bond length) as well as electronic effects. They are thought to weaken the adsorption of oxygen containing species, the lower coverage of which at the catalyst surface decrease the poisoning of active sites and improve the ORR activity. Alloying of Pt is also a way to prepare porous Pt based nanoobjects [109] with very high electrochemical activities, up to 36 times higher than classical Pt/C catalysts (figure 2.3.9).

Besides research on Pt-based catalysts, the development of non-Pt containing catalysts is featured by doped carbon nanostructures and nanoobjects. Tremendous literature and improvements have been provided by J.-P. Dodelet and co-workers researches over the three past decades, on so-called M/N/C catalysts, were *M* is a transition metal [108]. High ORR activity is established for high specific surface area materials containing carbon, iron (or cobalt) and nitrogen both of the latter at few percent level contents. The most active catalysts are obtained generally using multistep syntheses (figure 2.3.10) and involve high temperature treatment most often under ammonia [110].

However, if the carbon-doped nanostructures are very promising, they do not yet reach Pt performances. As a result, active layers based on this kind of catalysts have loading at the level of few mg per electrode square centimeter. This hinders active sites accessibility and induces mass transport limitations.

Due to highly significant progress recorded on the conductivity of Anion Membrane Exchange in the past decade, Anion Membrane Exchange Fuel Cells (AMEFCs) have recently gained significant interest in the fuel cell community [111]. Regarding HOR and ORR, for still unclear reasons, AMEFCs show a reverse situation compared to PEMFCs. In the former, the ORR is faster but the HOR is much slower than in PEMFCs. Noble metal catalyst activity, for HOR at pH 13 has been reported to be two orders of magnitude lower than at pH 0 [112]. It is worth nothing that the research performed on non-noble metal catalysts for the ORR in PEMFCs benefit to AMEFC cathodes. Indeed, some of the Fe/N/C catalysts actually show similar activity or even higher activity than Pt in basic media. In AMEFCs Pt is the most active catalyst for the HOR, this includes alloys with non-noble metal and core–shell structures. Other noble metal such as Pd, Ir, Ru, Rh provide significant activities towards the HOR. At present, Ni is the only non-noble metal showing a significant activity and many different nickel based catalyst incorporating other non-noble metal have been studied. The popular Raney nickel catalysts has been

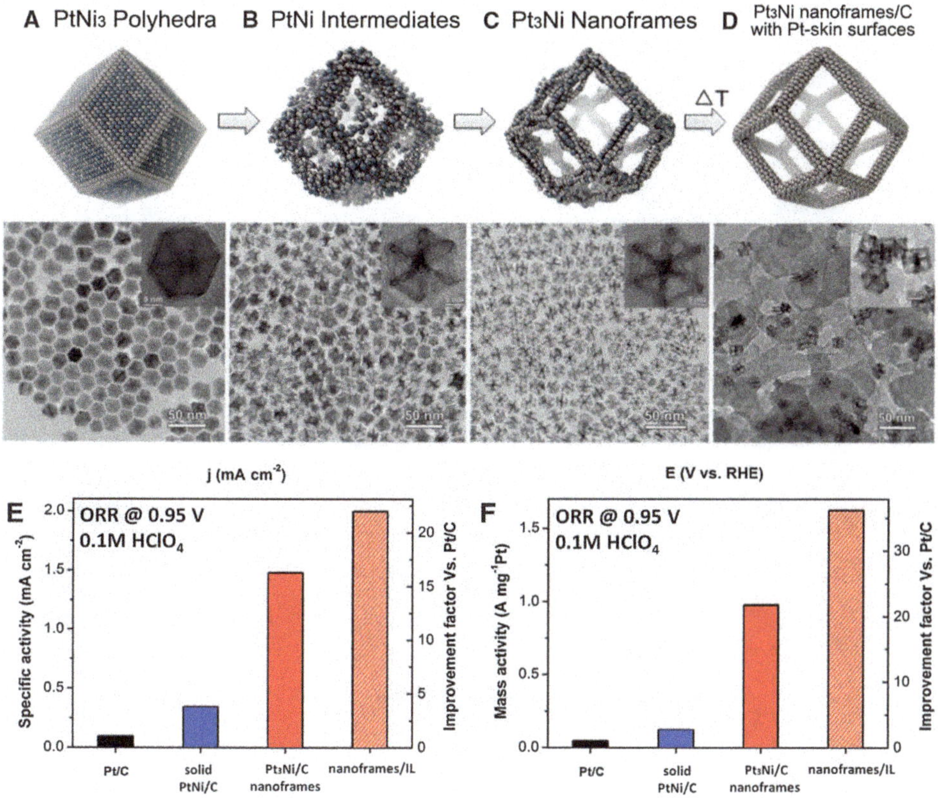

FIG. 2.3.9 – Scheme and Transmission Electron microscopy for Alloyed $PtNi_3$ catalyst and porous $PtNi_3$ nanoframe catalyst and their high electrochemical performances compared to classical Pt/C catalysts [109].

largely studied and used in liquid electrolyte (KOH) AFCs. Recent report by E.S. Davydova *et al.* provides detailed state of the art of knowledge and issues related to HOR in basic media and related catalysts [113].

Until recently, High temperature fuel cells working with an electrolyte carrying O^{2-} ions were the most frequent studied devices. O^{2-} conducting materials are typically metal oxides in which doping with a lower valence ion creates oxygen vacancies, resulting in a significant O^{2-} conductivity only effective at high temperature. For this reason, SOFCs are high temperature fuel cells, presenting both drawbacks and advantages. A certain time is required to reach the temperature at which the fuel cell works and the materials that form the cell must match in their thermal expansion coefficient. Phenomena like inter-diffusion and reaction between materials must be controlled. In contrast, high temperature is an advantage regarding catalysis, since it is possible to get rid of noble metal based catalysts use. A paper by W.H. Kan *et al.* provides a recent review on SOFCs [114]. Electrodes materials are ideally Mixed Ion-Electronic-Conductors while the O^{2-} electrolyte

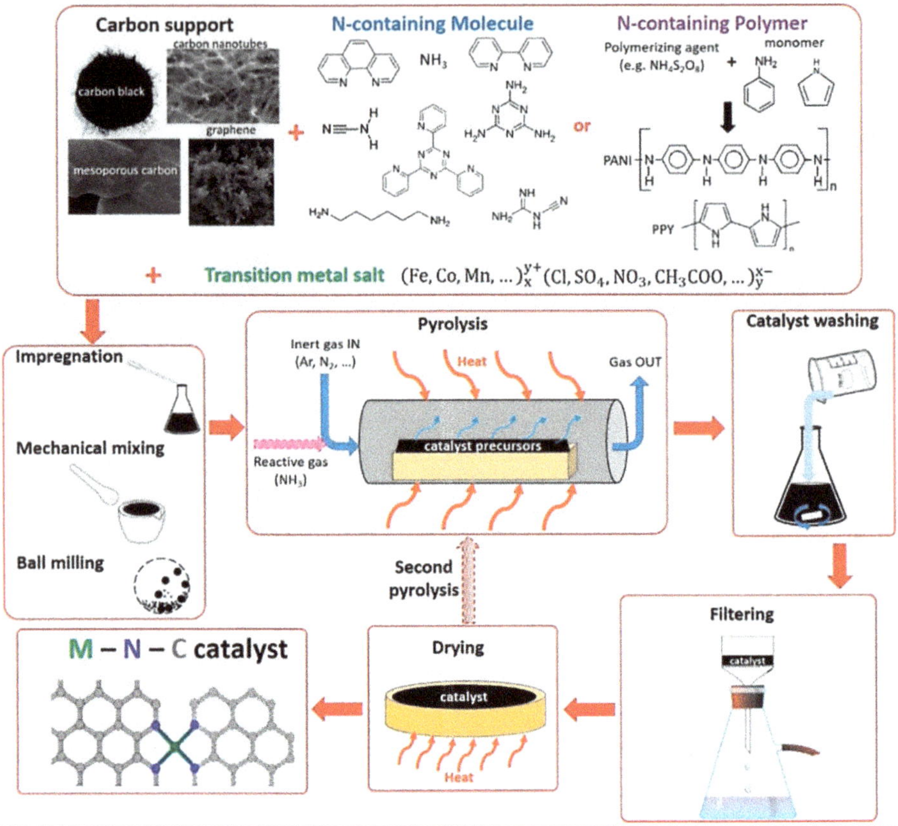

FIG. 2.3.10 – Illustration of multistep synthesis for non-noble M/N/C ORR electrocatalysts [110].

material has to be electron insulating. Most commonly used electrolyte materials are yttria-stabilized zirconia, gadolinium-doped ceria, and magnesium-doped lanthanum gallate. Cathode materials are perovskite family materials based on lanthanum strontium manganite and lanthanum strontium cobalt ferrite. NiO-yttrium Stabilized zirconia cermet is largely used as anode materials, Ni being the catalyst for HOR. Recent literature points to a strongly growing interest for high temperature fuel cells working with a proton exchange solid oxide electrolyte. These materials have been reported for the first time in the early eighties but suffer from low conductivity. Recently, important progress has been recorded on the synthesis of highly conductive materials, and impressive performances (higher than 1 W per cell square cm) have been reported for proton-conducting SOFCs [115]. A noticeable difference between O^{2-} and H^+ conducting SOFCs is the operating temperature, which is lower in the latter case (500–700 °C). A second particular aspect of proton conducting SOFCs is that the cathode has to be O^{2-}, H^+ and electron conducting while for O^{2-} SOFCs, proton conduction is not required (figure 2.3.11).

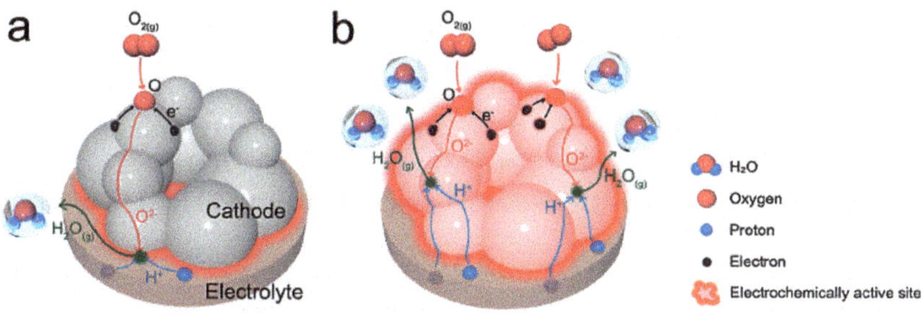

FIG. 2.3.11 – Illustration of water formation reaction at the H^+ SOFC cathode (a) using mixed ionic and electronic conduction (O^{2-}/e^-) and (b) a triple-conducting oxide (H^+, O^{2-}, e^-) [116].

Anion and proton conducting SOFCs are concerned more and more by nanoscale consideration since electrode materials involve mixing of ion and electron conducting materials for the first one, and mixing of mixed ionic-electronic conducting materials and proton conduction materials for the second one. Wet chemistry routes are now widely preferred in particular through infiltration processes [117]. Precursors with the desired stoichiometry are for example impregnated by ink-jet into a porous scaffold of one of the material that has to be combined with the other one, followed by thermal treatments [118]. Nanometer scale high area interface contacts between materials favor combination of different functions, here illustrated by O^{2-} or/and H^+ conduction and electron conduction properties. Cathode materials with triple conduction properties are materials with perovskite-related structures (figure 2.3.12).

2.3.3.2 Hydrogen Production

Paradoxically in the context of this paper, hydrogen production is currently responsible for about 830 million tons of carbon dioxide/year [107], about 2.6 times

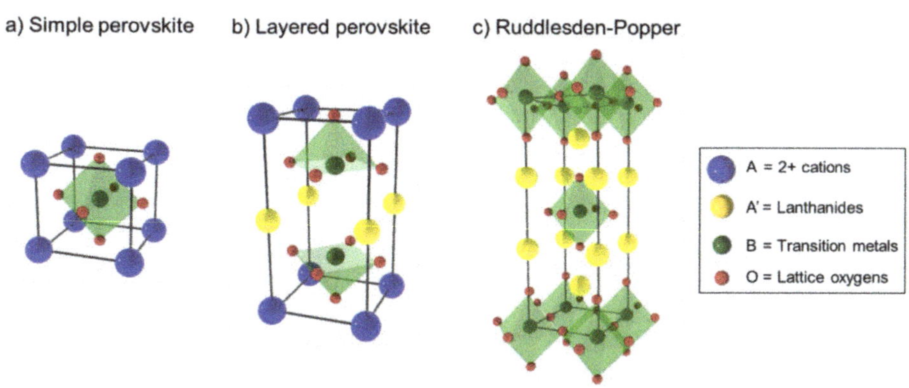

FIG. 2.3.12 – Perovskite related structures. https://doi.org/10.1016/j.rser.2019.04.042.

the CO_2 emission in France in 2018. Indeed, hydrogen is produced practically exclusively from fossil resources and mainly for ammonia production (fertilizers industry) and oil refining. They are different ways to produce "green hydrogen" without carbon emission or with low carbon emissions, among them: H_2 production from biomass, photocatalysis, photo-electrocatalysis, and electrolysis. These different approaches will be here briefly evocated and the reader will find in recent reviews detailed description.

Green hydrogen can be obtained from biomass gasification through thermo-catalytic conversion processes [119], which produces the so-called syngas, a mixture of hydrogen, carbon dioxide, carbon monoxide and methane, water as well as carbon and tars. Hydrogen production from biomass is carbon neutral. The H_2 content of syngas can be increased by additional process such as water gas shift reaction, which combines carbon monoxide and water into carbon dioxide and hydrogen (figure 2.3.13).

Catalysts involved in hydrogen production from biomass are essentially free of noble metal or uses low amounts of them. Alkali metal or oxides are used as well as aluminum and nickel, iron or a combination of these. The processes by which catalysts nanoparticles are prepared are an important factor on the efficiency of the process. Carbon conversion can be very high, reaching 80%–99%. Hydrogen production from biomass can be a low emission carbon process if coupled with CO_2 capture.

Photocatalysis and Photo-Electrocatalysis are processes to produce green hydrogen from water, both using sunlight for water splitting. Basically these two processes consist in using a semiconductor which upon photon absorption of higher energy than its band gap (Eg) will generate a pair of charge, *i.e.* an electron (e^-) is transferred to the conduction band, leaving a hole (h^+) in the valence band. Water may react with these charges to produce H_2 and O_2 (figure 2.3.14). In such processes nanoobjects are of particular interest since semiconductor electronic properties can be controlled through size control, and chemical tailoring.

Photocatalysis is the direct use of nanoparticle surface as shown in figure 2.3.14. It is conducted on nanoparticles in suspension or immobilized on a support. Different nanomaterials such as transition metal oxides, nitrides are studied as well as carbon-doped materials, generally combined with a co-catalyst which promotes either hydrogen evolution or oxygen evolution. The Solar to Hydrogen (STH) energy

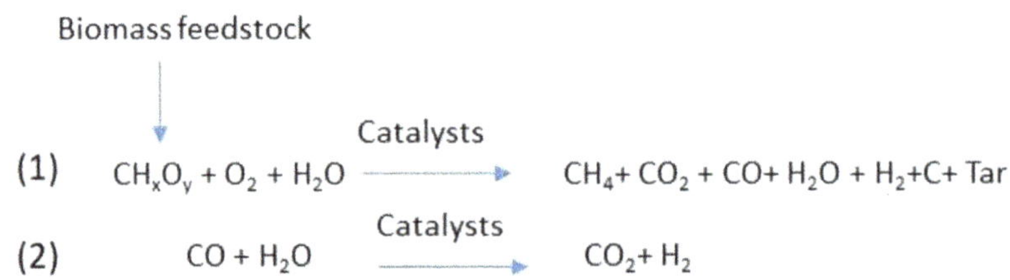

FIG. 2.3.13 – (1) Overall reaction for biomass gasification, (2) Water gas shift reaction for H_2 enrichment.

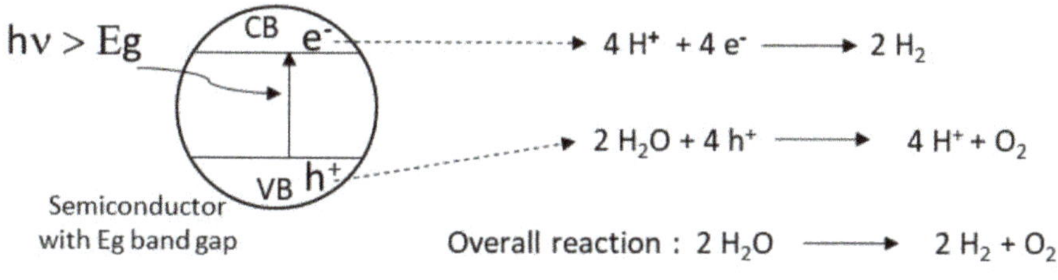

FIG. 2.3.14 – Illustration of H_2 and O_2 generation from H_2O using a semiconductor irradiated with light having an energy higher than its band gap.

conversion efficiency of these devices are extremely low (down to 1% generally), due to fast charge recombination and overall back reaction. B. Ipek *et al.* [120] reported a recent review on photocatalysis pointing at the present limits.

Photo-Electrochemical water splitting differs from photocatalysis as it generally involves a current flow in an external circuit between two electrodes. C. Jiang *et al.* [121] published a recent review on the subject that considers basics aspects, the different kind of devices, as well as the various semiconducting nanomaterials studied with their advantages and drawbacks. There are different configurations for Photo-Electrochemical Cells (PECs). The basic one is a photo-anode where water produces oxygen and hydrogen is formed at a counter electrode from protons (figure 2.3.15 left). A different configuration combines photo-anode and a photo-cathode (figure 2.3.15 right).

It is worth mentioning that most often PECs are externally biased to favor charge separation. Finally, the best STH conversion efficiencies are presently about 5%. 10% to few tens % of STH conversion efficiencies are actually obtained but by coupling a photovoltaic cell or an electrolyzer with PEC.

Electrolysis of water is extremely studied all over the world because it can be simply coupled with renewable electricity sources to produce high purity hydrogen with high yields. Electrolysers benefit from numerous developments performed in

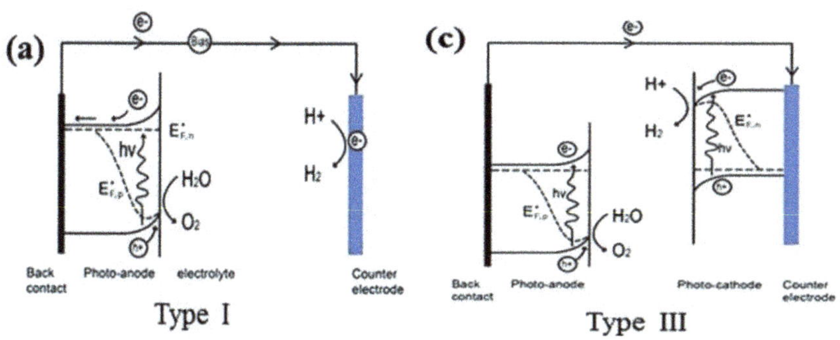

FIG. 2.3.15 – Example of configuration for photoelectrochemical cells dedicated to hydrogen production [121].

fuel cells technologies since reversed electrochemical reactions than those occurring in fuel cells stand in electrolysers. Some devices can even both work as fuel cell, or in the reverse mode of electrolyser. For Solid Oxide Electrolyser cells (SOEC high temperature), anion and proton conducting electrolyte can be used (figure 2.3.15). The materials involved for electrodes and electrolytes are similar in nature to those involved in corresponding fuel cells. An original development was recently reported by J. Kim *et al.* [122] in which an electrolyte with the complex formula $BaZr_{0.1}$-$Ce_{0.7}Y_{0.1}Yb_{0.1}O_{3-\delta}$ can conduct both O^{2-} and H^+. Thus, hydrogen can be simultaneously produced at the anode and at the cathode (figure 2.3.16).

For low temperatures, the Proton Exchange Membrane (PEM) Electrolysers have been commercially available devices for more than a decade. This technology produces highly pure hydrogen. It is also well adapted to coupling with renewable energies because it can withstand up to 300% of its nominal power with very short response time and exhibits long durability. As for PEMFCs, PEMs use noble metal based catalysts and the critical electrode is the anode at which the Oxygen Evolution Reaction (OER) occurs. Overpotential at the cathode results in actual potential between 1.4 and 2 V in acidic conditions, at which many materials corrode. This excludes the use of classical carbon based component, as it is the case in PEMFCs. Good catalysts for the OER in acids are RuO_2, IrO_2, PtO_2. IrO_2 is essentially used since it combines high activity, stability and significant conductivity. It is even more scarce than platinum, therefore strategies are developed to decrease catalyst cathode loading. Such strategies includes syntheses of nanosized particles, combination with non-noble metal (Sn, Co, Mn, Ta) oxides, combination with conductive (and corrosion resistant) nanosized supports to optimize catalyst utilization. At the cathode of PEMs where HER occurs, the catalyst is (as for PEMFCs) Pt (or Pd). Strategies bearing on alternative support development to decrease loading of these metal in the electrode are also developed. Alternative non-noble metal catalyst nanoparticles, such as CoP, have been recently reported [123]. For low temperature

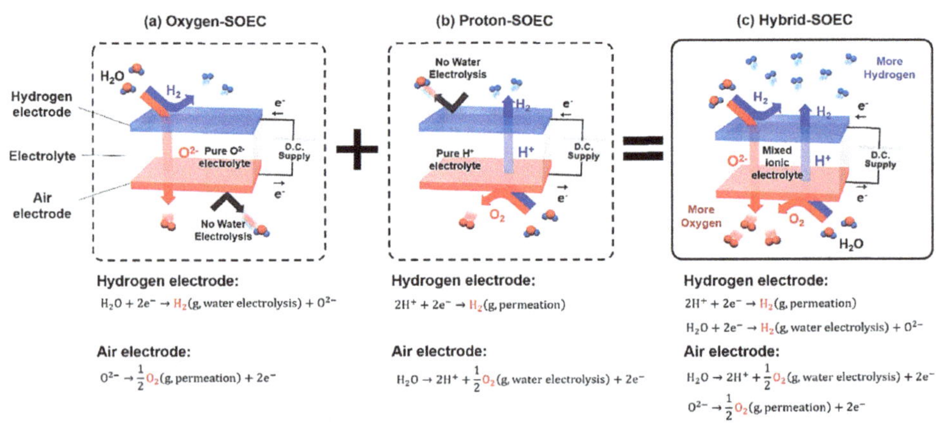

FIG. 2.3.16 – Schematic diagram of working principle for Solid Oxide Electrolysers (a) Oxygen-SOEC, (b) Proton-SOECs and (c) Hybrid-SOEC system [122].

anion conducting electrolysers, the question of catalysts is easier to manage as for Alkaline liquid electrolyte fuel cell or AEMFs. At the cathode Pt and Pd are used for HER but different kinds of non-noble catalyst nanoparticles are studied, based on Ni and Ni alloys with Co, Fe, Mo, as well as some of their phosphides, sulfides, or nitrides. These catalysts for low temperature electrolyser are practically exclusively developed for electrolysis in liquid NaOH or KOH. M.A. Khan *et al.* published a recent review on electrocatalysts for water electrolysis [124].

Hydrogen storage is currently essentially performed as compressed gas in high-pressure tanks. The study of nanoobjects for hydrogen storage issue is again motivated by the high specific surface they develop and the tailoring possibilities. At first sight, high specific surface area not only favors storage capacity but may also accelerate charging and delivering of hydrogen. In transportation application [125] important criteria among others are gravimetric density of the storage (wt% of hydrogen stored in the storing device), the speed at which hydrogen can be delivered, and the temperature at which storage and release take place. The fuel tank used in the fuel cell car mass-produced by Toyota has a gravimetric density of 5.7 wt%. Nanomaterials with very high specific surface area such as metal oxide framework, carbon based nanomaterials, metal hydride nanoparticles have been studied for hydrogen storage. Except for a recent report by L. Morris *et al.* [126] on Mg hydride nanostructured material showing 10.5% wt. storage performance, the gravimetric density of most nanomaterials is not sufficient, and the condition for hydrogen storage and release necessitates energy consuming temperature management.

References

[1] Cao G. (2004) Nanostructures & Nanomaterials, *Imperial College Press*, ISBN 9781783260881, P. 7.
[2] Alivisatos A.P. (1996) *Science* **271**, 5251, 933.
[3] Volmer M., Weber A. (1926) *Z. Phys. Chem.* **119**, 277.
[4] Katz J.L. Widersich H. (1977) *J. Coll. Inter. Sci.* **61**, 351.
[5] Haruta M., Delmon B. (1986) *J. Chim. Phys.* **83**, 859.
[6] Williams R., Yocom P.M., Stofko F.S. (1985) *J. Colloid Interf. Sci.* **106**, 388.
[7] Turkevich J., Stevenson P.C., Hillier J. (1951) *Discuss. Faraday Soc.* **11**, 55.
[8] Landman U., Hill R.N., Mosteller M. (1980) *Phys. Rev. B* **21**, 448.
[9] Hofmeister H., Huisken F., Kohn B. (1999) *Eur. Phys. J. D* **9**, 137.
[10] Benkoula S., Sublemontier O., Patanen M., Nicolas C., Sirotti F., Naitabdi A., Gaie-Levrel F., Antonsson E., Aureau D., Ouf F.X., Wada S.I., Etcheberry A., Ueda K., Miron C. (2015) *Sci. Rep.* **5**, 1.
[11] Antonsson E., Patanen M., Nicolas C., Neville J.J., Benkoula S., Goel A., Miron C. (2015) *Phys. Rev. X* **5**, 011025.
[12] Bryant B.N., Hirai A., Young E.C, Nakamura S., Speck J. (2013) *J. Cryst. Growth* **369**, 14.
[13] Langille M.R., Personick M.L., Zhang J., Mirkin C.A. (2012) *J. Am. Chem. Soc.* **134**, 14542.
[14] Zheng Y., Xiao M., Jiang S., Ding F., Wang J. (2013) *Nanoscale* **5**, 788.
[15] Teleki A., Wengler R., Wengler L., Nirschl H., Pratsinis S.E. (2007) *Powder Technol.* **179**, 131.
[16] Frens G. (1973) *Nat. Phys. Sci.* **241**, 22.
[17] Thouy R., Jullien R. (1996) *J. Phys. I* **6**, 1365.
[18] Lin M.Y., Lindsay H.M., Weitz D.A., Ball R.C., Klein R., Meaklin P. (1989) *Nature* **339**, 360.
[19] Faraday M. (1857) *Philos. Trans. R. Soc. London* **147**, 145.

[20] Schmid G. (1992) *Chem. Rev.* **92**, 1709.
[21] Brus M., Walker M., Bethell D., Schiffrin D.J., Whyman R. (1994) *J. Chem. Soc., Chem. Commun.* **801**.
[22] Murray C.B., Norris D.J., Bawendi M.G. (1993) *J. Am. Chem. Soc.* **115**, 8706.
[23] Pileni M.P. (2003) *Nat. Mater.* **2**, 145.
[24] Livage J., Henry M., Sanchez C. (1988) *Prog. Solid St. Chem.* **18**, 259.
[25] Niederberger M. (2007) *Accounts for Chemical Research* **40**, 793.
[26] Demazeau G. (2008) *J. Mater. Sci.* **43**, 2024.
[27] Li J., Wu Q., Wu J. (2015) Synthesis of nanoparticles via solvothermal and hydrothermal methods, *Handbook* of *Nanoparticles* (M. Aliofkhazraei, Ed). Springer International Publishing, Cham, pp. 1–28.
[28] Ji Z., Wang X., Zhang H., Lin S., Meng H., Sun B., George S., Xia T., Nel A.E., Zink J.I. (2012) *ACS Nano* **6**, 5366.
[29] Strobel R., Baiker A., Pratsinis S.E. (2006) *Adv. Powder Technol.* **17**, 457.
[30] Leconte Y., Leparoux M., Portier X., Herlin-Boime, Reynaud C. (2008) *Plasma Chem. Plasma Proc.* **28**, 233.
[31] Sublemontier O., Kintz H., Lacour F., Paquez X., Maurice V., Leconte Y., Porterat D., Herlin-Boime N., Reynaud C. (2011) *KONA Powder Particle J.* **29**, 236.
[32] Messing G.L., Zhang S.-C., Jayanthi G.V. (1993) *J. Am. Ceram. Soc.* **76**, 2707.
[33] Yan R.X., Gargas D., Yang P.D. (2009) *Nat. Photonics* **3**, 569.
[34] Goktas N.I., Wilson P., Ghukasyan A., Wagner D., McNamee S. (2018) *Appl. Phys. Rev.* **5**, 041305.
[35] Wagner R.S., Ellis W.C. (1964) *Appl. Phys. Lett.* **4**, 89.
[36] Lu W., Lieber C.M. (2006) *J. Phys. D* **39**, R387.
[37] Dick K.A., Deppert K., Martensson T., Mandl B., Samuelson L., Seifert W. (2005) *Nano Lett.* **5**, 761.
[38] Yang P., Lieber C.M. (1996) *Science* **271**, 1836.
[39] Wang N., Cai Y., Zhang R.Q. (2008) *Mater. Sci Eng. R* **60**, 1.
[40] Guniat L., Caroff P., Morral A.F.I. (2019) *Chem. Rev.* **119**, 8958.
[41] Dresselhaus M.S., Dresselhaus G., Eklund P.C., Rao A.M. (2000) Carbon nanotubes, *In The Physics Of Fullerene-Based And Fullerene-Related Materials* (W. Andreoni, Ed). Springer, Dordrecht, The Netherlands, pp. 331–379, ISBN 978-94-011-4038-6.
[42] Zhao Y.-L., Stoddart J.F. (2009) Noncovalent functionalization of single-walled carbon nanotubes, *Acc. Chem. Res.* **42**, 1161.
[43] Ying L.S., Bin M.A., Mohd Salleh H.B., Mohamed Yusoff S.B., Abdul Rashid J.B.,. Razak Abd (2011) *J. Ind. Eng. Chem.* **17**, 367.
[44] Pinault M., Pichot V., Khodja H., Launois P., Reynaud C., Mayne-L'Hermite M. (2005) *Nano Lett.* **5**, 2394.
[45] Mirzaei M., Sawan M. (2014) *Sensors* **14**, 17981.
[46] Evanoff K., Khan J., Balandin A.A., Magasinski A., Ready W.J., Fuller T.F., Yushin G. (2012) *Adv. Mater.* **24**, 533.
[47] Kreupl F., Graham A.P., Duesberg G.S., Steinhoegl W., Liebau M., Unger E., Hoenlein W. (2002) *Microelectron. Eng.* **64**, 399.
[48] Beigbeder A., Degee P., Conlan S.L., Mutton R.J., Clare A.S., Pettitt M.E., Callow M.E., Callow J.A., Dubois P. (2008) *Biofouling* **24**, 291.
[49] De La Zerda A., Zavaleta C., Keren S., Vaithilingam S., Bodapati S., Liu Z., Levi J., Smith B. R., Ma T.J., Oralkan O., Cheng Z., Chen X., Dai H., Khuri-Yakub B.T., Gambhir S.S. (2008) *Nat. Nano.* **3**, 557.
[50] Gao Z.L., Zhang K., Yuen M.M.F. (2011) *Nanotechnol.* **22**, 265611.
[51] Gao Z., Zhang X., Zhang K., Yuen M.M.F. (2015) *J. Phys. Chem. C* **119**, 15636.
[52] Mayne M., Grobert N., Terrones M., Kamalakaran R., Rühle M., Kroto H.W., Walton D.R. M. (2001) *Chem. Phys. Lett.* **338**, 101.
[53] Chaparro A.M., Folgado M.A., Feireira-Aparicio P., Martin A.J., Alonso-Alvarez I., Daza L. (2010) *J. Electrochem. Soc.* **157**, B993.
[54] Yashiro N., Usui T., Kikuta K. (2010) *J. Eur. Ceram.* **30**, 2093.

[55] Guillon O., Gonzalez-Julian J., Dargatz B., Kessel T., Schierning G., Räthel J., Herrmann M. (2014) *Adv. Engin. Mater.* **16**, 830.
[56] Lee Y.I., Lee J.-H., Hong S.-H., Kim D.-Y. (2003) *Mater. Res. Bull.* **38**, 925.
[57] Conibeer G., Green M., Cho E.-C., König D., Cho Y.-H., Fangsuwannarak T., Scardera G., Pink E., Huang Y., Puzzer T., Huang S., Song D., Flynn C., Park S., Hao X., Mansfield D. (2008) *Thin Solid Films* **516**, 6748.
[58] Ehbrecht M., Ferkel H., Huisken F. (1997) *Z. Phys. D* **40**, 88.
[59] Santaniello T., Migliorini L., Yan Y.S., Lenardi C., Milani P. (2018) *J. Nanopart. Res.* **20**, 250.
[60] Jaworek A., Sobczyk A.T., Krupa A. (2018) *J. Aerosol. Sci.* **125**, 57.
[61] Fauchais P., Heberlein J.V., Boulos M.I. (2014) *Thermal spray fundamentals: from powder to part.* Springer, New York.
[62] An A., Joshi B., Yarin A., Swihart M.T., Yoon S.S. (2020) *Adv. Mater.* **32**, 1905028.
[63] Hihara T., Sumiyama K. (1998) *J. Appl. Phys.* **84**, 5270.
[64] Wegner K., Barborini E., Piseri P., Milani P. (2006) *KONA* **24**, 54.
[65] O. Sublemontier, Y. Rousseau (2015) **FR1562865**.
[66] Caillard A., Cuynet S., Lecas T., Andreazza P., Mikikian M., Thomann A.-L., Brault P. (2015) *J. Phys. D: Appl. Phys.* **48**, 475302.
[67] Sublemontier O., Kintz H., Leconte Y. (2014) **US9925557B2**.
[68] Kiesler D., Bastick T., Kennedy M.K., Kruis F.E. (2019) *Aerosol. Sci. Technol.* **53**, 630.
[69] Sublemontier O., Rousseau Y., Leconte Y., Petit C., Monsifrot E., Perry F., Gaston J.-P., Chapon P., Stchakovsky M., Briois P., Billard A., Torrent F. (2018) Congrès Français sur les Aérosols, DOI: https://doi.org/10.25576/ASFERA-CFA2018-12545.
[70] Sublemontier O., Rousseau Y., Petit C., Muller C., Billard A., Briois P., Perry F., Gaston J.P. (2018) **FR 3077420**.
[71] Yan C., Zhao H., Perepichka D.F., Rosei F. (2016) *Small* **12**, 3888.
[72] Boyer J.-C., Vetrone F., Cuccia L.A., Capobianco J.A. (2006) *J. Amer. Chem. Soc.* **128**, 7444.
[73] Timmerman D., Izeddin I., Stallinga P., Yassievich I.N., Gregorkiewicz T. (2008) *Nat. Photonics* **2**, 105.
[74] Martí A., López N., Antolín E., Cánovas E., Stanley C., Farmer C., Cuadra L., Luque A. (2006) *Thin solid films* **638,** 511.
[75] Beard M.C., Midgett A.G., Hanna M.C., Luther J.M., Hughes B.K., Nozik A.J. (2010) *Nano Lett.* **10**, 3019.
[76] Aliberti P., Shrestha S.K., Teuscher R., Zhang B., Green M.A., Conibeer G.J. (2010) *Sol. Energy Mater. Sol. Cells* **94**, 1936.
[77] Green M.A., Dunlop E.D., Levi D.H., Hohl-Ebinger J., Yoshita M., Ho-Baillie A.W.Y. (2019) *Prog. Photovoltaics* **7**, 565.
[78] Hao X., Cho E.C., Scardera G., Shen Y., Bellet-Amalric E., Bellet D., Conibeer G.J., Green M.A. (2009) *Solar Energy Mater. Sol. Cells* **93**, 1524.
[79] Goktas N.I., Wilson P., Ghukasyan A., Wagner D., McNamee S., LaPierre R.R. (2018) *Appl. Phys. Rev.* **5**, 041305.
[80] O'Regan B., Grätzel M. (1991) *Nature* **353**, 737.
[81] Hardin B.E., Snaith H.J., McGehee M.D. (2012) *Nat. Photonics.* **6**, 162.
[82] Law M., Greene L.E., Johnson J.C., Saykally R., Yang P. (2005) *Nat. Mater.* **4**, 455.
[83] Jang Y.H., Jang Y.J., Kim S., Quan L.N., Chung K., Kim D.H. (2016) *Chem. Rev.* **116**, 14982.
[84] Wang P.H., Millard M., Brolo A.G. (2014) *J. Phys. Chem. C* **118**, 5889.
[85] Mahian O., Kolsi L., Amani M., Estellé P., Ahmadi G., Kleinstreuer C., Marshall J.S., Siavashi M., Taylor R.A., Niazmand H., Wongwises S., Hayat T., Kolanjiyil A., Kasaeian A., Pop I. (2019) *Phys. Rep.* **790**, 1.
[86] Olia H., Torabi M., Bahiraei M., Ahmadi M.H., Goodarzi M., Safaei M.R. (2019) *Appl. Sci.* **9**, 463.
[87] Martínez-Merino P., Alcántara R., Aguilar T., Gallardo J.J., Carrillo-Berdugo I., Gómez-Villarejo R., Rodríguez-Fernández M. (2019) *J. Navas, Energies* **12**, 4632.

[88] Poudel B., Hao Q., Ma Y., Lan Y., Minnich A., Yu B., Yan X., Wang D., Muto A., Vashaee D., Chen X., Liu J., Dresselhaus M.S., Chen G., Ren Z. (2008) *Science* **320**, 634.
[89] Harman T.C., Taylor P.J., Spears D.L., Walsh M.P. (2000) *J. Electronic Mater.* **29**, 1.
[90] Blackburn J.L., Ferguson A.J., Cho C., Grunlan J.C. (2018) *Adv. Mater.* **30**, 1704386.
[91] Liu M., Qin X.Y. (2012) *Appl. Phy. Lett.* **101**, 132103.
[92] Zhang Q., Uchaker E., Candelaria S.L., Cao G. (2013) *Chem. Soc. Rev.* **42**, 3127.
[93] Bruce P.G., Scrosati B., Tarascon J.-M. (2008) *Angew. Chem. Int. Ed.* **47**, 2930.
[94] Pomerantseva E., Bonaccorso F., Feng X., Cui Y., Gogotsi Y. (2019) *Science* **366**.
[95] Gibot P., Casas-Cabanas M., Laffont L., Levasseur S., Carlach P., Hamelet S., Tarascon J.-M., Masquelier C. (2008) *Nat. Mater.* **7,** 741.
[96] Okubo M., Hosono E., Kim J., Enomoto M., Kojima N., Kudo T., Zhou H., Honma I. (2007) *J. Am. Chem. Soc.* **129**, 7444.
[97] Magasinski A., Dixon P., Hertzberg B., Kvit A., Ayala J., Yushin G. (2010) *Nat. Mater.* **9**, 353.
[98] Liu N., Lu Z., Zhao J., McDowell M.T., Lee H.-W., Zhao W., Cui Y. (2014) *Nat. Nano.* **9**, 187.
[99] Ji X., Lee K.T., Nazar L.F. (2009) *Nat. Mater.* **8**, 500.
[100] Sourice J., Quinsac A., Leconte Y., Sublemontier O., Porcher W., Haon C., Bordes A., De Vito E., Boulineau A., Jouanneau Si Larbi S., Herlin-Boime N., Reynaud C. (2015) *ACS Appl. Mater. Interfaces* **7**, 6637.
[101] Charrier G., Desrues A., Barchasz C., Leroy J., Cornut R., Jousselme B., Campidelli S. (2016) *J. Mater. Chem. A* **4**, 15036.
[102] Wan J., Xie J., Mackanic D.G., Burke W., Bao Z., Cui Y. (2018) *Mater. Today Nano* **4**, 1.
[103] Yi E., Wang W., Kieffer J., Laine R.M. (2016) *J. Mater. Chem. A* **4**, 12947.
[104] Simon P., Gogotsi Y. (2008) *Nat. Mater.* **7**, 845.
[105] Lukatskaya M.R., Kota S., Lin Z., Zhao M.-Q., Shpigel N., Levi M.D., Halim J., Taberna P.-L., Barsoum M.W., Simon P., Gogotsi Y. (2017) *Nat. Energy* **2**, 1.
[106] Lagoutte S., Aubert P.-H., Pinault M., Tran-Van F., Mayne-L'Hermite M., Chevrot C. (2014) *Electrochim. Acta* **130**, 754.
[107] Technological report, International energy agency (2019), https://www.iea.org/reports/the-future-of-hydrogen.
[108] Shao M., Chang Q., Dodelet J-P., Cheniz R. (2016) *Chem. Rev.* **116**, 3594.
[109] Chen C., Kang Y., Huo Z., Zhu Z., Huang W., Xin H.L., Snyder J.D., Li D., Herron J.A., Mavrikakis M., Chi M., More K.L., Li Y., Markovic N.M., Somorjai G.A., Yang P., Stamenkovic V.R. (2014) *Science* **343**, 1339.
[110] Osmieri L. (2019) *Chem. Eng.* **3**, 16.
[111] Dekel D.R. (2019) *J. Power Sources* **375**, 158.
[112] Durst J., Siebel A., Simon C., Hasché F., Herranz J., Gasteiger H.A. (2014) *Energy Environ. Sci.* **7**, 2255.
[113] Davydova E.S., Mukerjee S., Jaouen F., Dekel D.R. (2018) *ACS Catal.* **8**, 6665.
[114] Kan W.H., Samson A.J., Thangadurai V. (2016) *J. Mater. Chem. A* **4**, 17913.
[115] Xu X., Wang H., Ma J., Liu W., Wang X., Fronzy M., Bi L. (2019) *J. Mater Chem. A* **7**, 18792.
[116] Kim J., Sendogan S., Kwon G., Ding D., Shin J., Liu M., Kim G. (2014) *Chem. Sus. Chem.* **7**, 2811.
[117] Ding D., Li X., Lai S.Y., Gerdes K., Liu M. (2014) *Energy Environ. Sci.* **4**, 552.
[118] Venezia E., Viviani M., Presto S., Kumar V., Tomov R.I. (2019) *Nanomaterials* **9**, 654.
[119] Li S., Zheng H., Zheng Y., Tian J., Jing T., Chang J.-S., Ho S.-H. (2019) *Int. J. Hydrogen Energy* **44**, 14266.
[120] Ipek B., Uzner D. (2019) IntecOpen publisher, https://doi.org/10.5772/intechopen.89235.
[121] Jiang C., Moniz S.J.A., Wang A., Zhang T., Tang J. (2017) *Chem. Rev. Soc.* **46**, 4665.
[122] Kim J., Jun A., Gwon O., Yoo S., Liu M., Shin J., Lim T-H., Kim G. (2018) *Nano Energy* **44**, 121.
[123] King L.A., Hubert M.A., Capuano C., Manco J., Danilovic N., Valle E., Hellstern T.R., Ayers K., Jaramillo T.F. (2019) *Nat. Nanotechnol.* **14**, 1071.

[124] Khan M.A., Zhao H., Zou W., Chen Z., Cao W., Fang J., Xu J., Zhang L., J. Zhang (2018) *Electrochem. Energy Rev.* **1**, 483.
[125] Rivard E., Trudeau M., Zaghib K. (2019) *Materials* **12**, 1973.
[126] Morris L., Hales J.J., Trudeau M.L., Georgiev P., Embs J.P., Eckert J., Kaltsoyannis N., Antonelli D.M. (2019) *Energy Environ. Sci.* **12**, 1580.

Chapter 3

Emerging Surface Engineering Processes

Abstract

This chapter covers the last developments regarding several processes dedicated to the deposition of coatings. A detailed description of each process is given. Cold spraying and suspension plasma spraying are investigated in two separate sections for the elaboration of thick coatings. The physics behind the powder particles or the liquid suspension behaviour governing the build-up of the coatings are precisely described. Gas-phase deposition processes for the deposition of thin films are then presented in three sections, physical vapor deposition, chemical vapor deposition and atomic layer deposition. Several variations of these conventional process families are introduced, keeping in mind the expected specifications of such coatings for various transportation and energy related applications. Ionized physical vapor deposition and metalorganic chemical vapor deposition processes are for instance explored, as they provide a better control over the performances of deposited materials.

Introduction

If some processes described in this chapter are known for a consequent amount of time by research teams, as they were principally developed in academic environments, their relevancy is today strategic for industrial companies wishing to improve drastically the performances of their materials or systems through the deposition of multifunctional coatings. The adjective "emerging" is here used regarding the spreading of each process at the industrial scale. This chapter focuses on emerging manufacturing processes for coating deposition, mainly through dry routes. The chapter presents first two thermal spraying techniques for the deposition of thick layers, by cold spraying of solid particles or plasma spraying of a liquid suspension. The rest of this chapter is dedicated to gas-phase deposition processes for thin films elaboration: physical vapor deposition (PVD), chemical vapor deposition (CVD) and atomic layer deposition (ALD). A subsection on ionized PVD is proposed, with high power impulse magnetron sputtering and cathodic arc deposition,

DOI: 10.1051/978-2-7598-2446-5.c003

that allows for the growth of layers with superior properties compared to conventional sputtering techniques. A part describing how processes can be more efficiently optimized through monitoring and modeling is proposed in the CVD section. This methodology is of course not specific to CVD processes and is extensively used for all deposition techniques. The issue of the influence of the precursor material design on grown coatings characteristics and properties is covered through the description of suspension plasma spraying compared to other thermal spraying processes, and through the tailoring of metalorganic molecules for efficient CVD processes.

Each process family is detailed with for example the main physical and chemical principles governing the growth of the coating, how each technology is practically implemented or their advantages and drawbacks. Examples of applications in the transportation and energy industries are also given where the relevancy of each deposition technique is highlighted in order to reach an optimal level of performance regarding the aimed application.

3.1 Thermal Spray – Cold Spray

Erick MEILLOT, Benjamin BERNARD, Samuel LETT, Léo KOVACS, Robin RIXAIN, Aurélie QUET, Raphaelle BERNARDIE, Céline CARUYER, Florent GIRARD and Vincent FOURNIER

CEA-DAM, Le Ripault, 37260 Monts, France

3.1.1 *Introduction*

The cold spray process has been developed for several decades in West industry and is now present in various important industry sectors all around the world. This paper aims to present some of its recent history, its principle with drawbacks and advantages, the way to build up coatings but also free-standing and net-shape pieces. Readers who would like to improve their knowledge upon cold spray can refer to the following book [1].

Industry requires more and more processes to apply coatings on pieces to protect them against environmental damages or to confer them new properties. Development axes are driven by coating performances and their cost as major considerations. Metallic coatings can be produced by different thermal spraying processes depending on deposit requirements (cold spray, plasma spraying, high velocity oxy-fuel, laser spraying...). Cold spray belongs to the family of thermal spray methods, but differs from other techniques in that it is a solid-state process, wherein the sprayed particles are deposited through supersonic-velocity impact with the substrate at temperatures far below the melting point of the sprayed material. As a result, the chemical nature of the initial material is not denatured and any significant grain growth is observed. Moreover, high particle velocities lead to very dense coatings. These main advantages explain the growing interest in cold spray.

Cold spray comes from investigations of Prof. Papyrin and colleagues at Institute of Theoretical and Applied Mechanics of Novosibirsk, in Russia, in the mid-1980s [2, 3]. Its principle will be discussed in the following chapter. Up until now, developments and improvement have been carried out without denaturing its basis. Figure 3.1.1 shows the improvement from a first homemade to one up-to-date commercial gun. These improvements enhance the use of different metal kinds, from ductile to harder ones.

3.1.2 *Principle: A Wide Range of Metallic Powder*

The basic principle of cold spray is shown in figure 3.1.2. It requires:

- A powder feeder.
- A gas heater with high-pressure gas supply.
- A system cooling the supersonic nozzle (not drawn in figure 3.1.2).
- A spray gun including a supersonic nozzle, from the "de Laval" type.

DOI: 10.1051/978-2-7598-2446-5.c013

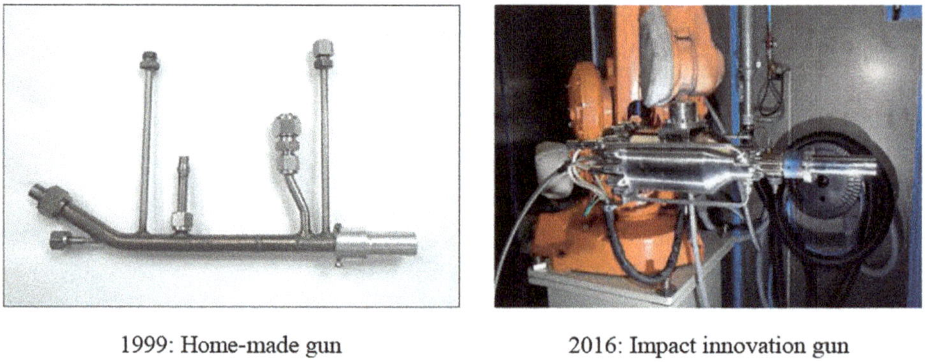

FIG. 3.1.1 – From old set-up to up-to-date one.

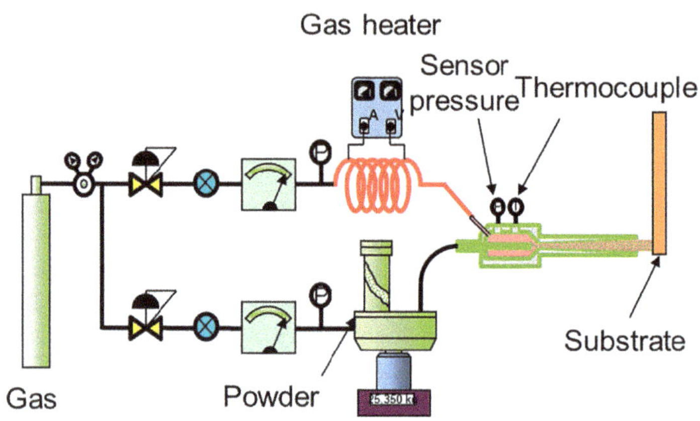

FIG. 3.1.2 – Schematic of cold spray process.

Generally, the spray gun is attached to a robotic arm in front of the substrate onto which the deposit is manufactured. This set-up stands in a closed spraying booth that helps to reduce noise and to contain dust from over-sprayed material. Only the supervision set-up is located outside with the operator.

Working gas, nitrogen, helium, or a mixture of these, is compressed at a high pressure of 1.0–5.0 MPa and introduced through the system in two different channels. The first one passes through the powder feeder in order to carry the particles (1–100 μm) to the gun. In the second one, an electric gas heater preheats the gas to temperatures between 100 and 1100 °C. Both channels converge to the convergent/divergent nozzle entrance (figure 3.1.3). A typical gun is equipped with a converging/diverging de Laval-type nozzle, which allows the gas and particles to reach supersonic velocities [3, 4]. The first stage of the nozzle carries the gas to the neck while the divergent expands and accelerates it. The higher temperature provides an additional increase in the gas velocity and, consequently, increases the

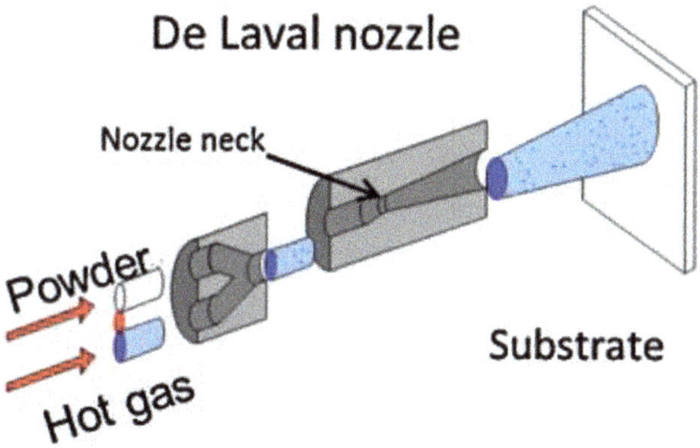

FIG. 3.1.3 – Principle of cold spray process.

particle velocity. Due to a short flight time of the particles in the rather low-temperature gas flow, the particles leave the nozzle with a high velocity but at low temperatures before colliding the substrate to form a coating by piling-up. As a result, the particles arriving at the substrate are still in solid state and then distort by plastic deformation on impact.

Although the principle of cold spray is basic, controlling the process is very difficult owing to the big amount of processing parameters. The type of gas, gas pressure and temperature, powder shape and size, nozzle design, standoff distance, traverse speed, are some of the parameters impacting the velocity of the particles and so all strongly affect the properties of the coating. As previously explained, the impacting particles are in solid state and so molten state can only appear during the impact on the substrate due to the shock between the two materials. Therefore, the particle velocity is the main macroscopic parameter that affects the microstructure of the coating. In this way, many studies have been carried out with a large range of powders to analyze the particle behavior and understand the deposit building.

3.1.2.1 Powder Supply

Various powder-manufacturing processes exist, however, regarding cold spray specifications, fused and crushed powder is usually preferred due to the resulting angular grain shape that causes high deformation depending on the way they impact the substrate. In the same order, a broader size distribution is preferred to enhance the final density, the heaviest particles generating a shot peening effect on the previous layer. The smallest particles must not be finer than a few micrometers or they could trouble the powder flowability (figure 3.1.4).

3.1.2.2 Velocity Analysis

Figure 3.1.5 shows different behaviors of copper particles depending on their velocity. A too low velocity leads to rebound upon the substrate surface. On the

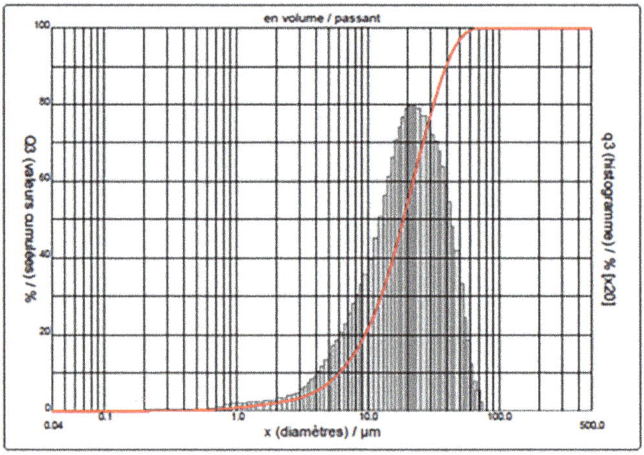

FIG. 3.1.4 – Size distribution of a typical cold spray powder feedstock.

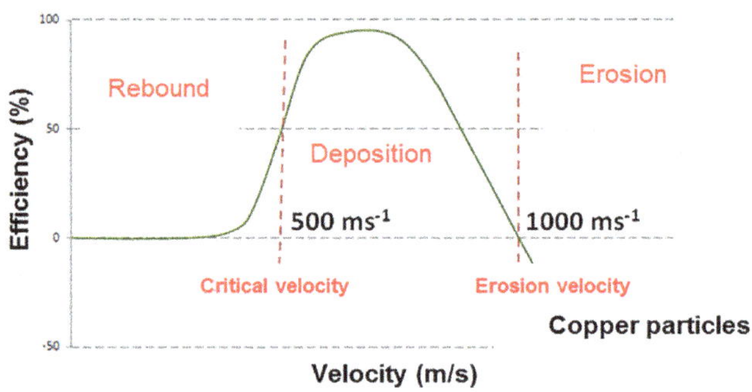

FIG. 3.1.5 – Spraying efficiency versus impact particles velocity.

opposite, a too high velocity erodes the material to be coated. The knowhow of the operator is to determine the window of deposition that exists between these two velocities. The critical velocity, a key parameter in cold spray, can be defined by a deposition efficiency of 50% (in weight).

The value of the critical velocity depends on many factors, such as the substrate material and feedstock powder, processing parameters, particle size and substrate surface properties, like roughness for example. Several authors proposed empirical expressions of this threshold in order to provide an estimation of the required velocity V_{crit} to achieve deposition such as:

$$\begin{aligned} V_{\mathrm{crit}} &= 667 - 0.014 * \rho + 0.08 * (T_m - T_{\mathrm{Ref}}) \\ &\quad + 10^{-7} * \sigma_{\mathrm{uts}} - 0.4 * (T_i - T_{\mathrm{Ref}}) \end{aligned} \tag{3.1.1}$$

where ρ is the particle density, T_m is the melting temperature, T_{Ref} is the reference temperature at which the ultimate tensile strength is determined, σ_{uts} is the ultimate tensile strength, and T_i is the initial particle temperature [5].

Or:

$$V_{\text{crit}} = \sqrt{\left[4 * F_1 \sigma_{\text{uts}} * \left(1 - \frac{T_i - T_{\text{Ref}}}{T_m - T_{\text{Ref}}}\right) * \rho^{-1}\right] + F_2 c_p * (T_m - T_i)} \tag{3.1.2}$$

where c_p is the particle specific heat capacity, and F_1 and F_2 are constants representing material-dependent calibration factors [6].

3.1.2.3 Bonding Conditions

As already discussed, the particle velocity is the main parameter for deposition. The first sprayed particles interact with the substrate. Due to their respective hardness, the first particles become embedded in the substrate as seen in figure 3.1.6. The kinetic energy is transformed into deformation energy and heat and the local temperature can rise close to the melting point of materials.

The process goes on: the following particles impact the first ones, deform themselves then stress the previous layers. Although many numerical simulations and experimental investigations have been performed to understand the bonding mechanism in cold spray, it remains poorly understood [4]. Yet, two different mechanisms have been proposed to explain the high bonding.

The first one, the most likely bonding mechanism in cold spray is associated with adiabatic shear instability [7] (ASI), as proposed by Assadi *et al.* [5] or Schmidt *et al.* [6]. The particle–particle or particle–substrate interfacial areas experience severe localized shear deformation during impact, which eventually disrupts the thin oxide surface films on the particles, permitting strong particle–material contact. This phenomenon combines with the development of high compressive stresses after the particles collide with the substrate.

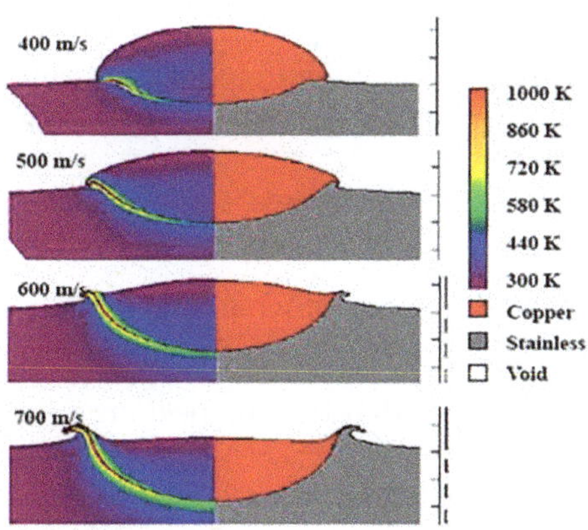

FIG. 3.1.6 – Simulation of a copper particle impacting onto stainless steel with a velocity greater than the critical one.

On a second route, Hussain *et al.* [8] proposed another bonding mechanism known as mechanical interlocking after cold spraying copper particles onto aluminum substrates. They suggested that the impaction of the copper particles on the aluminum substrate caused lips to form in the aluminum substrate, which partially enveloped the copper particles. This created a mechanical interlock between the substrate and impacting particles.

Whatever the mechanisms, particles at impact require high energy to deposit. On the next section, different ways to enhance their energy will be discussed.

Before that, let us have a look on the advantages and drawbacks of this tool.

3.1.2.4 Advantages and Drawbacks

In this paragraph, the advantages and drawbacks of cold spray are presented. More information is available from the following authors [9–11]. The cold spray process is of high interest due to the following advantages:

- It is well suited for the deposition of a large range of materials (temperature-sensitive, oxygen-sensitive and phase-sensitive materials) because of the low deposition temperature.
- Cold spraying of metals generally enhances fatigue resistance because of the micro "shot-peening" effect, generally inducing compressive residual stresses in the resulting deposits.
- Metal deposits have microstructures with high degrees of consolidation similar to wrought alloys due to intrinsic high-energy low-temperature features.
- Deposits present higher density and reduced presence of oxide phases.
- Cold spray has a significantly great deposition efficiency (DE), depending on the material properties.
- It presents small spray beam and so offers more precise control over the area of deposition onto the substrate (less need for masking).
- It generally increases the ability to join dissimilar materials because of less heat input into the substrate, which makes the substrate material to be of lesser importance.
- A high velocity of coating building can be obtained (up to 1 mm/pass).
- The ability to produce cermet coatings with a strong bonding between the metal matrix and the dispersant.
- The use of the process as a repair process is an efficient way to avoid the expense of producing a new part (*i.e.* cost saving selection) or environmental waste.

On the other hand, the cold spray process presents some drawbacks:

- Depending on both materials, powder and substrate, this couple can present a high level of residual stresses (due to shot peening), which requires heat treatment for stress relief.
- Not all materials can be deposited because of a too high hardness for example, leading to a too high critical velocity and so a weak deposition efficiency.

3.1.3 Investigation for Upgrading the Impact Particle Velocity

Deposition of hardest materials requires increasing the particle velocity through several ways. Modifying the nozzle design or increasing the gas pressure or temperature are commonly preferred. Numerical simulation is the best tool to predict the behavior of such enhancements. Any CFD (Computational Fluid Dynamic) code is relevant if it accepts 2D or 3D and compressive simulations.

3.1.3.1 Investigation on Nozzle Design and Flow Behavior

Figure 3.1.7 shows a schematic nozzle design with A_n representing its different areas. We have the convergent entrance for $n = i$, the neck surface for $n = *$ and the exit divergent surface for $n = e$. L_c and L_d stand, respectively, for the convergent and divergent lengths.

It is possible to establish a relationship linking the local cross-section surface $A(x)$ with the neck cross section surface and the local Mach number $M(x)$.

$$\frac{A(x)}{A^*} = \frac{1}{M(x)} * \left[\left(\frac{2}{\gamma+1}\right) * \left(1 + \frac{\gamma-1}{2} M(x)^2\right)\right]^{\frac{1}{\gamma-1}} \tag{3.1.3}$$

In this equation intervenes γ, the ratio of the constant-pressure- and the constant-volume-specific heats, which is approximately 1.66 for monoatomic gases (*e.g.* helium) and 1.4 for diatomic gases (*e.g.* nitrogen and oxygen). The dimensionless Mach number is defined by equation 3.1.4 where V_g is the local gas velocity and a is the speed of sound at the nozzle neck.

$$M = \frac{V_g}{a} \tag{3.1.4}$$

The local gas velocity V_g can be calculated from the following equation and depends on the nozzle design and on the gas nature *via* γ. In this equation $r = R/M_g$, with R the ideal gas constant, M_g the gas molar mass and T the gas temperature.

$$V_g = M\sqrt{\gamma r T} \tag{3.1.5}$$

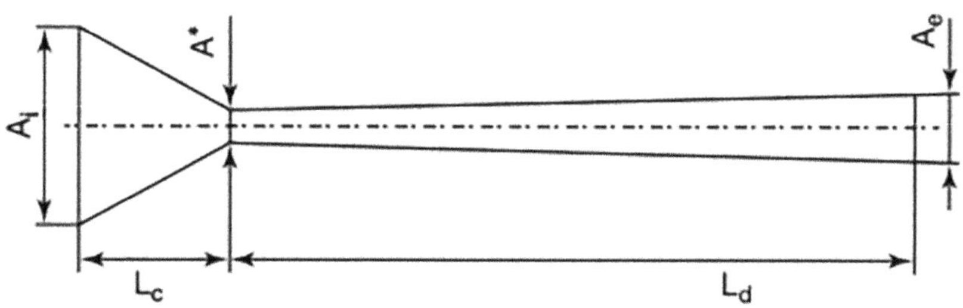

FIG. 3.1.7 – Schematic nozzle design [12].

Two types of nozzle design have been simulated: the first one (figure 3.1.8) has a short divergent leading to a sudden modification of the flow followed by a constant section extension,the second one (figure 3.1.9) has a continuous and long divergent from the neck to the exit.

Figure 3.1.10 shows the evolution of the Mach number *versus* x the nozzle axis of the two designs. The origin stands at the entrance of the convergent and theses values come from the centerline position.

The two-gas behaviors are the same when the gas goes out of the nozzle and flows in the atmosphere (fluctuations around 0.140 m position). For the first design, the gas attains quickly the theoretical velocity then this velocity drops slowly to reach the Mach number for pressure equilibrium. Just after the neck, adaptation of the pressure appears, as seen by the slow fluctuations of the curve. For the second design, there is a continuous increase in gas velocity to reach Mach number depending on the pressure balance at the exit.

For the second design, the gas temperature evolution has been calculated on the centerline with nitrogen feedstock (4 MPa and 800 °C). As seen in figure 3.1.11, the gas undergoes a drastic drop of temperature just after the nozzle neck due to the expansion phenomenon. This figure shows that the particles do not flow in a high temperature environment that the input gas temperature could predict.

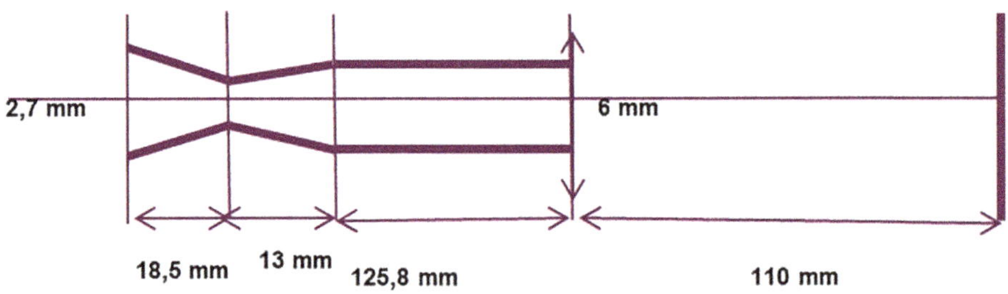

FIG. 3.1.8 – Short divergent with extension.

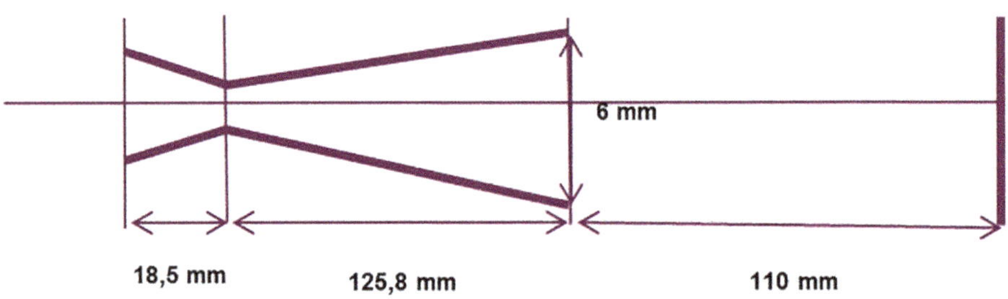

FIG. 3.1.9 – Continuous and long divergent.

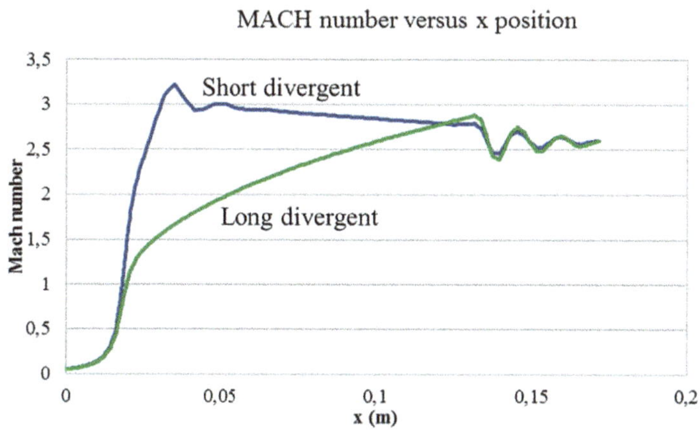

FIG. 3.1.10 – Evolution of the Mach number along the nozzle length for two nozzle designs.

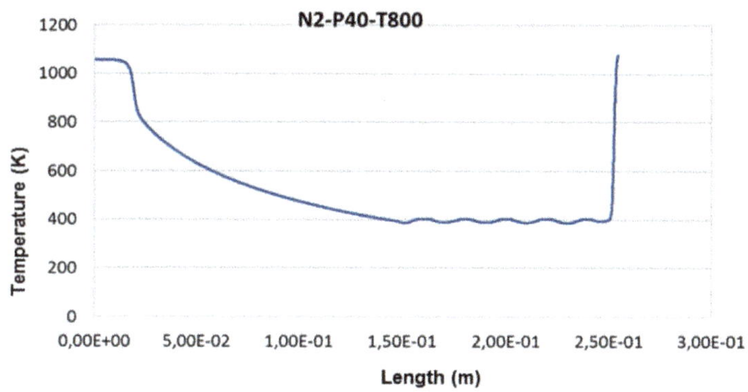

FIG. 3.1.11 – Evolution of the gas temperature along the nozzle length for the continuous and long divergent.

3.1.3.2 Investigation on Gas Nature and Flow Behavior

Depending on the deposited material properties (mechanical ones and hardness) and coating requirements (porosity), increasing the particle velocity at impact is mandatory. One way is the use of helium rather than nitrogen.

For the first nozzle design presented in the last section (see figure 3.1.8), calculations have been done with the same set-up of operating parameters (gas pressure of 2 MPa and gas temperature of 20 °C). Figure 3.1.12 represents the Mach number evolution *versus* the nozzle length for nitrogen and helium.

As expected, helium leads to a higher sound velocity at the nozzle neck and a higher gas velocity all along the extension. At the exit, the pressure balance imposes the same Mach number for the two gases.

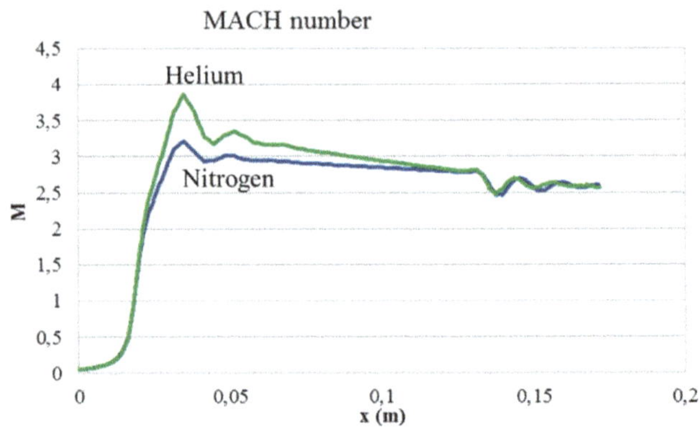

FIG. 3.1.12 – Evolution of the Mach number along the nozzle length for two gas natures.

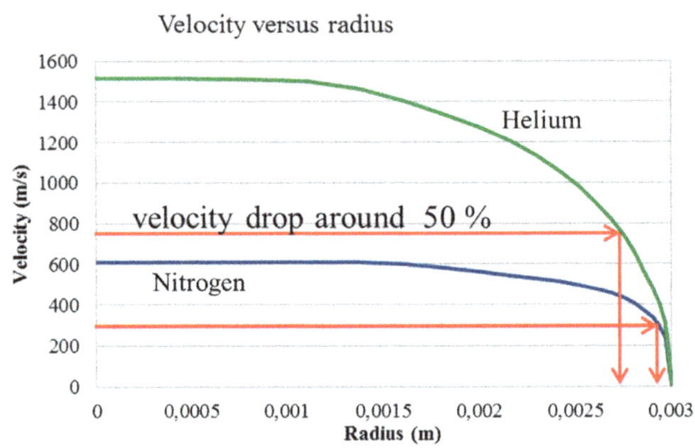

FIG. 3.1.13 – Evolution of the gas velocity versus the radius in the middle of the nozzle.

However, the difference between the two gases can be demonstrated when looking along the radius of the nozzle. Figure 3.1.13 shows the gas velocity evolution *versus* the nozzle radius in the middle of the nozzle extension. As seen, pointing out a velocity drop of 50% emphasizes the difference close to the nozzle wall, and of course the different velocity maxima between the two gases. A higher boundary layer is obtained for helium compared to nitrogen.

3.1.3.3 Investigation on Particle Behavior

Regarding the particle velocity, the theoretical equation of motion needs to be considered:

$$m_p V_p * \frac{dV_p}{dx} = \frac{C_D A_p \rho}{2} * \left(V_g - V_p\right) * \left|V_g - V_p\right| \tag{3.1.6}$$

Index p represents the particle while g refers to the gas. m stands for the mass, V the velocity and A the cross-section surface. ρ is the gas density while C_D is the drag coefficient (dimensionless number).

The drag coefficient generally comes from Henderson publication [13] for cold spray simulation because of Mach numbers over 1.75. The expression refers to the next equation:

$$C_{D\,\text{Henderson}} = \frac{0.9 + \frac{0.34}{M^2} + 1.86 * \left(\frac{M}{\text{Re}}\right)^{\frac{1}{2}} * \left[2 + \frac{2}{S^2} + \frac{1.058}{S} * \left(\frac{T_w}{T}\right)^{\frac{1}{2}} - \frac{1}{S^4}\right]}{1.86 * \left(\frac{M}{\text{Re}}\right)^{\frac{1}{2}}} \tag{3.1.7}$$

where Re is the Reynolds number, M the Mach number, T the gas temperature, T_w the particle temperature and S being defined as:

$$S = M\sqrt{\gamma/2} \tag{3.1.8}$$

3.1.4 Deposition

3.1.4.1 Build-Up of the Coatings

As previously detailed, deposition is the result of numerous particle deformations impacting on the substrate then the growing layers. This mechanism leads to coatings frequently presenting close microstructures as seen in the three pictures (figure 3.1.14). Optical and electron microscopy are generally used to characterize the coating/substrate interface and to evaluate the integrity of the deposit. As it is visible in figure 3.1.14, porosity decreases with higher pressures and as result, the particles are more deformed. The shot peening of the following particles arriving upon the deposit rises the deformation and so the residual stresses while the porosity decreases. In the next paragraph, discussion about the phenomenon of residual stress is proposed.

3.1.4.2 In-Situ Residual Stresses Measurement

One method to quantify residual stresses online is to monitor the deflection of the couple substrate/coating. It is commonly used for all deposition techniques, including PVD and CVD processes. The principle is presented in figure 3.1.15 for the *In-situ* Coating Properties (ICP) apparatus developed by RELIACOAT Inc (CANADA) where the substrate curvature is measured by three lasers. In the same way, two thermocouples get the backside temperature. The signal of all these captors goes to a computer that will determine on one side the stresses resulting on the shot peening and on the other side the cooling stresses. The shot peening stresses are calculated *via* Brenner and Senderoff relation [14] (equation 3.1.9) while the thermal stresses [15, 16] come from equation 3.1.10:

$$\sigma = \frac{E_s h_s * \left(h_s + \left(\frac{E_c}{E_s}\right)^{\frac{5}{4}} * h_c\right)}{6 h_c} * \Delta K \tag{3.1.9}$$

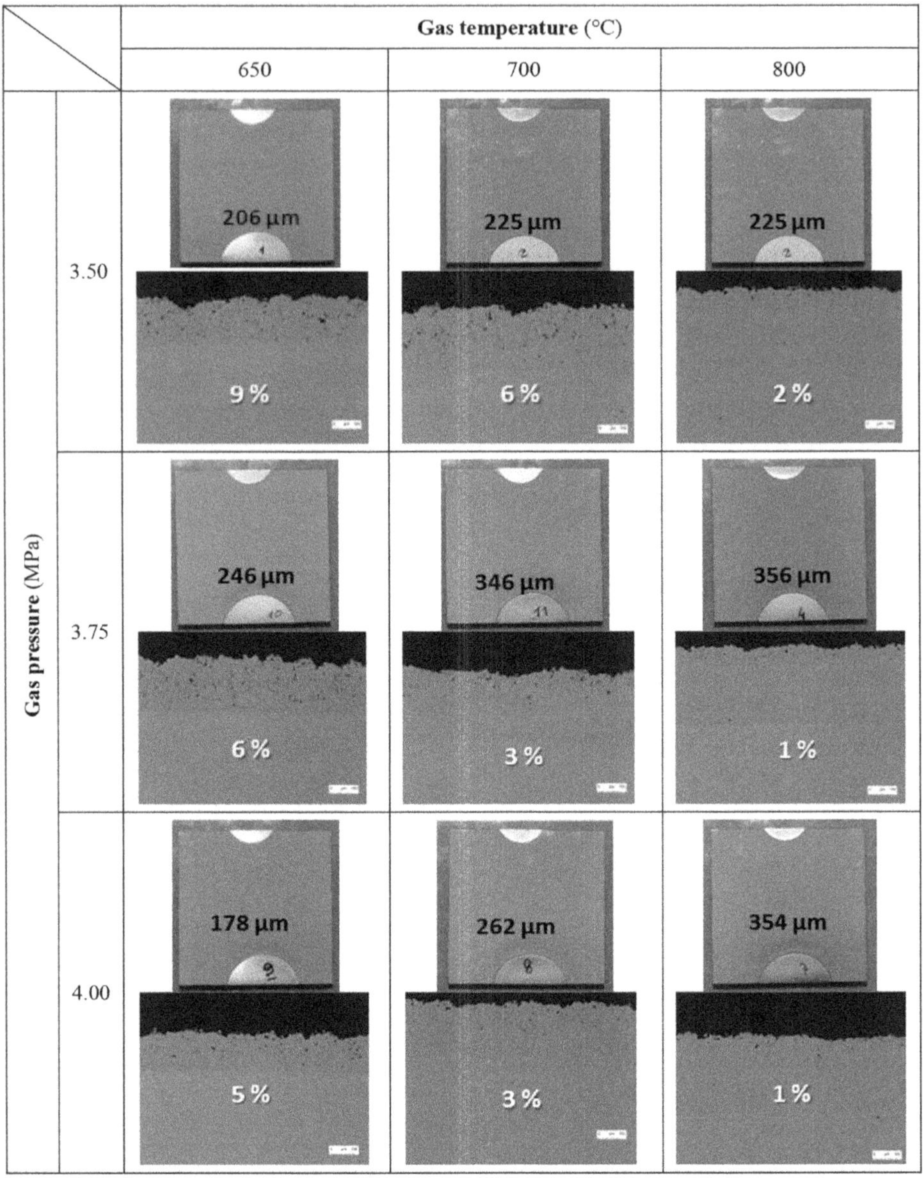

FIG. 3.1.14 – Scanning electron microscopy upper and cross-section of EM10-GVF coating versus different gas temperatures and pressures with mean thickness and porosity.

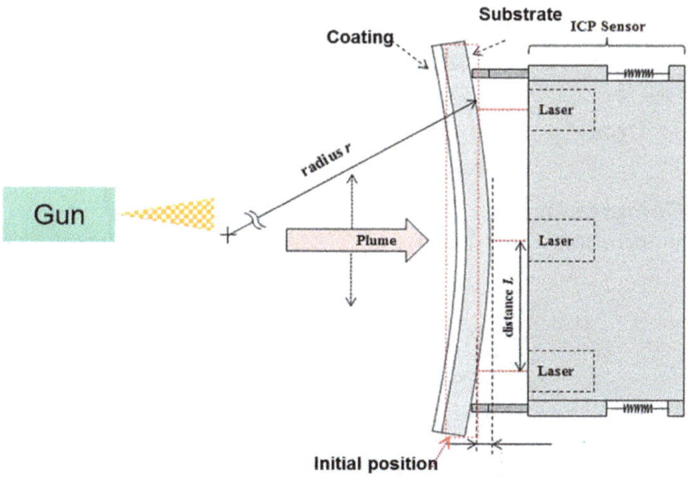

FIG. 3.1.15 – Deflection measurements principle.

$$\sigma_{\text{th}} = E_c \Delta T \Delta \alpha \tag{3.1.10}$$

where E is Young's modulus, h the thickness of the two materials, ΔK the curvature evolution, ΔT the temperature difference during deposition and post deposition and $\Delta\alpha$ the difference of thermal expansion coefficient between the coating and the substrate. The index s and c refer, respectively, to substrate and coating.

These equations are valid when the coating thickness is half the substrate one and when both Young's moduli are of the same order (the thicknesses ratio is less than 0.5 and the Young moduli ratio is less than 2) [14].

If shot peening stresses are generally compressive (equation 3.1.9), thermal stresses can be tensile or compressive depending on the values of the coating and substrate thermal expansion coefficients (equation 3.1.10). If the substrate coefficient is higher than the coating one, compressive stresses are generated while on the contrary, tensile stresses appear in the coating.

Two main mechanisms participate in the generation of stresses for cold spraying:

- The shot peening of the impacting particles (which can be the origin of the temperature rise).
- The cooling of the individual particles and of the global coatings and substrate.

When the curvature measurement is positive, the coating is in tension while when it turns negative, it integrates compressive stresses (figure 3.1.16). Metal coatings are usually in a compressive state.

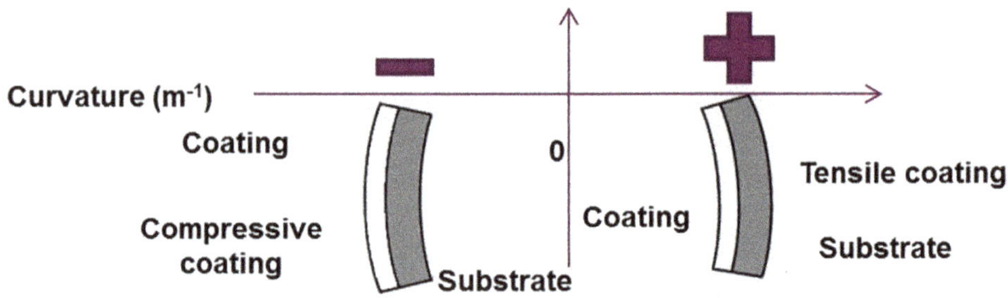

FIG. 3.1.16 – Compressive or tensile coatings.

3.1.4.3 Reducing the Residual Stresses

Controlling and reducing residual stresses in cold spraying is crucial for several applications like the realization of free-standing components (*i.e.* additive manufacturing). We will develop in this section the example of the manufacturing of a thick cylinder in tantalum by cold spray. First, the tantalum deposition on different substrate is investigated.

Several key parameters have been identified:

- The substrate nature which will host the particles.
- The thickness of the substrate.
- The transverse motion velocity of the cold spray gun in front of the substrate correlated with the powder feedstock.

Deposition has been carried out using always the same set of cold spraying tantalum parameters. First experience uses aluminum and stainless steel as substrates (table 3.1.1).

In table 3.1.1, the substrate presents higher rigidity from left to right due to its thickness and hardness. In this way, if the shot peening stresses remain constant or in the same range, with a weak decrease for aluminum, the main difference appears with the thermal stresses due to the high difference on the thermal conductivity of the substrates (see table 3.1.2). Therefore, aluminum deforms more easily under impact particles than stainless steel. Moreover, thermal stresses are compressive for both materials but are higher with the aluminum substrate. Stainless steel and

TAB. 3.1.1 – Influence of the substrate nature and thickness on stresses.

Stress/substrate	Aluminum	Stainless steel (316L)	Stainless steel (316L)
Substrate thickness (mm)	5	5	10
Shot peening stress (MPa)	−80	− 92	−93
Thermal stress (MPa)	−878	− 435	−480
Residual stress (MPa)	−958	−526	−573

TAB. 3.1.2 – Mechanical and physical properties of tested substrates.

Properties/nature	Tantalum	Stainless steel (316L)	Aluminum
Coefficient of thermal expansion ($10^{-6}.K^{-1}$)	6.5	16	23
Poisson ratio (Ø)	0.35	0.3	0.33
Elastic modulus (GPa)	190	200	70
Density (Ø)	16.6	7.9	2.7
Hardness (Hv)	90	160–190	35–48

aluminum receive the same heat amount but aluminum deforms more than stainless steel due to its higher thermal expansion coefficient.

As a result, lower compressive residual stresses are obtained with the stainless steel substrate. This is illustrated in figure 3.1.17 by the smaller curvature of the substrate after deposition.

In a second experience, two thicknesses of stainless steel substrate are studied: 5 and 10 mm. Raising the thickness not only increases the stiffness (the shot peening stress remains constant) but also the ability of the heat to diffuse and so the thermal stresses (table 3.1.1). As the residual stresses are compressive and reach a critical level, the coating entirely delaminates from the substrate as shown in figure 3.1.18.

A third experience investigates three transverse motion velocities (respectively 115, 230 and 460 $mm.s^{-1}$) linked with three different tantalum feedstock (respectively 40, 80 and 160 $g.min^{-1}$) in order to build the coating at the same growth rate. Figure 3.1.19 shows the evolution of the curvature *versus* the spraying time for the three motion velocities. For the lowest velocity, the substrate and the coating separate completely at the end of the cooling (blue curve). For the intermediate velocity, peeling appears once during the cooling (red line). The highest velocity keeps the integrity of the building.

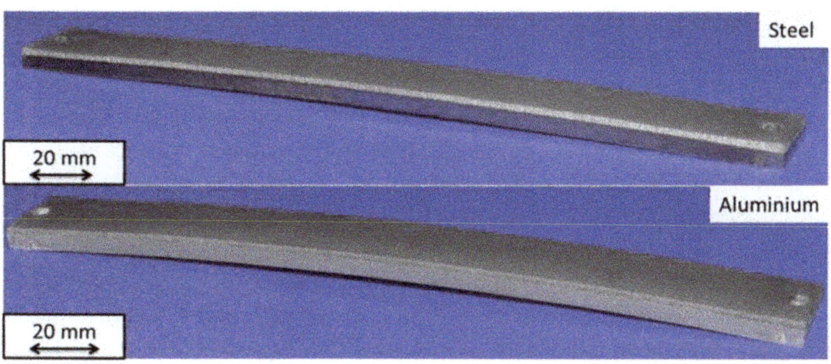

FIG. 3.1.17 – Deflection measurements of stainless steel and aluminum substrates after tantalum spraying.

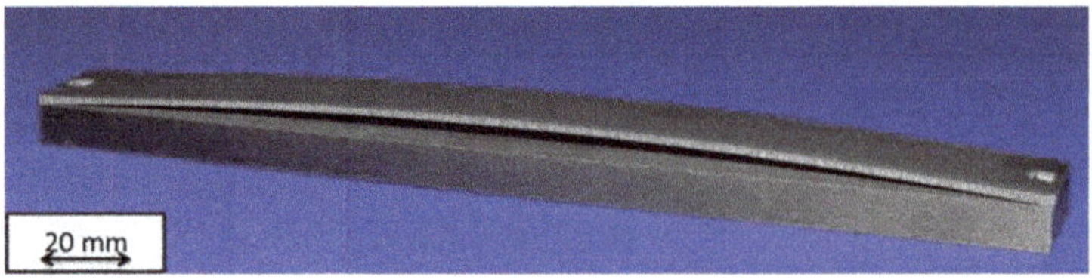

FIG. 3.1.18 – Deflection measurement of a tantalum deposit upon 10 mm stainless steel substrate.

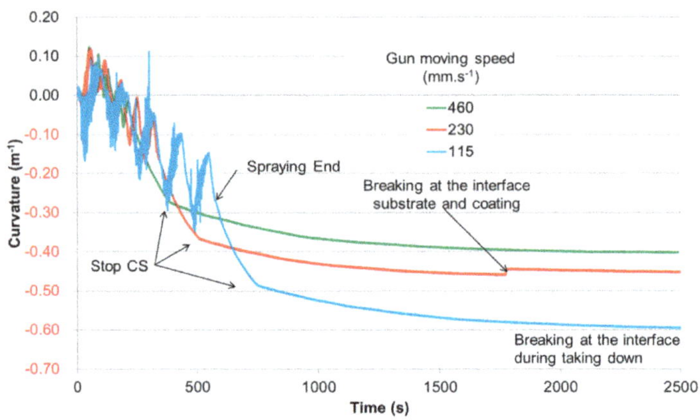

FIG. 3.1.19 – Deflection measurement of tantalum deposit upon 10 mm of stainless steel substrate.

The conclusion is that spraying tantalum powder with a higher transverse motion velocity and a lower feedstock on a rather thin stainless steel substrate prevents the apparition of residual stresses, *i.e.* maximum adhesion is reached for lower coating growth rates.

Another way of mitigating stresses is to play on thermal stresses. In this way, a cooling system has been proposed as described in figure 3.1.20. It consists of a cryogenic CO_2 spraying assisted by compressed air gas.

Two nozzles are placed on the same side as the cold spraying gun and are directed towards the substrate. The same set of spraying parameters for tantalum powder has been used with this cooling system except for the feedstock (160 $g.min^{-1}$) and the transverse motion velocity (115 $mm.s^{-1}$). These last parameters allow building up more rapidly a coating. As shown in figure 3.1.21, this succeeded.

When using the cooling system, one 5 mm thick coating is obtained without peeling or decohesion of the two materials while the entire delamination of the coating appears for only 1.6 mm without cooling. Additionally, using cooling surprisingly decreases the shot peening stress (see table 3.1.3). As expected, the thermal stresses go down significantly. Finally, compressive residual stresses can be reduced to approximately 300 MPa for tantalum spraying.

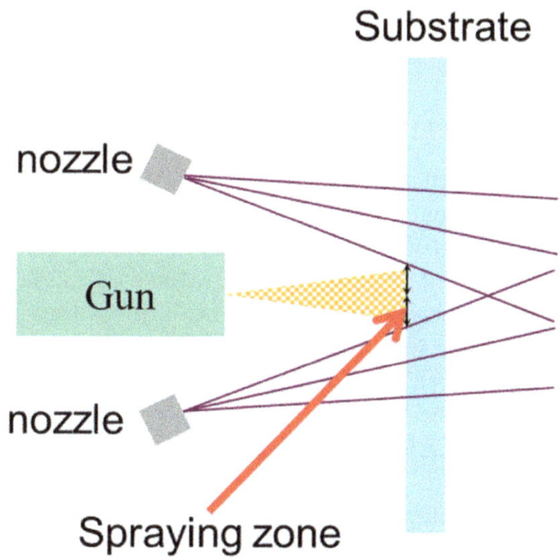

FIG. 3.1.20 – Cold spraying with cryogenic cooling principle.

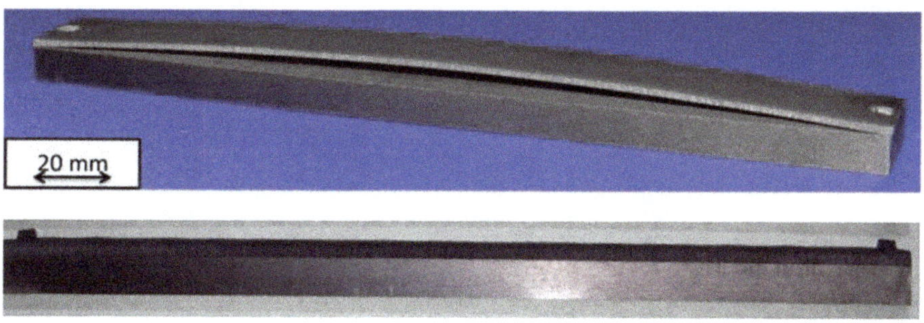

FIG. 3.1.21 – Tantalum cold spraying without (top) and with (bottom) cryogenic cooling system.

TAB. 3.1.3 – Stress measurement depending on the presence or not of a cooling system.

Stress/Substrate	Without cooling	With cooling
Aluminium substrate thickness (mm)	10	10
Shot peening stress (MPa)	−94	−48
Thermal stress (MPa)	−520	−253
Residual stress (MPa)	−614 (before delamination)	−301

The same methodology was successfully applied to Inconel, but cannot be directly transposed to all materials, such as TA6V.

3.1.4.4 What About TA6V?

Another experience has been carried out using TA6V alloy powder in order to investigate its behavior during cold spray process. Figure 3.1.22 shows the curvature evolution *versus* time for TA6V powder, compared to tantalum, using the same set or cold spraying parameters. The substrate is in this case made from aluminum. As seen in this figure, TA6V behaves differently and induces tensile stresses.

At this moment, further investigations are ongoing to determine which phenomenon is responsible for this behavior. In this case, adding a cooling system drastically enhances it and leads to higher tensile stresses.

The higher hardness of TA6V (table 3.1.4) implies a weaker deformation under impact compared to tantalum. Nevertheless, this leads also to a higher shot-peening and no response is available for the moment.

3.1.4.5 Mechanical Properties

Based on the results from the previous study concerning tantalum spraying, a thick free-standing cylinder has been built (figure 3.1.23). This cylinder authorizes mechanical properties introspection of the raw material.

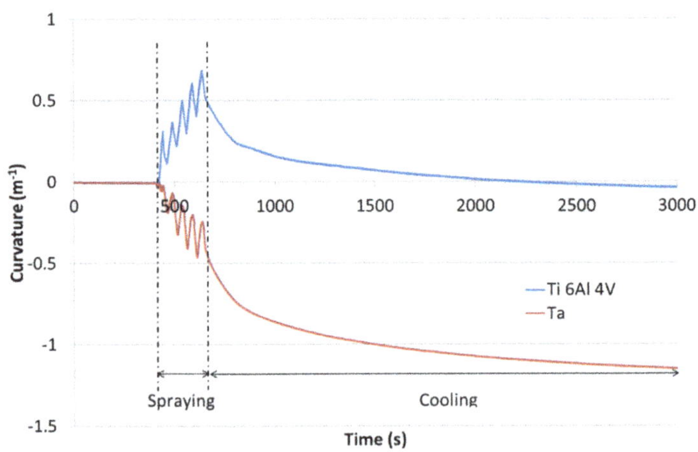

FIG. 3.1.22 – Curvature evolution versus time for tantalum and TA6V cold spraying.

TAB. 3.1.4 – Properties of the sprayed materials.

Properties/material	TA6V	Tantalum	Aluminum
Hardness (Hv)	346	76	48
Young's modulus (GPa)	115	190	70
Thermal expansion coefficient (10^{-6} K^{-1})	9	6	23

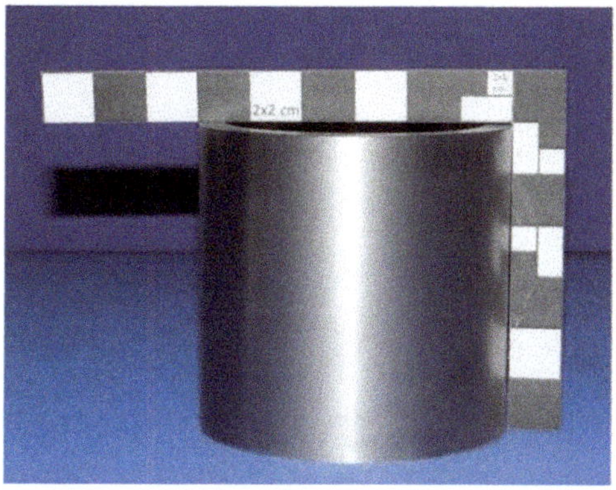

FIG. 3.1.23 – Tantalum free-standing cylinder built by cold spray additive manufacturing.

As expected, the density seems homogenous with 0.7% of porosity. SEM observations after chemical attack show an important hardening work of the material. Figure 3.1.24 illustrates the global material deformations and the *in-situ* grain deformations. Grains could deform as splat shape, indicating a high local temperature at impact. The Vickers hardness measurements (table 3.1.5) show a large dispersion depending on the position from where the samples were taken in the cylinder.

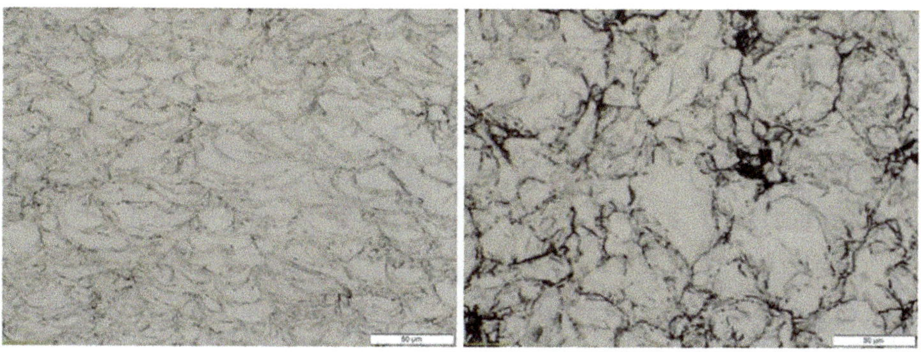

FIG. 3.1.24 – Scanning electron microscopy observations of the global tantalum coating and of the local tantalum grain deformations.

TAB. 3.1.5 – Vickers hardness measurements (minimum, maximum, average and dispersion) of as-sprayed tantalum.

Vickers hardness measurements	Sample 1	Sample 2	Sample 3
Mean value	239	242	239
Maximum value	268	268	260
Minimum value	196	221	220
Dispersion	30	24	23

Figure 3.1.25 details the behavior of the as-sprayed material with compressive and tensile tests for different samples taken from the free-standing cylinder.

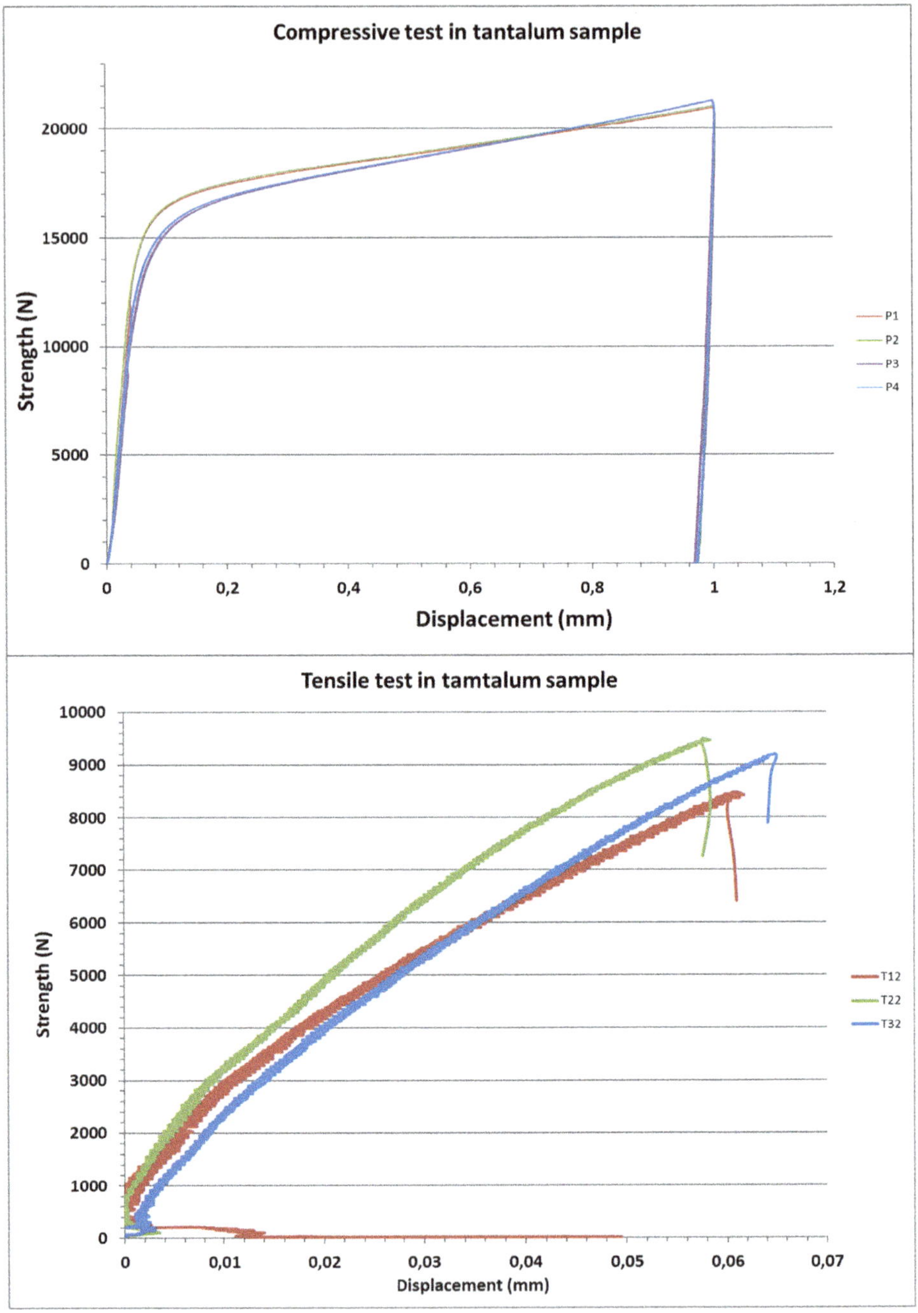

FIG. 3.1.25 – Tantalum compressive and tensile tests.

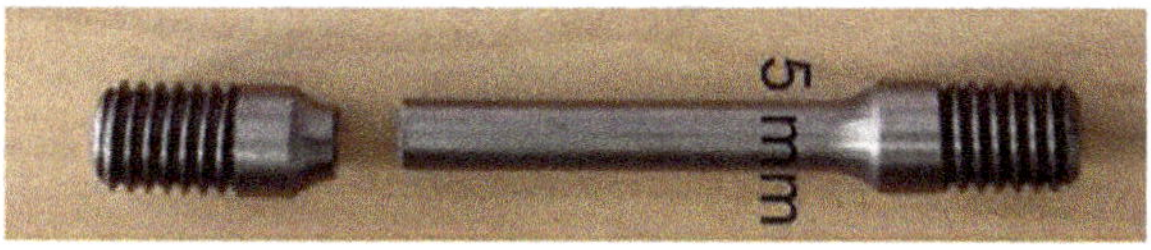

FIG. 3.1.26 – Tantalum tensile test breakdown.

For the compressive strength test, the material presents a breakdown threshold at high intensity: 1% deformation is reached at 700 MPa. Moreover, the behavior is different depending on the position of the samples in the cylinder. For the tensile strength test, the breakdown arrives in the elastic zone (around 400 MPa) with a weak deformation (6/100) and a breaking close to the neck of the samples (figure 3.1.26).

In conclusion, raw and as-sprayed tantalum is brittle and requires heat treatment to enhance its mechanical properties for using it as self-standing components. Investigations are ongoing at the moment to improve these properties for additive manufacturing.

3.1.5 Applications

Numerous studies have claimed that cold spray provides coatings with superior corrosion, wear resistance, and mechanical integrity in the as-sprayed condition [17]. These coatings can be used in many sectors, including the aerospace, automotive, transportation, die casting, petrochemical, mineral and metal processing, electronics, marine, and ceramics and glass manufacturing industries [18, 19].

Recently, a review concerning additive manufacturing [20] included repairing, remanufacturing and building up free-standing parts.

A complete review concerning cold spray applications can be read in [21].

3.1.6 Conclusion

In summary, the cold spray process differs from the other techniques of surface treatment deposition by the fact that the powder mainly remains in solid state. It is available both for metal deposition and for additive manufacturing of net-shape parts.

The impacting particle velocity is the main parameter governing the coating growth and it must be adapted depending on the material properties and on the final material requirements. Generally, the residual stresses are compressive.

Increasing the particle velocity implies an increase in the shot peening stresses with work hardening of the final material. For additive manufacturing, the end-user material is brittle and requires heat treatment in order to enhance its mechanical properties.

Coatings from cold spraying could be used in many industrial sectors but end-user materials are still under development.

3.2 Thermal Spray – Suspension Plasma Spraying

Erick MEILLOT, Benjamin BERNARD, Samuel LETT, Léo KOVACS, Robin RIXAIN, Aurélie QUET, Raphaelle BERNARDIE, Céline CARUYER, Florent GIRARD and Vincent FOURNIER

CEA-DAM, Le Ripault, 37260 Monts, France

3.2.1 Introduction

This process comes from the classical Atmospheric Plasma Spraying (APS) one, which is an industrial tool to achieve thick ceramic coatings (thicker than 100 μm). In the following section, some historical elements are proposed before a presentation, on one side, of the classical APS process and, on the other side, the developments to integrate nanomaterials and submicronical materials spraying.

Plasma spraying process emerged at the beginning of the 20th century. It took years to achieve its current status but ceramic coatings built by APS are everywhere, *e.g.* for thermal barriers or wear resistance applications. Whatever the type of material feedstock, plasma spraying belongs to the family of surface treatment.

It is at the end of the 20th century that the idea of using nanomaterials with this kind of process has been proposed [22]. Nanomaterial coatings present specific properties, quite different from micrometric one. This will be discussed in the following paragraphs.

The aircraft industry intends to increase operating temperatures of gas turbine engines to improve efficiency by developing more effective Thermal Barriers Coatings (TBCs) to protect metallic alloys. Atmospheric Plasma Spraying (APS) is also widely used to perform TBCs for aero engines, particularly for combustion chambers.

Readers who want to go further can refer to the "bible" of spraying [23].

3.2.2 Principle: From Micrometer to Nanometer Powder

Figure 3.2.1 shows the principle of the atmospheric plasma spraying which is a classical thermal spray process producing ceramic coatings. On the one hand, a control-command unit regulates various influxes (gas, water for cooling, powder and electricity). On the other hand, these different networks are connected to a plasma gun. It consists of two water-cooled electrodes, a cone-shaped cathode and a cylindrical-shaped anode, between which an electrical arc is created. A gas flow (generally, argon or gas mixtures such as argon/helium or argon/hydrogen or argon/helium/hydrogen) is injected between the electrode into the arc to transform the electrical energy into thermal and kinetic energies. As a result, the flow goes out of the anode at high temperature (10 000–15 000 K) and velocity (1000–2000 m.s^{-1}). In the created plasma plume close to the anode exit, the powder (from 5 to 60 μm diameter) to deposit is injected with a carrier gas (generally argon). Particles are heated and accelerated at different thermal kinetic states before impacting the

DOI: 10.1051/978-2-7598-2446-5.c023

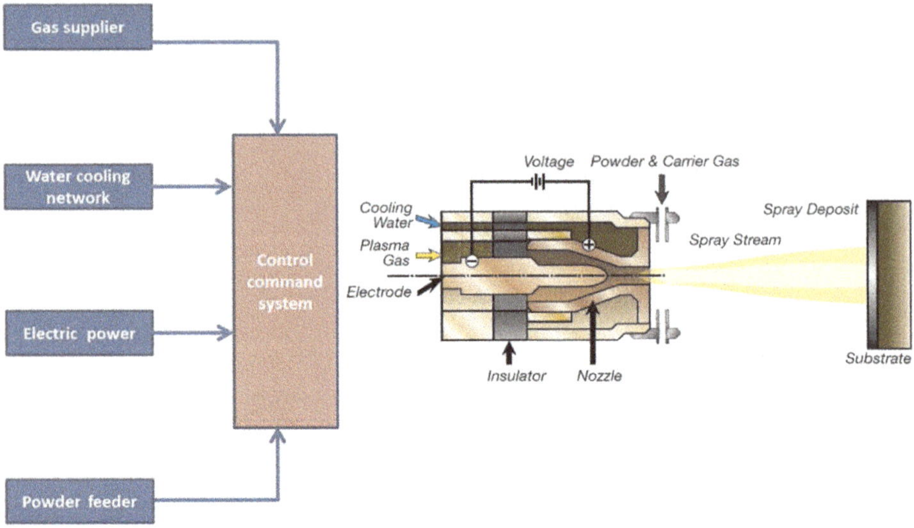

FIG. 3.2.1 – Principle of atmospheric plasma spraying system.

substrate, as disk or splashing lamellae [24, 25]. The impacting particles create the coating by piling-up successively. Coating properties depend strongly on the particles thermal kinetic states but also on the surface properties of the substrate (roughness, temperature, nature...) [26, 27].

As explained before, the idea to introduce submicron materials (<1 μm) into the plasma plume has been imagined at the end of 80s. The main issue is the momentum difference between the plasma plume and the nanomaterial injected with a carrier gas. Getting an adequate balance requires a too high gas flowrate, which deviates the gas flow. Resolving this issue implies to change the medium. The nanomaterial is thus mixed in a liquid (water, organic fluid) in order to get a suspension, a solution or a sol. Then this mixture is introduced in a pressurized feeder which carries it to the injector close to the plasma plume.

This concept implies strong modifications of the plasma/input interactions [28], as seen in figure 3.2.2. Another modification concerns the spraying distance: from around 100 mm for micrometric particle, the distance decreases to around 30–50 mm for nanometer particles.

The main points to keep in mind are the following. Suspension plasma spraying decomposes itself in several steps involving very small entities over very short time scales. After injecting the suspension, this latter undergoes first a hydrodynamic interaction fragmentation approximately during 1 μs into droplets smaller than a hundredth of a micron. When the fragmentation is completed, thermal effects can develop with the diminution of the droplet size and of its thermal inertia. During the step consisting in the vaporization of the carrier liquid, the nanoparticle behavior is not well known: do the particles leave the droplet or stay inside? Pawlowski [29] proposed a theoretical explanation. When the particles are free, they undergo

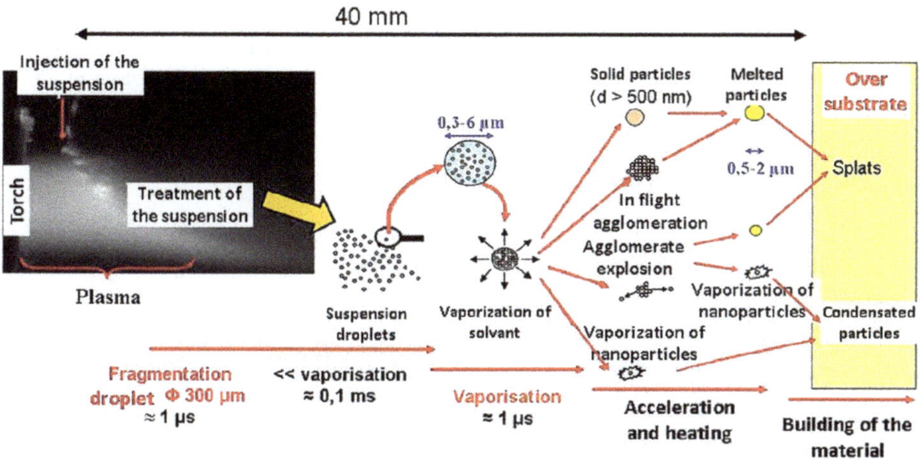

FIG. 3.2.2 – Principle of suspension plasma spraying.

thermal and kinetic effects of the gas flow, such as acceleration, heating, vaporization, in-flight agglomeration or agglomerate explosion before impacting the substrate to create the coating. This complexity is increased by the gas flow from the torch exit which presents instabilities due to the motions of the electrical arc inside the torch [30]. These movements depend on the working conditions, one of which is the gas nature. For example, a mixture containing a small percentage of hydrogen leads to the restrike mode with a saw tooth voltage signal, as seen in figure 3.2.3 [31]. Monoatomic gas mixtures, such as helium/argon, generate takeover mode, which is much more regular. According to this, the behavior and the atomization of the liquid jet depend on the time during which the interaction takes place in a long or a short plasma jet.

Moreover, although the fragmentation theory is well defined with various breakup modes depending on the Weber number [32], these investigations have been

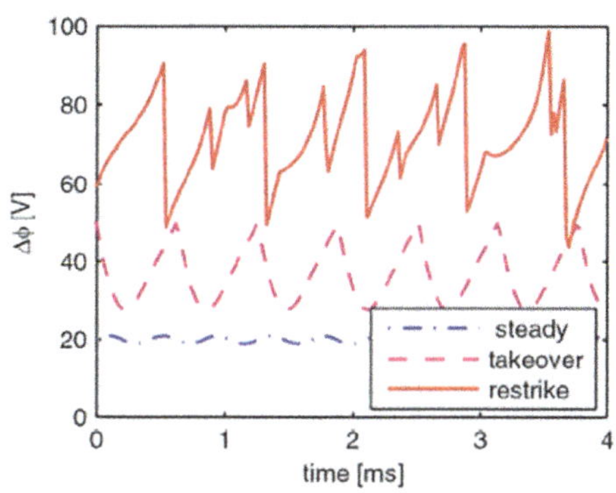

FIG. 3.2.3 – Different modes of electrical arc instabilities inside the nozzle.

carried out in steady modes when the jets do not present high velocity gradients. So another query is: what happens with high-gradient (when it comes to velocity but also temperature and transport property gradients) flows, such as plasma, which give rise to a fast evolution in Weber number during interaction with a liquid?

3.2.3 SPS: A Process to Control

Controlling the process requires investigations step by step from the gun behavior to the plasma/particles interactions and the substrate/particles interactions. Each step is discussed in the following paragraphs.

3.2.3.1 The Gun Behavior

For suspension plasma spraying, the torch remains similar to the one of conventional plasma spraying. Depending on the plasma operating parameters (gas nature and flowrate, intensity, old or new anodes, anode diameter...), the three modes described in figure 3.2.3 can be seen. They appear during the fragmentation step depending on the Weber number We, defined in equation 3.2.1:

$$We = \rho_G * (V_G - V_L)^2 * \frac{D_L^2}{\sigma} \tag{3.2.1}$$

where ρ_G and V_G represent, respectively, the density and velocity of the gas, V_L and D_L the velocity and the diameter of the liquid drops while σ is the superficial tension of the liquid.

Figure 3.2.4 presents different types of fragmentation depending on the operating plasma parameters. These experimental images [33, 34] come from a F4 OERLIKON METCO gun by laser shadowgraphy with water introduced in the flow.

The Weber number modification comes from the gas velocity interacting with the liquid shape. This behavior has been reproduced by numerical simulation in order to calculate the different as-fragmented droplet distributions: size, positions (x, y, z) and velocity (v_x, v_y, v_z).

3.2.3.2 Droplet Generation – Continuous Train of Single Droplets

With 3D CFD code using only Navier–Stokes equations, the injection of several water droplets (diameter of 360 μm and injection velocity of 22 m.s^{-1}) has been simulated in an isothermal plasma flow (argon – hydrogen mixture at 3000 K) in order to validate the numerical models. This assumption implies that all the gas properties are calculated at this temperature. Figure 3.2.5 represents the interaction between these two fluids.

From equation 3.2.1 the Weber number is integrated over the interval [100, 350], depending on the gas velocity and on the position in the flow, which defines a sheet stripping breaking mode. This mode is shown in figure 3.2.5. It reveals the appearance of filaments at the droplet surface before breaking into numerous smaller droplets.

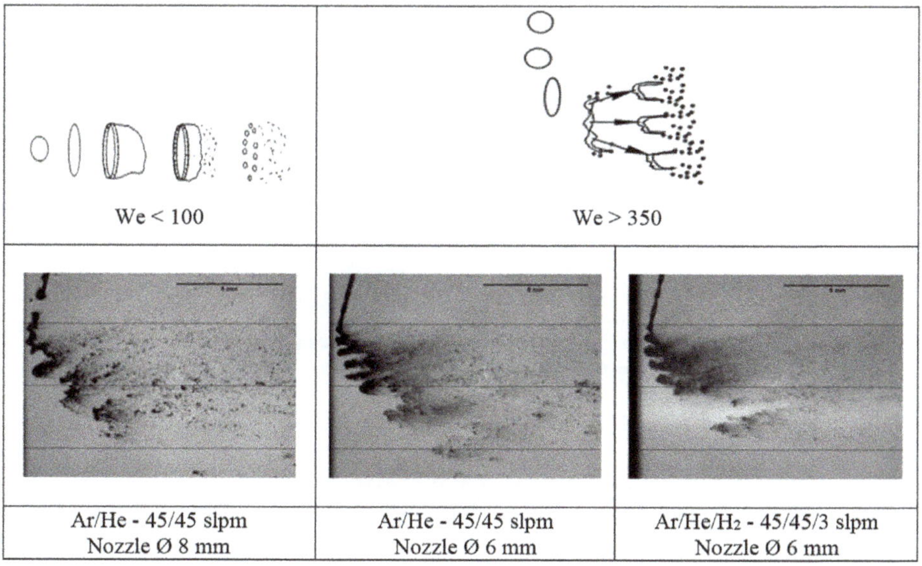

FIG. 3.2.4 – Different kinds of liquid fragmentation depending on the operating plasma parameters.

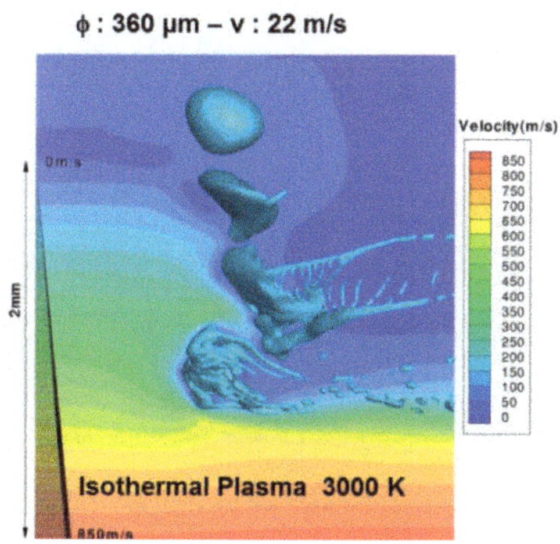

FIG. 3.2.5 – Injection of several liquid water droplets in an isothermal plasma jet.

Next step was the simulation of the same water injection in a realistic plasma flow (figure 3.2.6).

In this case, high temperature gradients imply high transport coefficients and density gradients of the gas. As a result, the filaments are rapidly integrated into smaller droplets leading to a quicker breaking of the water jet.

Shadowgraphy images corroborate this behavior as seen in figure 3.2.7.

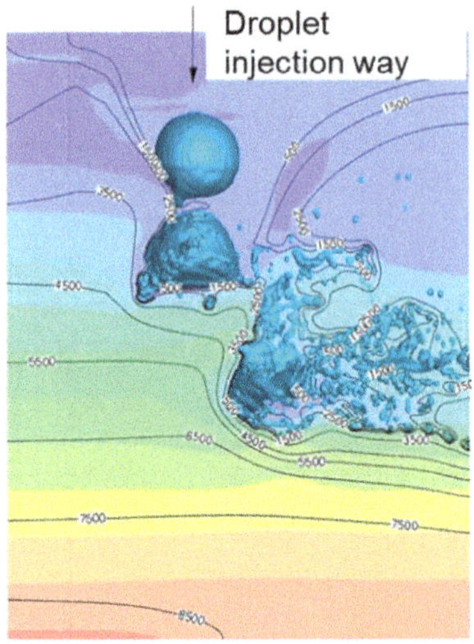

FIG. 3.2.6 – Injection of several liquid water droplets in a realistic plasma jet.

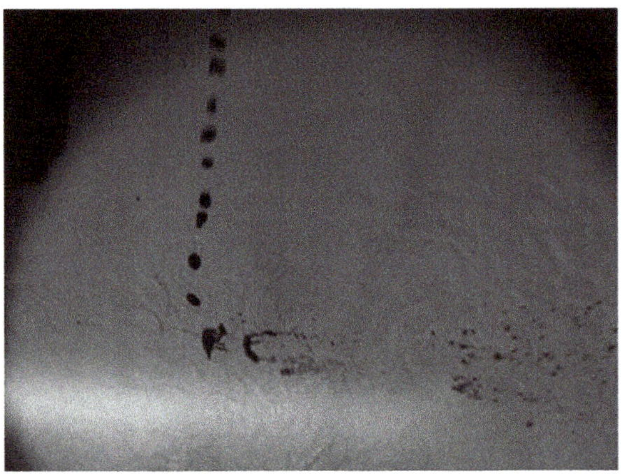

FIG. 3.2.7 – Injection of several liquid water droplets in a realistic plasma jet.

3.2.3.3 Droplet Generation – Continuous Injection of a Liquid Shape

After this validation, similar simulations have been carried out with a continuous water shape injection in an unstable plasma flow of hydrogen/argon [35]. Figure 3.2.8 presents the evolution of the liquid shape in interaction with the plasma flow. A similar behavior of the shape interacting with the flow is present: appearance of filaments at the shape surface, growth and disintegration of the filaments into

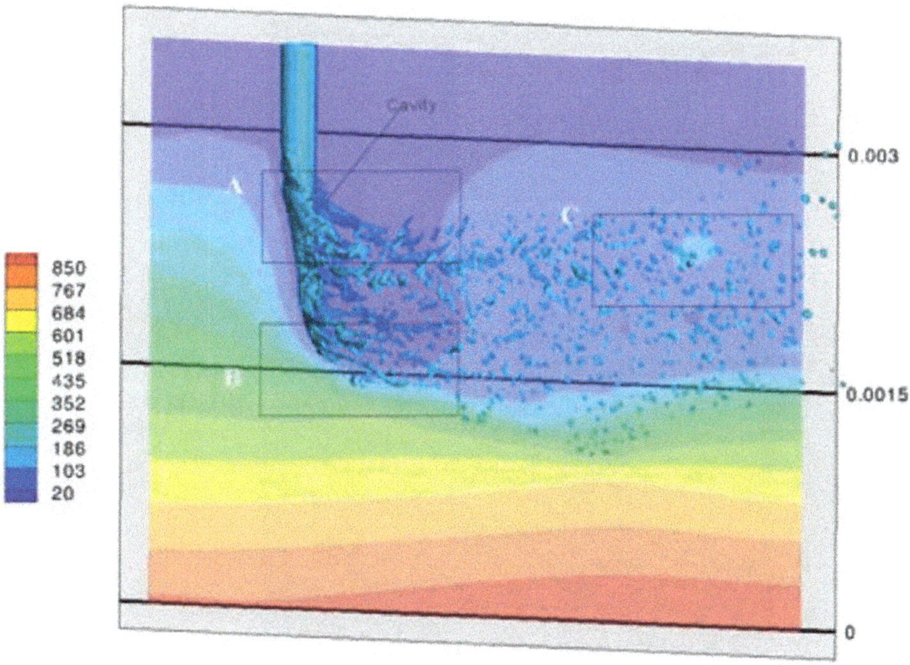

FIG. 3.2.8 – Continuous injection of a liquid in a realistic plasma jet by numerical simulation.

smaller droplets in a short time. Three zones (A, B and C) can be defined from figure 3.2.9.

The rapid growth of the Weber number in the first layer was responsible for the cavity between the threads visible in zone A. This is characteristic of the bag breakup mode of fragmentation. The variation in the Weber number in one direction and then the other in a very short space led to this mode with a mixture of fragmentations. In zone B, where the Weber number did not move very rapidly, the filaments had the possibility to spread before being destroyed. The values described a mix of bag and stamen breakup and sheet stripping. Zone C shows a large variety of droplet shapes according to the mix of breakup modes, while the droplet diameters seemed more homogeneous close to the 0.0015 m radius line.

Consequently, the fragmentation is assumed to end 10 mm from the torch exit and close to the centerline. At this position, the liquid should be completely fragmented into small droplets following the size distribution shown in figure 3.2.10.

Figure 3.2.10 shows that the main population of the fragmented droplets (defined by their near-spherical shape) has a diameter close to 20 μm regardless of the time [36]. Still-existing ligaments are not included in these distributions. The dwell time of 20-μm droplet in the surrounding atmosphere is estimated to be less than 100 μs before impacting the substrate. The following heat treatment analysis is based on this diameter.

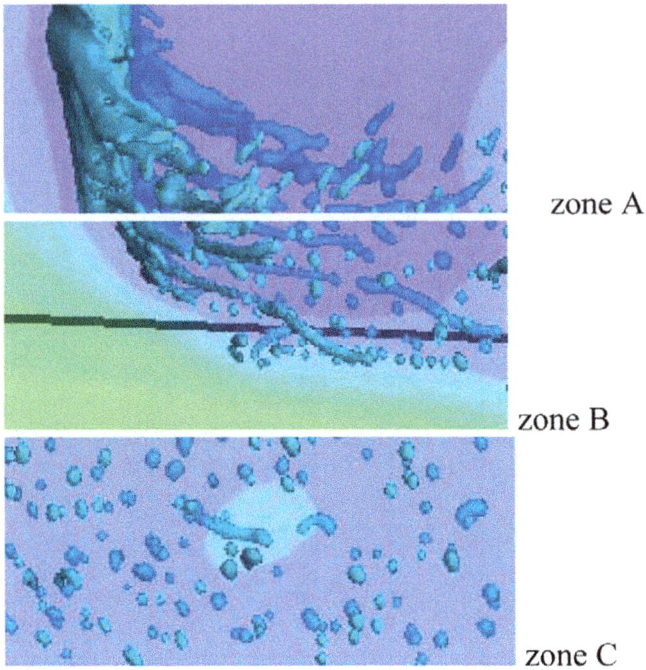

FIG. 3.2.9 – Details of three zones (A, B and C) from previous image at 185 μs.

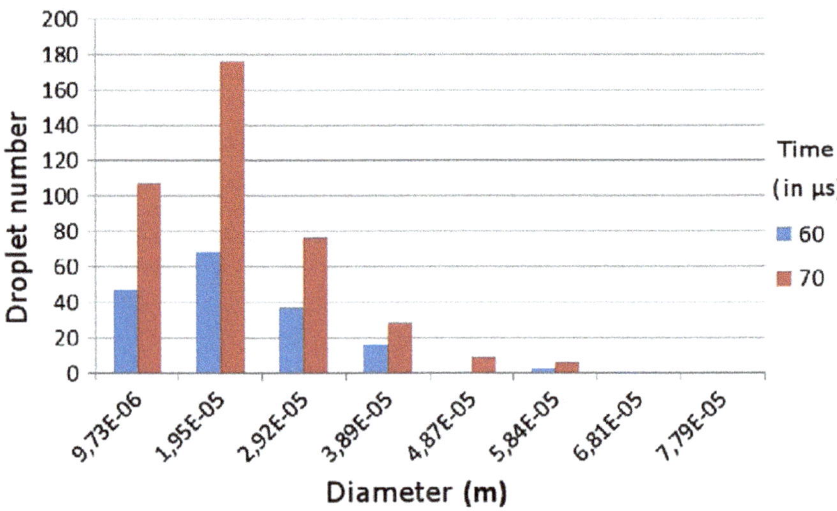

FIG. 3.2.10 – Size distribution of the droplets at different times in the unsteady plasma flow.

3.2.3.4 Heat Treatment of the Droplets

Analysis of the heat treatment undergone by the droplets during the projection was published in [37] including radiation and convection assuming a spherical droplet. It shows two things:

- Radiation. It appears that the available radiation energy of the plasma is not high enough to heat and completely evaporate the droplet. Radiation is only one activation mode to trigger the process. However, Basu *et al.* explained [38] that it takes more than 200 μs to start the significant vaporization of the fuel droplet. In SPS, the dwell time is only 100 μs for the 20 μm water droplet before reaching the substrate. Consequently, radiation is not the main mechanism responsible for heating the droplet.
- Convection. If the droplet remains spherical during its flight, heating by convection cannot be considered to be the main mechanism due to the steam sphere around it. Calculations have determined a spherical layer of steam around the droplet of 8.3 μm thick.

In consequence, the droplets have to undergo a second breaking. This has been detailed by simulation in the plasma flow of two different singular droplet sizes, a smaller one (10 μm) and a bigger one (60 μm), as seen in figures 3.2.11 and 3.2.12.

The smaller droplets flatten completely in a very short time increasing the exchange surface between them and the flow, thus increasing the heat transfer by convection. The bigger ones present a different behavior due to their mass and size: they deform in a peanut shape with a singular shrinkage leading to a second breaking into two different sizes droplets. If this phenomenon is slower than the previous one for smaller droplets, the secondary breaking provides a full evaporation of droplets before impact.

3.2.3.5 Before Impacting

After evaporation of the liquid, the nanoparticles can flow in the plasma surrounding atmosphere in the same way than the micro-particles with acceleration and heat transfer.

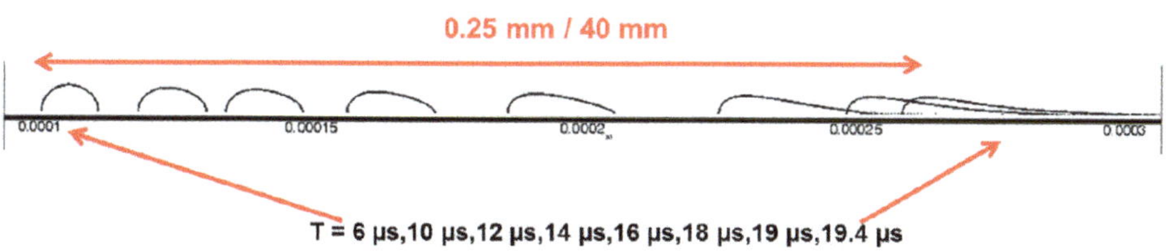

FIG. 3.2.11 – Deformation and second breaking of smaller droplet in the flow.

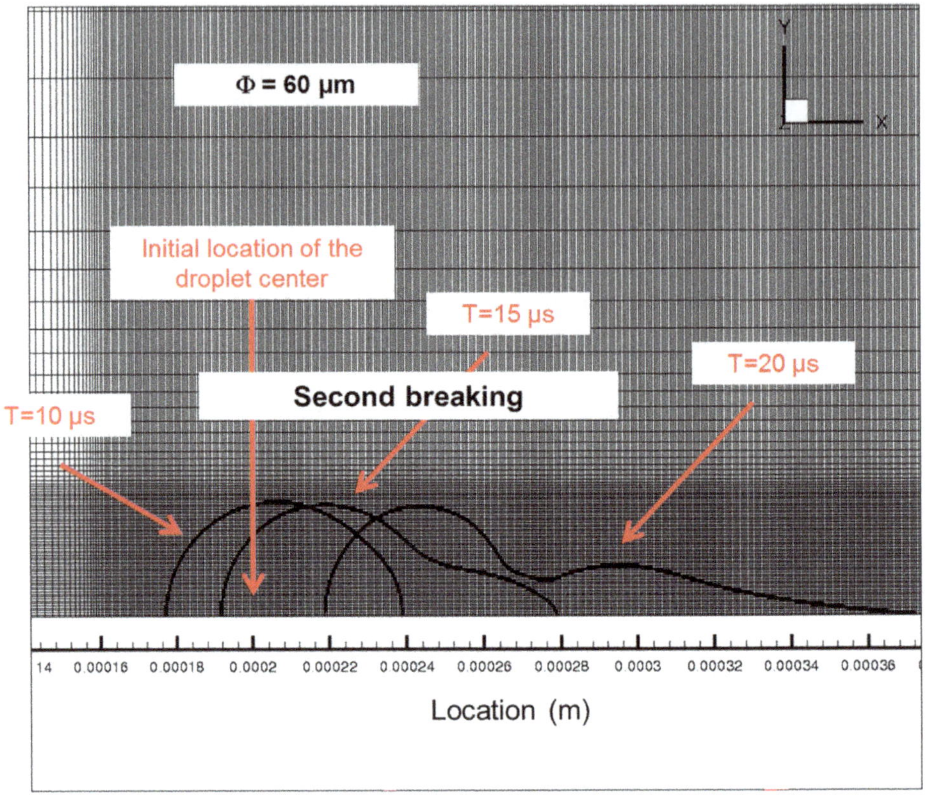

FIG. 3.2.12 – Deformation and second breaking of bigger droplet in the flow.

3.2.4 Build-Up of the Coating

The SPS process allows building of coatings with a columnar microstructure, which is quite different from APS deposition. K. Van Every *et al.* [39] proposed a build-up mechanism to explain the formation of columns by SPS mainly attributed to deviation of small particle at the vicinity of the substrate, while B. Bernard *et al.* [40, 41] identified key parameters to control the morphology of columns based on three types of zirconia coatings (figure 3.2.13). The column compaction is the main observed variation. Coatings A and B show typical SPS columnar structures with inter-columnar voids. Coating C presents a compact columnar structure where inter-columnar voids tend to disappear. The SPS microstructural modulation and the impact of spraying parameters were widely described in [41] and were only summarized here.

Coating A made using a lower transverse speed than coating B has an enlarged column size distribution, particularly for higher coating thicknesses. The lower transverse speed induced a higher amount of deposited material at each pass of the torch in front of the substrate. Deposition velocity was calculated at 1.15 μm/pass

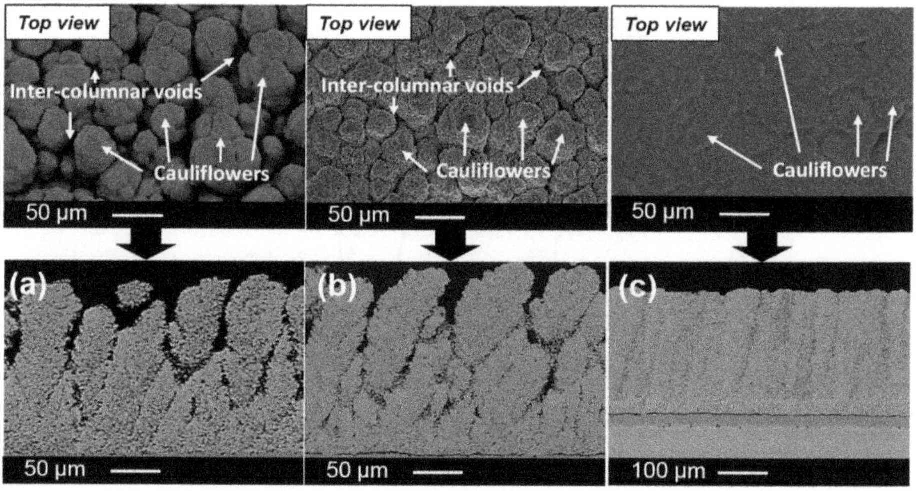

FIG. 3.2.13 – SEM pictures (cross sections and top views) of SPS coatings A (a), B (b) and C (c).

compared to 0.70 μm/pass for coating B. It allowed for coating B a longer available time for column organization leading to a better compaction and in fine attenuated inter-columnar voids.

The compact structure of coating C in figure 3.2.13c could be attributed to the use of a lower plasma enthalpy and a higher suspension feed rate. It was proposed that the lower enthalpy of plasma was not able to perfectly melt large particles leading to an increase in proportion of well-molten fine particle. These little particles were more sensitive to deviation by the plasma flow at the vicinity of the substrate. It led to a more pronounced lateral development of columns. At the end, it induced overlapping of columns, accelerated by the higher suspension feed rate.

B. Bernard *et al.* [40] proposed to divide the growth rate in two contributions, one transverse and the other parallel to the substrate surface. Bigger particles contribute principally to the transverse growth rate because they are not deviated by the flow when impacting the substrate. Parallel growth rate results from the smaller particles, which can be deviated close to the substrate. Therefore, the column grows in the shape of a tie (figure 3.2.14). This assumption can be validated by the evolution of the mean column diameter *versus* the coating thickness (figure 3.2.15).

3.2.5 Potential Applications for Aeronautics

In aeronautics [42] and over many years, thermal barrier coatings (TBCs) were studied in order to improve gas turbine efficiency. The increase in the thermal insulation performances of the ceramic top layer of TBCs, mainly composed of yttria-stabilized zirconia (YSZ), was identified as a main issue for the increase in

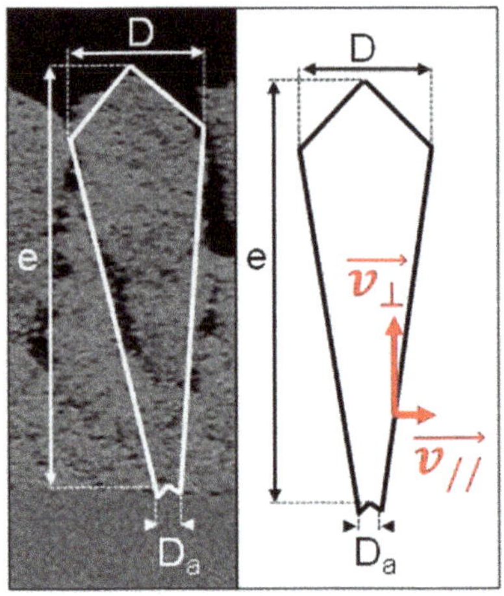

FIG. 3.2.14 – Tie shape column growing for coating build-up.

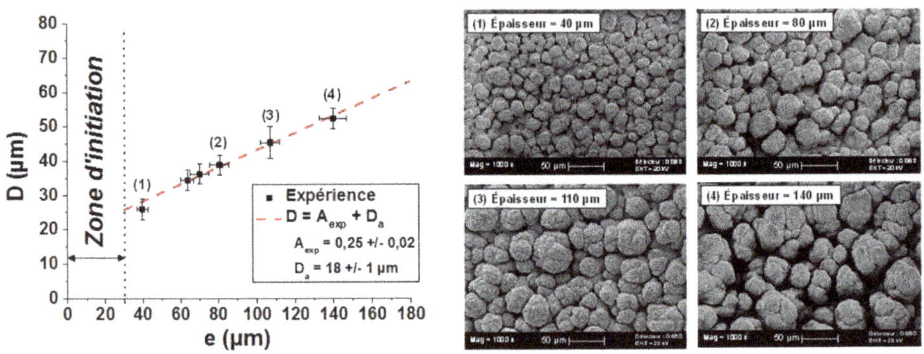

FIG. 3.2.15 – Evolution of the mean column diameter versus the coating thickness.

operating temperatures of the next generation of turbine engines [43]. Since the 80s, the most efficient deposition process regarding operating conditions of high-pressure parts (mainly thermal cyclic resistance) is electron beam physical vapor deposition (EBPVD). A high-energy electron beam is used to vaporize an ingot of YSZ. Then, vapors condense on the surface of the metallic substrate leading to the coating growth. A highly thermal-compliant columnar structure is obtained, granting a high lifetime.

On the other hand, atmospheric plasma spraying (APS), widely used before the 80s, consists of injecting powder through a plasma jet in order to be melted and accelerated. Then, molten particles are spread onto a metallic substrate as individual splats, stacking to form a lamellar microstructure. This typical microstructure favors low thermal conductivity. It also causes thermal expansion mismatch between the ceramic and metallic part of the TBC, leading to a relatively low lifetime on the high pressure components of the turbine engine, like mobile blades, also considered as a main issue [44, 45]. However, this kind of coating is successfully and commonly applied on combustion chamber components, which is not exposed to similar thermal mismatch [46].

B. Bernard *et al.* [42] have explored the microstructure of SPS coatings, such as the previous one in order to diminish the thermal conductivity and to enhance the lifetime of the YSZ deposit. The main conclusions are reported here.

For comparison, two optimized SPS columnar structures and a typical YSZ columnar EB-PVD were characterized. The thermal conductivity of the SPS YSZ coating was found to be lower than 1 $W.m^{-1}\ K^{-1}$ from room temperature up to 1100 °C regardless of the column morphology, while the EB-PVD coating was recorded at 1.3 $W.m^{-1}\ K^{-1}$ at 1100 °C. These kinds of SPS columnar structures meet the thermal insulation requirements of the next generation of gas turbine engines.

In addition, the lifetime estimated by thermal cycling fatigue tests appears promising for a TBC application regardless of the microstructure of SPS coatings, or the bond coat used on AM1 substrate (β-(Ni,Pt)Al) and platinum-diffused one (γ-Ni/γ'-Ni_3Al) commonly applied for EB-PVD coating.

However, a superior thermal cycling resistance was mainly observed for SPS columnar structure with well-separated columns compared to compact ones. The thermal compliance offered by inter-columnar voids is more efficient than vertical cracks that appeared *in situ* during thermal cycling fatigue tests for the compact columnar structure. Thermal cyclic lifetime of YSZ SPS columnar structure equals EB-PVD TBC system and are suitable for the next generation of gas turbine.

The preparation of the bond coat surface was also a relevant parameter for the optimization of the lifetime. A high substrate roughness induced a lower organization of the YSZ SPS top coat columns leading to lower thermal compliance.

3.2.6 Conclusion

Suspension plasma spraying (SPS) is identified as promising for the enhancement of thermal barrier coating (TBC) systems used in gas turbines. Particularly, the emerging columnar microstructure enabled by the SPS process is likely to bring about an interesting TBC lifetime. At the same time, the SPS process opens the way to a decrease in thermal conductivity, one of the main issues for the next generation of gas turbines, compared to the state-of-the-art deposition technique, so-called electron beam physical vapor deposition.

The interest of such process comes from it characteristics as an industrial set-up with a high efficient spraying rate.

This paper has described different steps to control in order to manage coating deposition, from the gun behavior to nanomaterial injection into the plasma flow with the different interaction steps.

It took a long way to introduce APS process in mass-scale industry. With only twenty years of investigations in academic institutions, this process presents now high-interest in industry, especially for aeronautics.

3.3 Physical Vapor Deposition: Principles, Ionized PVD and Examples of Application

Frédéric SANCHETTE[1,2] and Alain BILLARD[3]

[1] *ICD LASMIS, Université de Technologie de Troyes, Antenne de Nogent, Pôle Technologique de Haute-Champagne, 52800 Nogent, France*
[2] *Nogent International Center for CVD Innovation, LRC CEA-ICD LASMIS, UTT, Antenne de Nogent-52, Pôle Technologique de Haute-Champagne, 52800 Nogent, France*
[3] *Institut FEMTO-ST, CNRS, UTBM, Univ. Bourgogne Franche-Comté, Site de Montbéliard, 90010 Belfort Cedex, France*

3.3.1 Introduction

The physical vapor deposition methods (PVD Physical Vapor Deposition) are a set of coating synthesis techniques of metallurgical or ceramic films for various applications as in mechanics, optics, microelectronics, chemical and aeronautical industries, etc... In a general way, coatings are deposited at low pressure (<10 Pa) following mainly three stages: creation of a metallic vapor from a source (or target), its transport in the target – substrate space and its condensation on the surface of a substrate to be coated. If the name PVD includes a large quantity of processes, it is advantageous to classify them into three main categories according to the method of obtaining the metallic vapor: by thermal effect; evaporation or cathodic arc technology (in this case, electron emission results from the combination of a field effect and a thermionic effect), or by mechanical effect; sputtering of a solid source (magnetron sputtering). PVD technologies allow obtaining a high control of the physico-chemical properties of the deposited coatings (density, roughness, structure, microstructure, chemical composition...).

Using conventional direct current magnetron sputtering, the majority of the sputtered particles remains mainly neutral, which prevents the control of energy and direction of the species impinging the surface of the growing film. For many applications and, for example, in the case of integrated circuits manufacturing, it is better to get a flow of particles, coming from the plasma, perpendicular to the surface of the substrate, favoring the growth of compact films.

Concerning barrier coatings for protection of components against severe environments, ionized metallic vapor enhances the ability to independently control ion flux and ion energy at the substrate. It also allows ion cleaning with ionized particles to be deposited, together with better control of the *in situ* ion bombardment of the growing film and better thickness homogeneity on biased substrates with complex morphologies. Moreover, ionized metallic species in the vicinity of a biased substrate improve control of thin film structure and properties compared to conventional DC (Direct Current) magnetron sputtering for instance.

DOI: 10.1051/978-2-7598-2446-5.c033

Nowadays, several techniques make it possible to produce plasmas characterized by a large fraction of ionized species. One of these techniques is cathodic arc deposition, known to produce dense coatings. The main problem of this technology, which is widely used to coat machining and forming tools, lies in the creation and deposition of micro droplets reducing the quality of the films. Other sources make it possible to ionize the plasma to a greater extent (typically a few tens of percent) compared to a conventional magnetron discharge with direct current (<1%). It is the case, for example, of magnetron discharges assisted by RF loop or microwave.

Concerning magnetron sputtering, it is also possible to increase the density of plasma by increasing the average power applied to the cathode. The problem is then the target's heating. A recent solution to this problem lies in the application of high power pulses with low duty cycle (HiPIMS: High Power Impulse Magnetron Sputtering). While the target's heating is limited by the average power applied to it, the peak power applied during short pulses can be very high. Applying this technique to a magnetron cathode achieves ionization rates above fifty per cent.

The PVD main principles are first presented, including basic details on plasmas as well as the different methods to obtain a metallic vapor. Then conventional magnetron sputtering, HiPIMS technology and Cathodic arc deposition are successively briefly described. Examples of deposited materials for protection of components against extreme environments are also given.

3.3.2 PVD Main Principles

3.3.2.1 Glow Discharge

Electric glow discharge is a type of plasma formed by passing a current between two electrodes biased in the range 100 V to several kV through a gas, usually argon or another noble gas, under reduced pressure. The glow discharge owes its name to the fact that plasma is luminous, the luminosity being produced because the electrons gain sufficient energy to generate visible light by excitation or ionization collisions with the residual gas atoms, which generate photons. By abuse of language, a partially or fully ionized gas is called plasma, which is electrically neutral and conductive of electricity.

3.3.2.2 PVD – 3 Steps

Figure 3.3.1 shows the 3 main steps of PVD processes. It is possible to split PVD processes into 3 distinct stages,vapor generation, transfer at low pressure and condensation on the substrates. The different PVD processes are distinguished by means of obtaining the vapor of the material to be deposited,thermal evaporation (conventional or from a high energy electron beam), magnetron sputtering (conventional or HiPIMS) and cathodic arc deposition. In this chapter, only magnetron sputtering (conventional or HiPIMS) and cathode arc deposition are described.

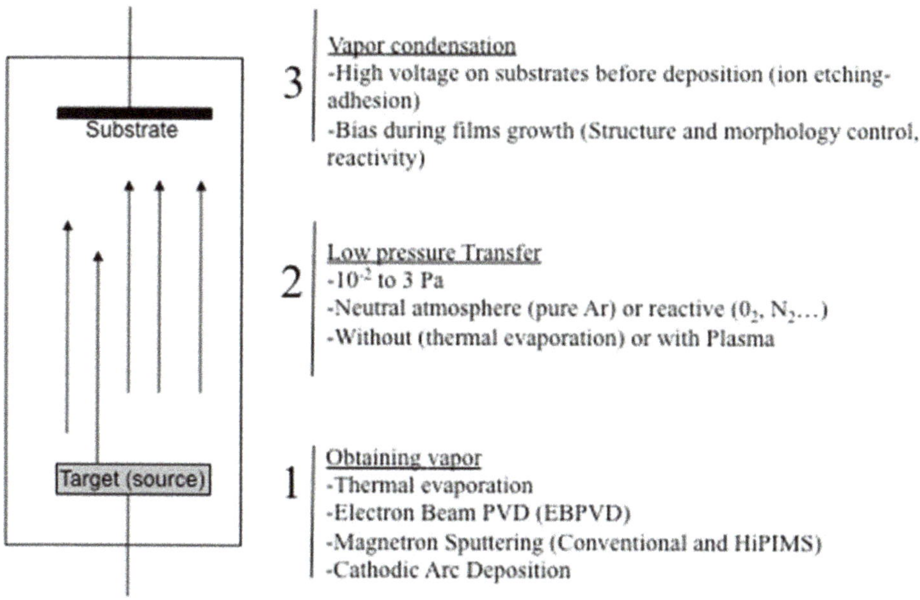

FIG. 3.3.1 – 3 steps of PVD process.

3.3.3 Standard Direct Current (DC) Magnetron Sputtering (MS)

An abnormal glow discharge is produced by applying a voltage in the order of 1 –3 kV between the cathode (the target, material to be deposited) and the anode (the walls of the grounded reactor) in a rarefied atmosphere of argon at a pressure reduced generally between 0.1 and 10 Pa (figure 3.3.2).

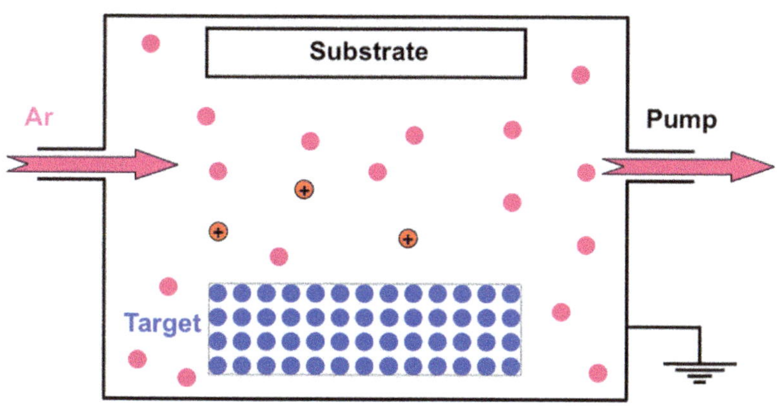

FIG. 3.3.2 – Schematic diagram of the principle of magnetron sputtering deposition process.

As a result, partially ionized plasma composed of electrons, ions and neutral species is formed between the two electrodes. Positive ions are accelerated towards the negatively polarized target under the effect of an electric field and these ions bombard the target's surface. Several mechanisms can occur on the surface of the target (figure 3.3.3):

- Reflection of the incident ion neutralized by charge transfer (1 of figure 3.3.3).
- Implantation of the incident ion (2 of figure 3.3.3).
- Ejection of species from the target by momentum transfer (3 of figure 3.3.3).
- Emission of secondary electrons which ensure the self-maintenance of the discharge (4 of figure 3.3.3).

The ejected atoms acquire enough energy to cross plasma and finally condense on the first surface encountered. The growth of a thin layer takes place in many stages:

- Adsorption of atoms on the substrate surface.
- Interaction between deposited species and/or those of the substrate forming a stable chemical bond.
- Nucleation.
- Volume growth.
- Diffusion of atoms in volume in the layer and/or in the substrate.

Close to the surface, the incident atom is subjected to a potential energy formed by the atoms of the surface. The interactions involved are Van der Waals interactions. This atom diffuses on the substrate surface during a time that depends on the interactions between the adatoms, the substrate, the temperature and the deposition rate. During its diffusion, it chemically binds with the substrate on a favorable site (Chemisorption). This type of site is called nucleation site. When the adatom arrives on one of the sites already occupied, it binds to the nucleus in the course of formation. Nucleation of nuclei is preferably carried out at the level of the defects present on the substrate surface.

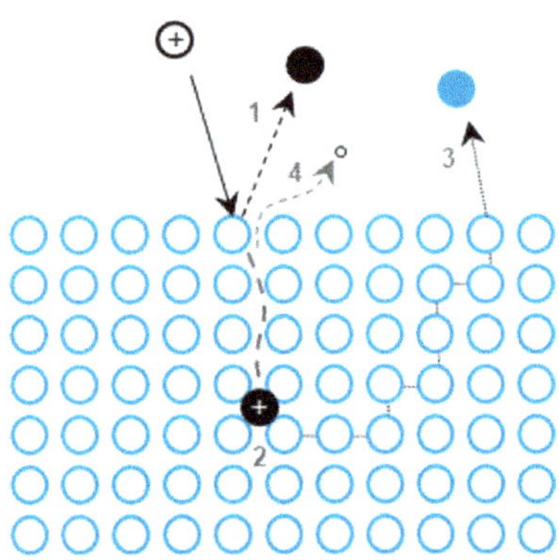

FIG. 3.3.3 – Main mechanisms resulting from the interaction of an energetic ion and a surface.

The growth of films deposited by PVD methods can be divided into different stages of nucleation and growth. These processes are determined by thermodynamic parameters as well as kinetic parameters. The actual growth mode is determined mostly by the conditions such as deposition rate, temperature and pressure.

A thermodynamic criterion developed by Bauer [47] for the growth mode in equilibrium conditions is given by the following equation:

$$\Delta\sigma = \sigma_f + \sigma_i - \sigma_s$$

where σ_f is the free enthalpy of the film surface, σ_i the free energy of the film-substrate interface and σ_s the free energy of the substrate surface.

It is possible to distinguish three growth modes from this criterion (figure 3.3.4).

- If $\Delta\sigma \leq 0$, the atoms are bonded more strongly to the substrate than between each others and the films tend to extend over the substrate to minimize total energy. In this condition, a full recovery of the substrate by the films is favorable and the growth mode of Frank van der Merwe (layer-by-layer type) is observed (figure 3.3.4a).
- If $\Delta\sigma \geq 0$, the atoms are bonded to each other more strongly than they are bound to substrate. In this case, the Volmer-Weber growth mode (of the island type) is obtained (figure 3.3.4b).
- There is a third possibility: $\Delta\sigma$ can change from a negative value to a positive one depending on the thickness of the film. This comes from the strain energy, proportional to the thickness of the film, which can bring additional energy to $\Delta\sigma$. In in this case, the growth mode of Stranski–Krastanov (mixed type) is obtained (figure 3.3.4b).

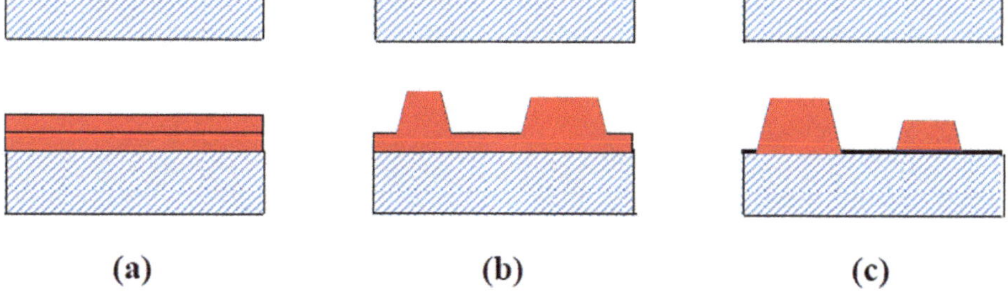

FIG. 3.3.4 – Diagram of the three growth modes of thin films: Morphology of a film by the mechanisms of (a) Frank van der Merwe, (b) Stranski–Krastanov and (c) Volmer–Weber.

In most practical cases, the growth of the films takes place in conditions out of thermodynamic equilibrium, and is strongly influenced by kinetic processes. Therefore, the final macroscopic state of the film is not necessarily the most stable thermodynamically but the most favorable from the kinetics point of view. Morphology of thin layers strongly depends on the deposition conditions and is the subject of a so-called structure zone model (the SZMs). The first model of structure zone for metallic thin films was proposed by Thornton in 1977 [48]. In this model, the most important deposition parameters are the working pressure and the homologous temperature (the ratio of the substrate temperature T_s over the melting temperature T_m of the deposited material was used: T_s/T_m). The SZM's concept was extended to a more generalized situation by Anders in 2010 [48], which considers an energy-homologous temperature diagram.

During deposition, the substrate temperature influences the mobility of the surface adatoms. The increase in the substrate temperature leads to increase in the mobility of species adsorbed on the substrate surface and their bulk diffusion, which leads to densify the deposited layer. The deposition pressure impacts the energy and impinging angle of the incident particles. Decreasing the deposition pressure has a similar effect as increasing the deposition temperature. Thus, low working pressure and high deposition temperature promote the growth of compact layers (figure 3.3.5). This compact morphology is often associated with compressive stresses. According to Hoffman's model [49], stress evolution with changing the deposition pressure is related to the gap size involved by the incident particles energy impinging on the growing film. In fact, the inter-columnar distance decreases with the increase in the bombardment energy of the film surface, related to the decrease in deposition pressure. As a result, the attractive interatomic forces increase, resulting in an increase in tensile stress values. At low deposition pressure, the film has a dense structure, which leads to a compressive macrostress state in the film as a consequence of repulsive forces and the film atoms implantation by knock-on process [49]. The rise in compressive stress with increasing negative bias voltage could be explained by film densification and the increase in lattice defect densities due to the increase in ion bombardment energy [50].

FIG. 3.3.5 – TiO_2 coatings deposited by magnetron sputtering at (a) 0.3 Pa and (b) 1.5 Pa.

Deposition of compounds as oxides, carbides or nitrides can be achieved by magnetron sputtering of metal targets in reactive plasmas consisting of a mixture of argon and chemically active gas (O_2, CH_4, N_2...). Physical and chemical mechanisms of reactive magnetron sputtering are developed in [51].

3.3.4 From DCMS to High Power Impulse Magnetron Sputtering (HiPIMS)

High power impulse magnetron sputtering (HiPIMS) technology is a kind of ionized physical vapor deposition technology based on magnetron sputtering developed in recent years [52]. It exploits high power pulses as the power supply mode to provide peak power density up to the order of $kW.cm^{-2}$ for magnetron sputtering cathodes [53], while simultaneously, current density up to 3–4 $A.cm^{-2}$ is obtained on the target's surface (figure 3.3.6). A frequency in the range of 50–5000 Hz and a low duty cycle in the range of 0.5%–5% are applied [54, 55] to ensure that the magnetron sputtering cathode does not overheat since the average power is comparable to that of conventional direct current magnetron sputtering. This mode of power supply results in high energy density and high ionization fraction of the sputtered particles [56, 57], which facilitates a more effective regulation of the film growth by controlling the energy and direction of the sputtered species. It is a significant advantage over conventional DC magnetron sputtering where the sputtered vapor consists mainly of neutral species [58, 59]. Moreover, the pulsing of the discharge also allows for much greater flexibility in the HiPIMS process in film preparation due to additional control parameters such as pulse width, duty cycle, and pulse voltage. Based on the above discharge characteristics, HiPIMS technology has become a hot spot in international research and the studies on (i) the characteristics of high power impulse discharge [60, 61], (ii) the physical mechanism of plasma [62, 63], and (iii) the deposited thin films [64, 65] has progressed rapidly. The high ionization rates (few tens %) allow deposition of conform coating with good thickness homogeneity on biased complex substrates. Moreover, metallic ions can be used for substrates etching before the deposition stage in order to improve adhesion. Last but not least, metallic ions allow the control of morphology, microstructure, structure and texture (preferential orientations for instance) for optimizing functional properties of films (figure 3.3.7). It should be noted that target ions M^+ may be easily attracted back to the cathode direction and then accelerated across the high voltage sheath to take part in the re-sputtering mechanism. This re-sputtering caused by back attracted target ions M^+ promotes an increase in the ionization rate but impairs the deposition rate. Thus, deposition rates are much lower with HiPIMS than with conventional DC magnetron sputtering.

3.3.5 Cathodic Arc Deposition (CAD)

The vacuum Cathodic Arc Deposition (CAD) [66] process is now widely used on an industrial scale to prepare protective hard coatings on cutting tools, metal moulds, etc... It has been recognized that high ionization levels of cathodic arc discharges

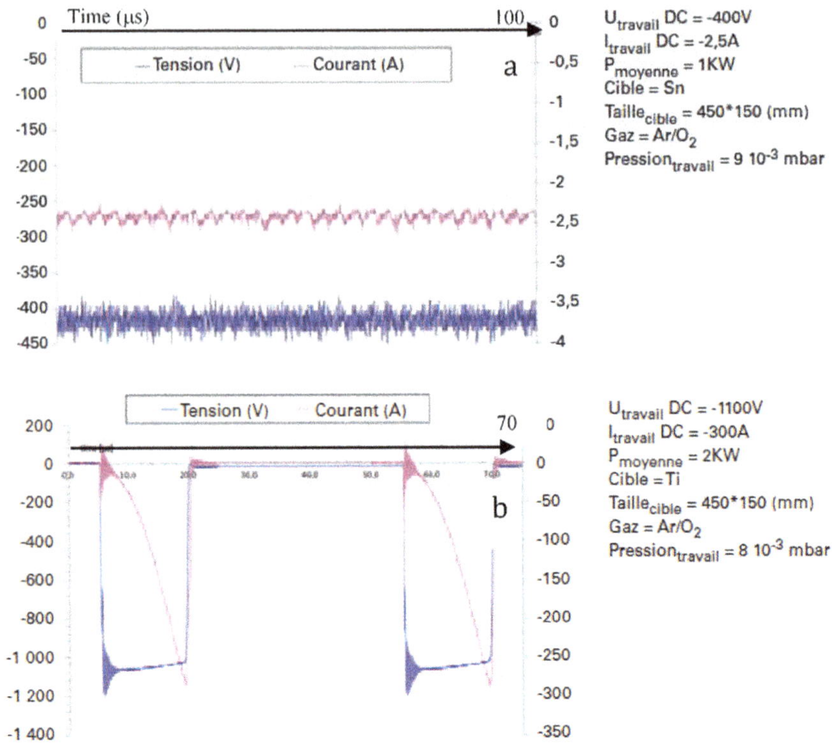

FIG. 3.3.6 – Two modes of power supply of a magnetron cathode, from reference [52]. (a) Cathode (target) voltage (from 400 to 600 V) and direct current. (b) HiPIMS: pulses of high peak on the cathode (250 kW here). In some cases the peak power can reach a few megawatts.

and high ion energy can provide advantages, as for HiPIMS technology, such as enhanced adhesion required for mechanical applications involving high loads. CAD allows deposition of a wide range of hard compounds as nitrides or carbonitrides. However, CAD is also often associated to microdroplet generation that degrades the surface roughness of coatings. The recent trend is to develop advanced nanostructured hard coatings in order to enhance properties as hardness, toughness, and oxidation resistance. This paragraph recalls the basic aspects of CAD technology.

The cathodic arc discharges are generally produced from cold cathodes. They are self-sustained at low-pressure (10^{-3} Pa typically) in the vapor from the cathode, which is the material to be deposited, and does not require any carrier gas. The geometry and configuration of cold cathodes allow vertical positioning in the deposition chamber to coat parts with complex geometry. In the case of cold cathodes, arc is called discontinuous and electron emission results from the

	Pulsed DC	**HiPIMS**
	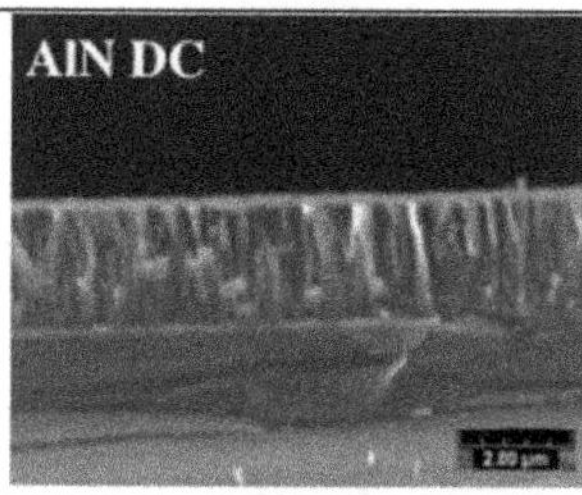	
H (GPa)	195 ± 6	252 ± 9
E (GPa)	8.2 ± 0.5	17.9 ± 1.0
e (µm)	2.1	2.5

FIG. 3.3.7 – Cross sectional SEM images of AlN films deposited by conventional DC magnetron sputtering and by HiPIMS. Associated mechanical properties: hardness and Young's modulus.

combination of a field effect and a thermionic effect. This process leads to ion energy in the range 10–100 eV. Like HiPIMS, ionized metallic vapor, created in cathodic arc discharges, enhances the ability to independently control ion flux and ion energy at the substrate. It also allows ion cleaning with ionized particles to be deposited, together with better control of the *in situ* ion bombardment of the growing film and better thickness homogeneity on biased substrates with complex morphologies. Moreover, ionized metallic particles in the vicinity of a biased substrate improve control of thin film structure and properties compared to conventional magnetron sputtering for instance.

Phenomena, associated with the material's emission by discontinuous cathodic arc process, are often qualitatively described by a stationary or near stationary model. Figure 3.3.8 describes schematically the physical processes involved according to this interpretation. The breaking contact between the trigger and the cathode surface leads to the initiation of an arc associated with a spot that is a source of intense emission of electrons. Inelastic collisions between electrons and atoms of the material constituting the source (target) lead to the formation of positive ions in a region near the surface of this source. This positive space charge comes from the faster flow of electrons, which are more mobile than ions. The cloud of ions plays the role of a potential barrier of a few tens of volts, which slows down electrons and accelerates ions. The electric field generated by this barrier (of the order of 10^{10} V/m), amplifies electron emission owing to a decrease in their work output. Moreover, some ions (about 10%) are accelerated towards the cathode, contributing to the creation of new cathode spots. The other ions, strongly accelerated from the cloud, allow the synthesis of coatings [67–70] that grows under the effect of ion bombardment of the biased substrates. The material flux emitted by explosion or by evaporation is composed of ions, neutral atoms, electrons and

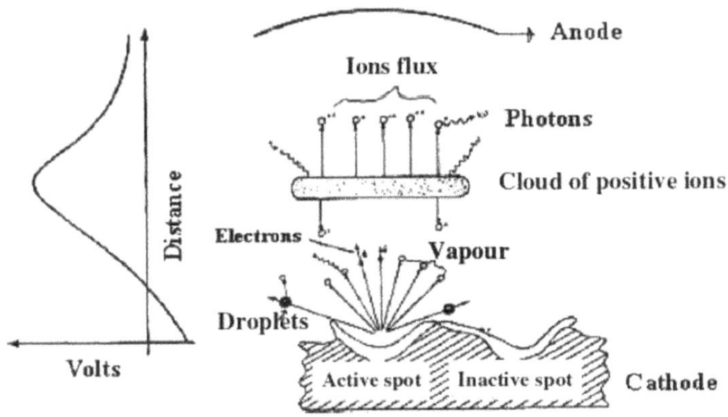

FIG. 3.3.8 – Evaporation process in a quasi-stationary configuration and potential barrier generated by creating a positive charge space near the cathode surface, from reference [69].

microdroplets of the material to be deposited. The vapor's density near the surface is between 10^{17} and 10^{19} cm^{-3} [71, 72].

Microdroplets, or macroparticles, have sizes ranging between 0.1 and 10 μm and are mainly emitted within [0, 30°] angle from the cathode's surface with speeds around 1700 m.s^{-1}. These microdroplets are the direct consequence of either explosion of the matter, or of high plasma pressures within a cathode spot on the locally molten material. These microdroplets can significantly increase the density of defects (figure 3.3.9).

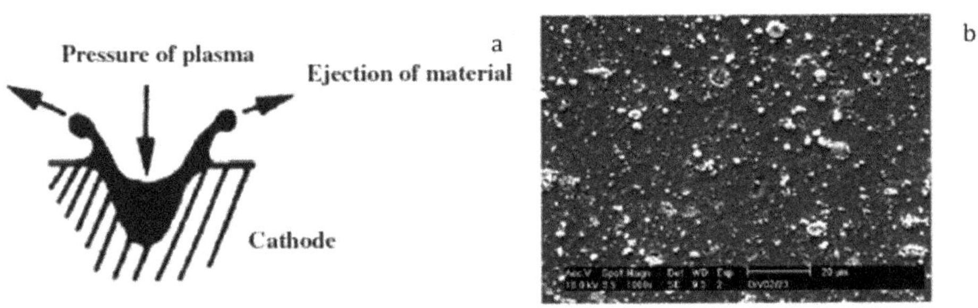

FIG. 3.3.9 – (a) Droplet emission process as a result of the plasma pressure on the cathode molten material [73]. (b) Scanning electron micrographs of multilayered TiN/AlTiN surface deposited by CAD [74].

3.3.6 Examples of Application

3.3.6.1 Cr-Coated Zircaloy-4 as Enhanced Accident Tolerant Nuclear Fuel Claddings for Light Water Reactors [75]

Coatings with thicknesses between a few microns and $\sim 10\ \mu m$ deposited on a Zircaloy-4 substrate have been studied with the objective to provide a significant reduction in the oxidation-induced embrittlement of the nuclear fuel cladding, especially in accidental conditions, such as LOss-of-Coolant-Accident (LOCA) conditions [75].

The studied coatings, obtained either by DCMS or by HiPIMS, included nitride and metallic multi-layered ones. The results of this screening analysis showed that the first generation of chromium-based coatings (figure 3.3.10) exhibited the most promising behaviour: good compromise between oxidation resistance and adhesion to the metallic substrate, good fretting resistance and improved resistance to oxidation in steam at high temperature (HT).

Figure 3.3.11 shows the weight gain and typical optical micrographs obtained on the different coated Zircaloy-4 sheets samples after steam oxidation for 850 s at 1100 °C and direct water quenching down to RT (LOCA conditions). This HT oxidation time–temperature condition is representative to LOCA conditions. From these results, the best behavior (lower weight gain) is observed for ("PVD multi-pass") Cr/Cr coatings, while PVD monolayer Cr coatings are not fully protective [76]. Indeed, for the HT oxidation conditions tested here, Cr/Cr coatings exhibit a continuous residual metallic chromium layer beneath the outer chromia (Cr_2O_3) layer with no indication of significant oxygen ingress into the Zircaloy-4 substrate. Thus, they can be considered as fully protective. A poor oxidation behavior is observed for most of the ceramic (nitrides) coatings. For one extreme case, *i.e.* for Cr/AlTiN coating, the final weight gain is even slightly higher than the one of

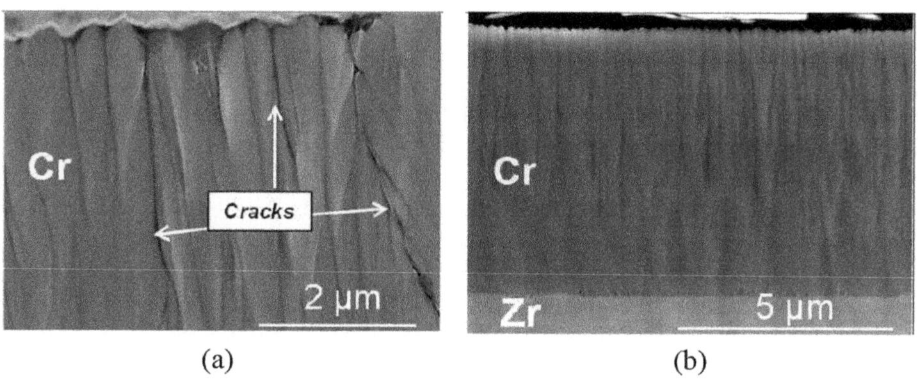

FIG. 3.3.10 – Cross sectional SEM images of pure Cr coatings on Zircaloy-4 substrate. Cr layer obtained by (a) conventional DC magnetron sputtering and by (b) HiPIMS, from reference [75].

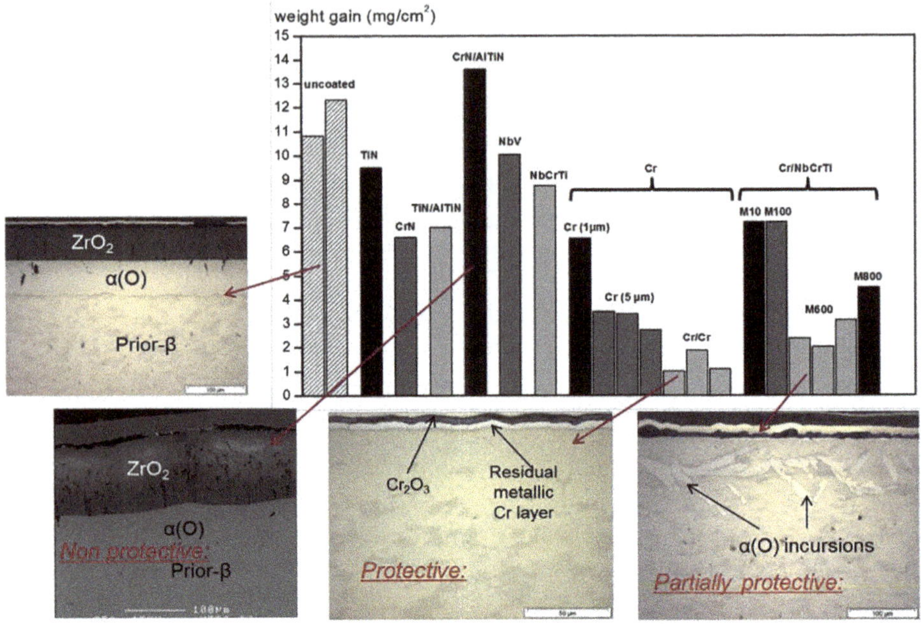

FIG. 3.3.11 – Weight gain and typical optical micrographs of a transverse cross-section obtained on several coated Zircaloy-4 sheets samples after high-temperature steam oxidation for 850 s at 1100 °C and direct water quenching down to room temperature, from reference [75].

reference uncoated Zircaloy-4, consistently with the formation of a thicker outer zirconium oxide.

3.3.6.2 Nanostructured Hard Coatings Deposited by Cathodic Arc Deposition on Carbide Inserts for Machining of Inconel 718 (Aeronautical Application)

In spite of being the most used superalloy for aerospace applications [77], Inconel 718 is still complex to process. Indeed, its machining causes many productivity troubles because of high wear rates of cemented carbides cutting tools. This nickel-based alloy is very difficult to machine because of its elevated temperature strength, high hardness and high work hardening rates leading to high machining forces [78, 79]. Superalloys contain also abrasive carbide particles and as a consequence very high wear rates are observed on the flank face of cuttings tools [80]. Several concepts have been proposed to overcome these challenges, such as coating having a superlattice structure.

A multilayer with a single crystal growth and with a single layer thickness in the size of the lattice parameter is named "superlattice" (figure 3.3.12). The superlattice concept is based on the deposition of a layered structure of two materials with the

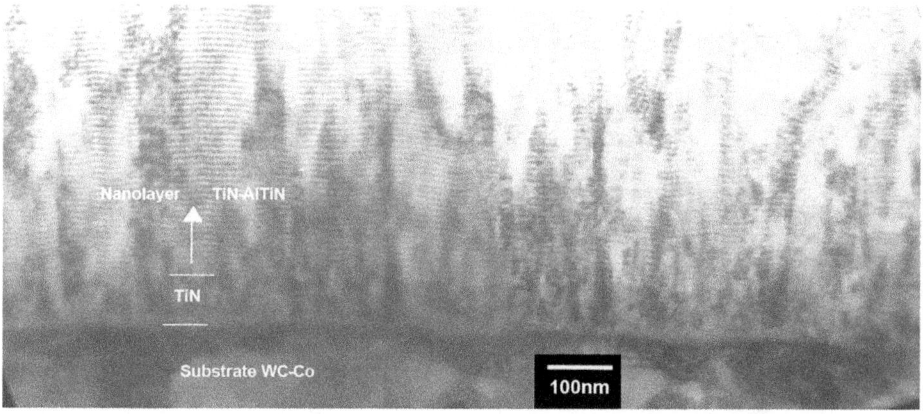

FIG. 3.3.12 – Cross-section bright field Transmission Electron Microscope (TEM) image of a TiN/AlTiN superlattice coating. Growth through interfaced from the TiN layer underneath, from reference [74].

same crystal structure where interfaces between the layers provide energy barriers to the motion of dislocations. It generally shows very good mechanical and tribological properties, distinct from individual layer material ones. Several explanations have been proposed including dislocation blocking by layer interfaces, Hall–Petch strengthening, strain effects at layer interfaces and supermodulus effect [81]. The superhardness has been explained by a model developed by Chu and Barnett [82] based on the dislocation motion within a superlattice structure, promoting the role of the real interface structure on the dislocation mobility.

It is possible, by layering two transition element nitrides, to synthesize superlattice coatings from two compounds, which have similar crystallographic cells, a similar lattice constant. These types of hard coating are deposited by CAD (see section 3.3.5) because of the advantages of high-ionized plasma, which allow optimizing substrate-coating interface by metallic ion bombardment leading to high adhesion levels.

Figure 3.3.13 illustrates the influence of the composition and nanolayer period (bilayer's thickness) on mechanical properties (nanohardness H and Young's modulus E). Values are scattered due to the high roughness of coatings deposited by cathodic arc deposition caused by microdroplets (see section 3.3.5). The evaluated nanohardness of the nanolayer coatings (between 47 and 55 GPa) were found to be considerably higher than that of multilayer coatings (between 35 and 39 GPa) and single layer as TiN (35 GPa) and CrN (29 GPa). A multilayer with a period in the size of lattice dimensions is a recent material (a superlattice), and generally shows properties not directly related to the individual layer materials.

Standard cutting tests (cutting force and flank wear measurements) were carried out on Inconel 718 by means of an instrumented tool holder in order to evaluate the performances of several hard coatings deposited on an 890 carbide inserts (figure 3.3.14). Nanolayered TiN/TiAlN coating is the best solution for, as an

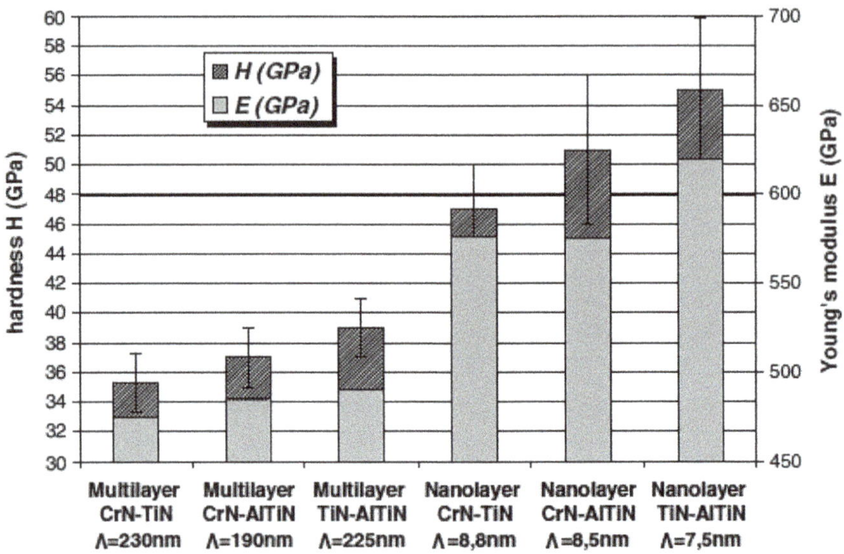

FIG. 3.3.13 – Comparison of hardness and Young's modulus between multilayer coatings for CrN–TiN, CrN–AlTiN and TiN–AlTiN associations. Multilayer: period (bilayer thickness) >100 nm; nanolayer: period (bilayer thickness) <100 nm, from reference [74].

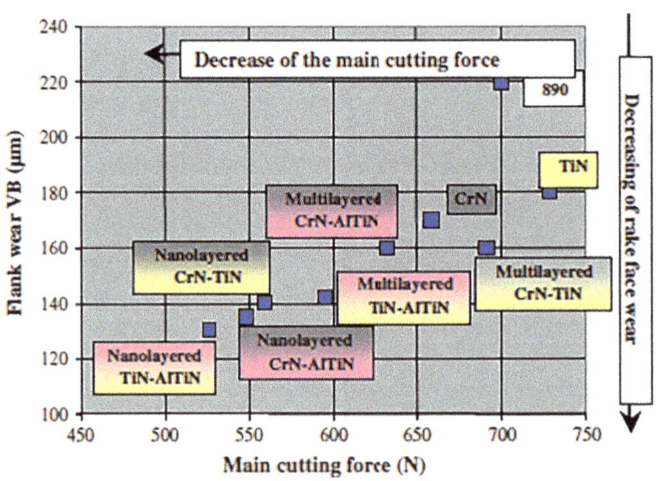

FIG. 3.3.14 – Evolution of flank wear of coated cutting tools with cutting force for various compositions and periods for finishing steps. Substrate: carbide insert grade 890. Multilayer: period (bilayer thickness) >100 nm; nanolayer: period (bilayer thickness) <100 nm. Finishing operation on the Inconel 718: 40 m/min, 0.2 m/rnd, 1 mm depth of cut.

example, a finishing step. This coating induces the lowest cutting force on the tool and limits built-up edge phenomenon. It is also very effective for avoiding strong abrasive wear (low flank wear) due to the very high hardness of the material. High temperature resistance of AlTiN coating and good tribological behavior also prevent sticking phenomena on rake face. As a consequence, this coating gave the best result not only in terms of cutting lifetime but also in terms of machining quality.

3.3.7 Conclusions

PVD technologies are today widely used in many fields because the deposited thin layers often have much better properties than those of the bulk materials from which they come. Indeed, the condensation of metallic vapor on a "cold" substrate is very far from thermodynamic equilibrium conditions and deposition reactor technologies make it possible to obtain multi-functional architectured coatings.

In this short course, the general principles of PVD technologies have been discussed and three processes have been described: conventional magnetron sputtering, HiPIMS and the CAD process. These last two processes, based on highly ionized plasmas, make it possible to obtain barrier coatings for extreme environments (high temperatures, aggressive environments, severe mechanical stresses...).

3.4 Chemical Vapor Deposition

Francis MAURY[1] and Alexandre MICHAU[2]

[1] *CIRIMAT, CNRS/INPT/UPS, 4 allée E. Monso, 31030 Toulouse cedex 4, France*
[2] *Des – Service d'Etudes Analytiques et de Réactivité des Surfaces (SEARS), CEA, Université Paris-Saclay, 91191 Gif-sur-Yvette, France*

3.4.1 Introduction

Since the pioneer monograph on "Vapor Plating" in 1955 [83], many comprehensive books have been published on Chemical Vapor Deposition (CVD) as a result of the important and rapid growth of this technology for many applications, essentially in microelectronics and metallurgical coating industries. The short list of representative generic books given in references will be useful for readers for a good overview and especially for finding more details [83–95]. Some of them cover all aspects including fundamentals, applications, materials and special processes [83, 84, 86, 87], others are dedicated to variants of the basic CVD process as Epitaxy (MOVPE) [85, 89, 90], Atomic Layer Deposition (ALD) [94] or Chemical Vapor Infiltration (CVI) [93], while others review specific CVD processes for different classes of materials as metals [88], nonmetals [91], semiconductors [92] and polymers [95].

During the past decades two families of CVD processes are at the origin of the great development of CVD techniques: (i) Plasma Enhanced CVD (PECVD) and (ii) Metal–Organic CVD (MOCVD). A specific course is dedicated to the first while this one gives essentially a focus on the second. The use of metalorganic precursors lowers the deposition temperatures and makes it possible to deposit a wide variety of materials and microstructures. This lecture updates knowledge based on recent advances obtained in particular by the authors' laboratories. The different categories of CVD reactions will not be reviewed because they can be found in the cited references [83, 84, 86, 87], and also because using metalorganic precursors the chemical mechanism proceeds mainly by pyrolysis reactions.

An important originality of CVD processes is that it is possible to deposit multifunctional thin films and thick protective coatings but also monolithic or nanostructured coatings, nanotubes, fibers, as well as powders by precipitation from the vapor phase. Also the process has a high throwing power, which leads to a conformal coverage of complex surfaces as fibers, tubes or other 3D things. The growth of composite coatings and their nanostructuration as multilayers or nanocomposites can be controlled. CVD can be utilized in continuous strip coating, *e.g. in situ* deposition of solar control coatings on float glass. Thanks to the different activation methods and the diversity of its operating modes, the CVD process is versatile, flexible and economically competitive.

DOI: 10.1051/978-2-7598-2446-5.c043

3.4.2 Principles and Fundamentals

All CVD processes rely on similar mechanisms based on several main steps (figure 3.4.1). The first one consists in generating a reactive vapor containing the precursor to decompose. A precursor is usually a molecule comprising the main element to be incorporated into the growing film. After being generated, the reactive vapor has to be transferred to the activation area where as its name suggests, a reaction set will be activated in order for the precursor to decompose or to react with other reagents, and subsequently grow a thin film. In the majority of CVD chambers, the vapor phase is transported by a carrier gas in a viscous regime. This carrier gas is chemically inert towards the precursor to avoid premature decomposition. The activation triggers a set of chemical reactions, homogeneous (in gas phase) or heterogeneous (involving at least one surface), that will ultimately lead to the nucleation and growth of a film. Volatile products from surface reactions then desorb and are evacuated through a pumping system including traps to retain by-products before the exhaust and effluent disposal. Processes without carrier gas are rare but we have to keep in mind that pack-cementation is an industrial isothermal CVD process operating in quasi-static regime [96].

Although gases are ideal sources to control mole fractions, commercially available precursors are supplied mainly as powders or liquids, which enable them to be easily stored and transported. Powders can be directly sublimated or dissolved into a compatible solvent and then evaporated. Liquids are directly vaporized with a thermostated bubbler or indirectly in the case of an aerosol system working under a continuous or pulsed mode, under atmospheric or reduced pressure.

CVD processes are operated in two different regimes, named diffusive and kinetic (or reactional) regime, which govern the growth rate (figure 3.4.2). In the diffusive regime, diffusion and transport phenomena are limiting the coating growth because

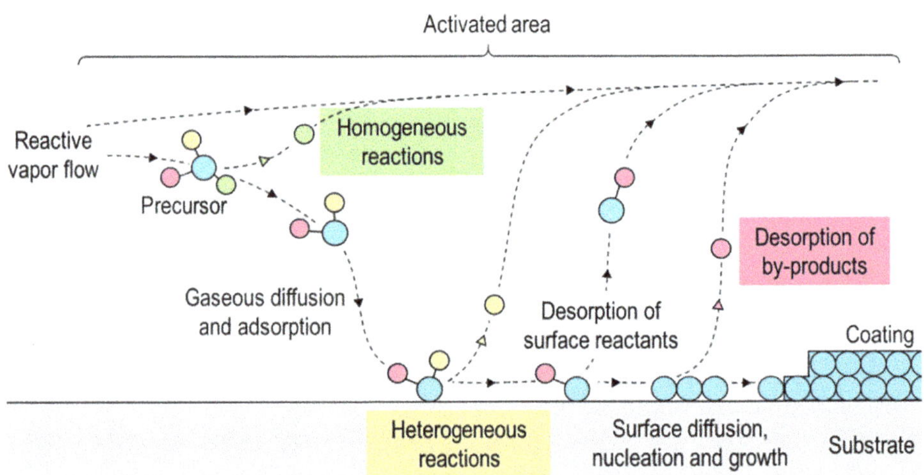

Fig. 3.4.1 – Main CVD mechanisms.

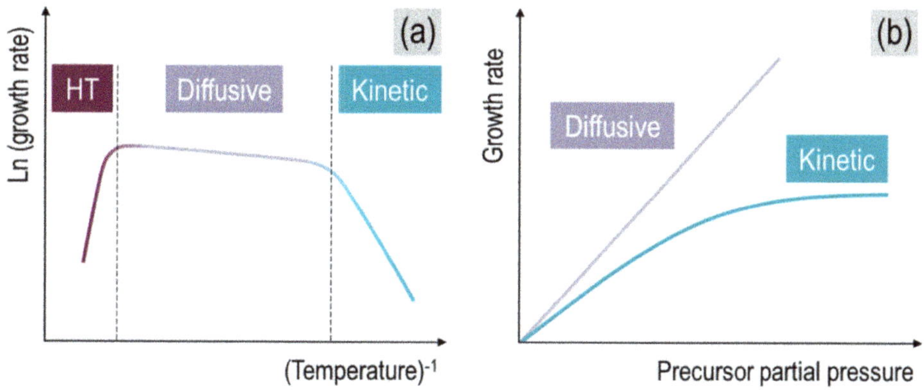

FIG. 3.4.2 – (a) Definition of CVD regimes versus deposition temperature with the growth rate and (b) influence of the precursor partial pressure for each growth regime.

of an insufficient matter influx. The deposition temperature is generally high and is expected to have a small influence on the growth rate because activation energies of physical phenomena as diffusion are low. This is the typical regime for epitaxial growth of functional layers. The second regime, the kinetic one, is not limited by transport phenomena but by the reaction kinetics of the precursor decomposition. The amount of reagent necessary for the surface reaction is in excess. Thus, the growth rate is strongly influenced by the deposition temperature and associated activation energies are high (of the order of chemical bond strengths that must be broken). This last regime is preferred to uniformly cover objects with complex 3D geometries or high aspect ratio. Combined with a sufficient enough reactive vapor flow, homogeneous thicknesses are achieved on objects impossible to properly coat with other deposition processes as those depositing line-of-sight. There is in fact another regime at higher temperatures, noted HT in figure 3.4.2, promoting nucleation in gas-phase. It is industrially exploited for pigments and powders production [97].

If the temperature is a key parameter for the growth kinetics, the total pressure is also a rate limiting factor since the diffusivity of a gas is inversely related to the pressure. At low pressure, surface reactions are the rate determining step and the mass-transfer parameters are far less critical than under higher pressure (figure 3.4.2).

3.4.3 Thermal CVD: Hot-Wall vs. Cold-Wall

Thermal CVD is the generic term of CVD processes where the gas phase is activated by thermal energy. Thanks to a simple implementation, thermal CVD is one of the most common CVD variants. Two types of reactors can be distinguished depending on the heating mode: hot-wall and cold-wall. In hot-wall reactors, the whole deposition chamber is heated, including the walls of the chamber and the substrate-holder. This is usually done by placing the chamber inside an oven and the useful volume of the reactor is considered isothermal. The gas-phase is then activated in the entire chamber and both homogeneous and heterogeneous reactions contribute to the decomposition mechanism of the precursor. On the other hand, only the restricted area of the sample holder is heated in cold-wall reactors. The use of an induction coil or IR lamps and a susceptor serving as a substrate-holder allows for a very confined heating. As the gas-phase is activated only very locally, heterogeneous reactions play this time a predominant role in the decomposition mechanism.

These two configurations have very distinctive attributes. Precursor consumption is for example exacerbated in hot-wall reactors since the whole deposition chamber is activated. It means that a coating is obtained not only on the substrates but also on the walls of the chamber. Hot-wall reactors allow the deposition on a larger amount of substrates with various shapes and distribution, since a bigger volume is activated. Cold-wall reactors are typically designed to coat flat substrates as for epitaxial growth of Si and III–V semiconductors. Additionally, a more homogeneous thermal field is expected in hot-wall reactors whereas strong gradients are observed in cold-wall reactors. However, it is worth noticing that achieving an optimal thermal field in a CVD chamber is a difficult matter, especially in large volumes, often eased by the use of appropriate Proportional–Integral–Derivative controllers (namely PID), thermocouples and multiple heating components. Finally, the choice of the heating mode depends on the type of substrate and on the desired material to deposit, *i.e.* the chemical system used.

Thermal CVD operates classically with halide precursors. Despite their relatively well-known decompositions mechanisms, their main inconvenience is usually their decomposition temperature around 1000 °C and the corrosive atmosphere produced by their decomposition, which strongly narrows potential substrate materials. Also the choice of volatile precursors in metal halide chemistry is limited given the increasing complexity of the materials to be deposited. These limitations are overcome by using metalorganic precursors (see section 3.4.6).

3.4.4 Activation in CVD Technologies

The diversity of energetic activation methods brings great flexibility to CVD processes. Their implementation generally aims at a significant drop in deposition temperatures. Activation consists in supplying different sources of energy to the chemical system so that it produces the growth of the desired film at lower temperature. The examples presented here are therefore to be distinguished from specific operating modes of CVD reactors that aim the growth of very particular quality of the material, and which are described as specific CVD processes in the section 3.4.5.

3.4.4.1 Plasma Enhanced-CVD

Talking about activation of a gas phase leads immediately to think of the use of a plasma. Indeed a plasma makes it possible to considerably lower the deposition temperature and thus overcome the main constraint of thermal CVD processes. The developments and applications of PECVD techniques are considerable and they will be presented in a dedicated course.

3.4.4.2 Laser and Photo-CVD

Laser CVD is the result of the thermal energy from the laser impinging and heating the absorbing substrate (thermolysis). As for cold-wall reactors, the growth is restricted to the heated area because absorption by the gas phase is negligible, and the mechanisms are similar to thermal CVD. Applications include direct writing of micron size patterns. In Photo-CVD, activation is induced by UV photons, which are sufficiently energetic to break the chemical bonds of reactants in the gas phase (photolysis). A good matching of the absorption wavelength of precursors and the radiation used (lamp or UV laser) is required. Deposition temperature is significantly reduced with this technique. To illustrate with an innovative result, certain polymers can be deposited near ambient temperature by Photo-CVD as a continuous film on the surface of functional liquids of low volatility for a tight encapsulation necessary for optical micro-systems (figure 3.4.3) [98].

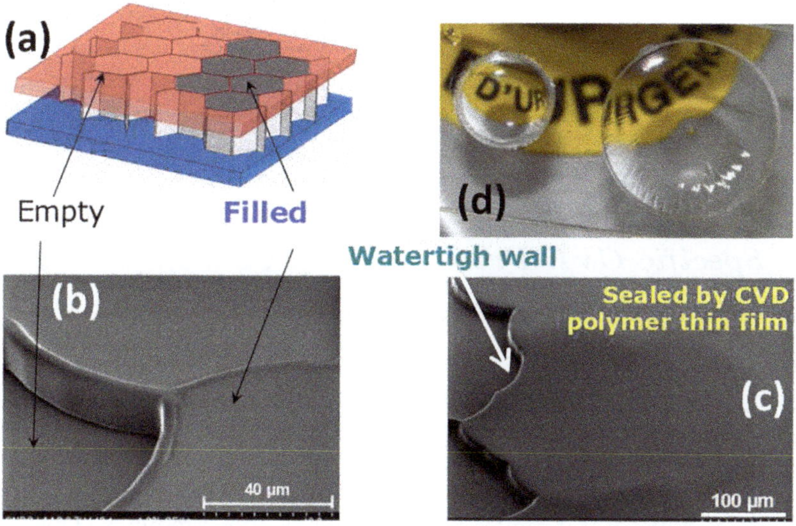

FIG. 3.4.3 – Watertight sealing by a CVD polymer film of microcells filled with functional liquid (a–c); encapsulation with a PMMA film by Photo-CVD of liquid droplets 10 and 20 mm in diameter acting as optical lens (d). Pulsed Direct Liquid Injection (DLI) was used to vaporize the monomer in the UV activated zone.

3.4.4.3 Hot Wire Chemical Vapor Deposition (HWCVD) & Initiated Chemical Vapor Deposition (iCVD)

HWCVD is a method in which the reactive vapor dissociates at the surface of resistively heated filament to produce radicals that form intermediate species and adsorb on the heated substrate to deposit a film. Depending on the chemical system, a catalytic decomposition occurs on the refractory filament (mainly W or Ta) and the technique is sometimes called Catalytic Chemical Vapor Deposition (Cat-CVD). Thanks to the high efficiency of H2 dissociation on W wire, the technique is mature for optoelectronic applications to produce device-quality *a*-Si:H films. Technology is making rapid progress and implementation in roll-to-roll reactors was investigated for mass production [99].

Also polymer layers used as water vapor barriers are deposited by a variant of this Cat-CVD process named iCVD where a thermally labile initiator is added in the feed stream to start the polymerization. Thus the initiator decomposes at the heated filament into radicals and a free-radical polymerization begins on the substrate near room temperature. Oxidative-CVD is another variant for polymer deposition where a volatile oxidizing molecule, such as Fe(III) derivatives, is simultaneously injected with the monomer to participate in step growth polymerization on the substrate. Here no filament is required, the chemistry is a spontaneous reaction, and it is considered as a chemically activated process [95].

3.4.4.4 Flame-CVD & Combustion Chemical Vapor Deposition (CCVD)

In Flame-CVD, also called Combustion CVD (CCVD), the metalorganic precursor is added to the burning gas and the flame is moved just above the surface to be coated. The high energy within the flame decomposes the precursors into highly reactive intermediates, which readily react with the substrate to form a coating. The CCVD process is well adapted to deposit thin films in ambient atmosphere. Consequently, several oxides have been deposited using inexpensive precursors. The technique is also efficient for the production of C nanomaterials.

3.4.5 Specific CVD Processes

3.4.5.1 Vapor Phase Epitaxy (MOVPE & CBE)

Epitaxy is defined as the growth of a single crystal film on a single crystal substrate with which lattice matching is established; the latter then serves as a seed crystal. The film and the substrate can be the same material (homoepitaxy) or not (heteroepitaxy); epitaxial growth occurs only if the lattice mismatch is not too high. Metal–Organic Vapor Phase Epitaxy (MOVPE) is a very important process in semiconductor industry to produce Si, III–V and II–VI compounds. This process is implemented in specially designed cold-wall CVD reactors operating in the diffusive regime and generally under low pressure using a viscous flow of reactants [85, 89, 90]. Alternatively, Chemical Beam Epitaxy (CBE) uses ballistic beams of molecular precursors that impinge on the surface of a heated substrate, where they decompose to incorporate atoms into lattice sites of a growing epitaxial film. In contrast to MOVPE, homogeneous gas phase reactions and diffusive transport in the gas phase do not play a role in the overall reaction kinetics of CBE. This technique requires an UHV reactor, which facilitates the use of suitable *in situ* control of the growth (*cf.* section 3.4.7).

3.4.5.2 Selective-CVD

Selective deposition has received much attention in CVD for microelectronics applications to avoid the use of masks and simplify manufacturing of devices. Selective Deposition is obtained when the reactivity of the adsorbing gases differs between different surfaces. Surface chemistry including nucleation is a key step in these processes because the growth starts immediately on the substrate material whereas the nucleation on dielectrics is retarded because chemisorption is more difficult (the valence band is full). Selective growth of epitaxial Si, SiGe, III–V and II–VI compounds as well as W and Cu was investigated [100].

3.4.5.3 Homo-CVD & Catalytic CVD (CatCVD)

HOMO-CVD is a non-isothermal process where radicals are created in homogenous reactions by cracking of the gas phase at a temperature significantly higher than that of the substrate. The average energy of this activation process is significantly lower than that of plasma deposition (0.1 eV compared to 1–20 eV) and there is no damaging effects induced by energetic ions and radiation. It was developed in the early 80's for the growth of *a*-Si:H with photovoltaic quality [101].

3.4.5.4 Fluidized Bed-CVD

CVD is an important technique for surface modification of powders in fluidized beds through either grafting or deposition of thin films. Beyond pioneering applications in the nuclear power industry, the surface modified powders find applications in the fabrication of monolithic or composite structural parts, heterogeneous catalysis and other domains. FBCVD processes combine the constraints of two elemental processes occurring simultaneously: (i) mixing of powders in such a way that their entire surface is exposed to the reactive vapor phase and (ii) gas–solid chemical reactions as in conventional CVD process. Significant progress results from modeling of FBCVD processes [102].

3.4.5.5 Chemical Vapor Infiltration (CVI)

CVI is a specific CVD process where the reactive vapor phase infiltrates a porous structure as fibrous weaves. A conformal coverage occurs on the fibers and the structure is gradually densified to form a fiber/matrix composite. The kinetics is different because reactants have to diffuse inward through the porous structure and the byproducts have to diffuse out. The process operates under low pressure to facilitate gas diffusion and in the kinetically limited regime to achieve a maximum densification before having an occlusion of open porosities. It is a very slow process that can take several weeks to reach maximum densification. In order to increase the densification rate, different procedures have been developed but, depending on the shape of the preform to be densified, they cannot all be used. Among them, in isothermal isobaric CVI (I-CVI) developed in Bordeaux, the vapor phase surrounds the porous preform and enters by diffusion. In forced CVI thermal-gradient (F-CVI) developed in Oak Ridge National Lab., the precursors enter under pressure ($\approx$2 MPa) into the cool side of the preform and flow through it up to the hot zone where the deposition occurs. CVI is used to manufacture ceramic/ceramic composites as SiC/SiC for structural parts resistant to high temperatures in aerospace industry such as nozzles of rockets or aircraft fighters and C/C for aircraft brakes [93, 103, 104].

3.4.5.6 Atomic Layer Deposition (ALD)

When thin films of excellent conformity and high crystallographic quality are desired, it is very important to produce them by well-controlled surface chemistry which involves layer-by-layer growth mode resulting from the alternating injection of reactants separated by a purge time. This variant was named AL-CVD in the pioneering patent of Suntola in 1974. This process, now known as Atomic Layer Deposition, is undergoing rapid development for thin layer applications [94], and is the subject of a specific lecture in this series (see section 3.5).

3.4.6 Precursor Tailoring for Metalorganic CVD

Metalorganic compounds are increasingly important because they decompose at low temperature and metalorganic chemistry offers a wide choice of volatile molecules. The chemist can develop an engineering of molecular precursors to adapt their design and properties for specific MOCVD processes to deposit desired compound materials. Also, tailor-made metalorganic sources are more suitable for activated CVD techniques as photo-CVD or PECVD processes. Thus MOCVD strongly contributes to the great flexibility of the generic parent processes. Using a metalorganic precursor the reaction pathway is generally based on its thermal decomposition (pyrolysis), which forms the main element or more elementary fragments as surface nutrient species for the growth. The criteria for choosing precursors have been reviewed elsewhere [105] and they are summarized in table 3.4.1.

3.4.7 Process Optimization Through Monitoring and Process Modelling

3.4.7.1 Diagnostics in CVD

In situ monitoring benefits from several advantages despite the constraints imposed by their implementation inside a CVD chamber. Usually non-invasive, they provide real-time knowledge about mechanisms and species that are impossible to study

TAB. 3.4.1 – General and specific characteristics of molecular precursors for efficient CVD processes.

Physico-chemical properties	– Good volatility (>10^{-1} torr at ~100 °C). – High purity. – Good long-term stability in its container at RT. – Liquid (low m.p.) rather than solid (or gas). – Soluble in solvents suitable for the process. – Compatible with a multi-source process.
Thermal properties	– Good thermal stability during vaporization and transport. – Ability to be cleanly decomposed to deposit the desired material.
Safety requirements	– Non-toxic/non-pyrophoric/non-corrosive (including by-products).
Economic requirements	– Readily available. – Low cost. – Recyclability to make the process cost-effective.
Specific criteria imposed by the process	– Nature of the desired material. – Nature of the substrate. – Processing conditions (low temperature, epitaxy, activation method, etc.). – Simplicity of the process (single-source approach).

ex situ. However, online monitoring supposes to sample a representative surface area or vapor phase near the substrate and associated techniques can be invasive and are complex to implement [106]. Kinetics related information or identification of intermediate reactants are extremely valuable in the objective of process modelling and optimization. With the awareness of the precursor decomposition pathway or the contribution of gas-phase over surface reactions, it is possible to adjust pressure and thermal fields in a CVD reactor in order to promote or inhibit the formation of chemical species and grow a material with the desired structure and phase.

Online monitoring for CVD processes relies on two main approaches. First, the identification and quantitative analysis of chemical species serve for the determination of reaction mechanisms. Classical mass spectrometry and optical (UV, IR, FTIR...) spectroscopies are frequently used and well suited. Optical emission spectrometry is also widely used for PECVD processes. These techniques will not be discussed further here and we will focus on the growth control during the process.

3.4.7.2 Growth and Kinetics Control in CVD

The second approach concerns process monitoring and growth control. All CVD reactors are equipped with classical temperature and pressure gauges, sometimes with more advanced equipment such as a quartz crystal balance or an ellipsometer for growth rate monitoring [106]. This approach is illustrated with two original examples.

Example 1. Optical pyrometry is a sensitive, simple and low-cost technique commonly used for the measurement of deposition temperature in CVD processes. It is much less known that it can be used as diagnostic tool to provide fruitful information at the nanometric scale during the growth of thin films. Thus, monochromatic IR pyrometry ($\lambda = 1{,}6$ μm) was used for *in-situ* and real time control of the early stages of the growth of various CVD films. Figure 3.4.4 illustrates two examples: (a) the growth at 500 °C of a transparent coating as TiO_2 on Si(100) shows oscillations of the pyrometric signal *versus* deposition time due to interferences induced by multi-reflections at the interfaces. Simulation of these data gives a thickness of 2760 nm (by SEM: 2800 nm),$n_{\text{film}} = 1{,}97$, $k_{\text{film}} = 0{,}001$ (by ellipsometry: $n = 2{,}43$, $k = 0{,}0009$) [107], (b) the growth of metallic Fe at 750 °C by MOCVD using Fe $(Cp)_2/H_2O$ showing a rapid nucleation but the growth of a dense and pure Fe film requires in addition a step of solid state reactions with secondary phases for about 10 min related to those of blast furnace chemistry (reduction of oxides by C) [108].

Example 2. If gas phase chemistry plays an undeniable role in MOCVD, it is the surface chemistry that controls the quality of the films deposited. The near threshold photoemission is very sensitive to all events on a surface and allows for instance analysing the dynamics of epitaxial growth from organometallics. It is based on the detection of the total photoemission current produced by the excitation of the surface by UV radiation whose energy is close to the photoemission threshold of this surface. When a layer-by-layer growth mode is established (epitaxy), the extreme surface is alternately rich in Ga and As which regularly modifies the work function and therefore generates oscillations of the total photoemission current. Their period therefore corresponds to the formation of a monolayer, which makes it possible to measure the growth rate in real time as shown in figure 3.4.5 [109]. In fact, all gas-surface interactions affect the intensity of the total photoemission current. It is then possible to obtain experimental kinetic data on the elementary steps of adsorption, desorption and heterogeneous decomposition of precursors which can then be used in modeling the CVD process, assuming that the surface chemistry

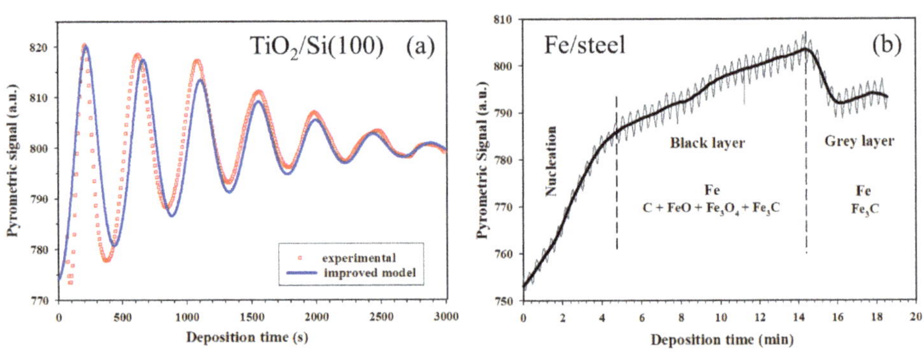

FIG. 3.4.4 – Temporal variation of the pyrometric signal during the early stage of MOCVD: (a) of TiO_2 on Si(100) substrate at 500 °C and (b) of metallic Fe on stainless steel at 750 °C (secondary phases are in parentheses); more details in the text.

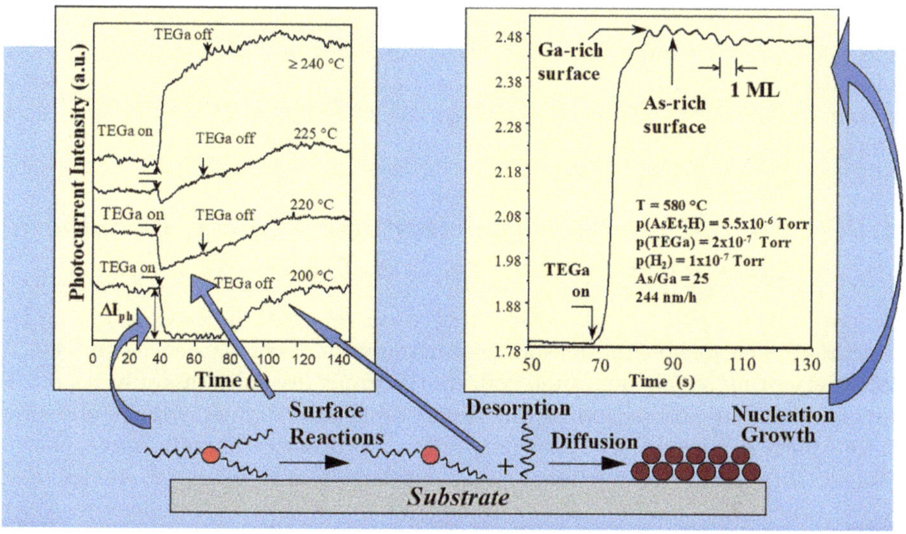

FIG. 3.4.5 – Temporal variation in the intensity of the total photoemission current during the interaction of $GaEt_3$ on GaAs(100) (left) and during the first stages of the epitaxial growth of GaAs from $GaEt_3/AsEt_2H/H_2$. The sketch symbolizes the elemental steps of vapor phase growth of a thin film.

investigated in such high vacuum environment is representative of a more conventional CVD process (figure 3.4.5) [110].

3.4.7.3 Other Diagnostics and Process Modeling

In addition to thickness control, the knowledge of the stresses in film is a powerful information as the stress level partially governs major properties of thin films, whether they are structural (adhesion, toughness...) and functional (optical, electrical...). The most common method consists in evaluating the substrate curvature induced by the growth of a coating directly resulting from its stress level, frequently with the help of an optical system measuring the deflection of light.

Apart from additional knowledge and comprehension of phenomena occurring in a CVD deposition chamber, online monitoring generates data sometimes used to adjust and feed models, from very simple ones describing a phenomenological law to more complex ones involving multiphysics and multiscale modelling. Models help to better seize and understand CVD processes but also to optimize them, mostly to reach homogeneous coverages and thicknesses [111]. Several aspects are to be considered, the thermochemistry and kinetics of the entire chemical system that allows for the prediction at the equilibrium (or not) of the formed species in heterogeneous or homogeneous phases and the modelling of the deposition chamber. This latter includes the modelling of thermal and pressure fields, reactive and unreactive flow rates and finally the geometry of the reactor. The modelling of the decomposition of aluminum tri-isopropoxide on the inner surface of a glass bottle improved for instance the homogeneity of the coverage by the amorphous alumina coating [112]. In another case, the simulation of the MOCVD deposition of amorphous chromium carbide from bis(ethylbenzene)chromium into long and narrow tubes assisted efficiently in the scale-up process applied to the internal protection of nuclear fuel claddings [113].

3.4.8 Concluding Remarks: Historical and Contemporary Applications in Energy and Transport Fields

Some applications of CVD processes have been reported in sections 3.4 and 3.5, industrial applications being the source of many variants of the generic process. The wide variety of materials deposited reflects the great diversity of applications. This includes industrial production of functional films for microelectronics and photovoltaics, polymer layers for packaging, hard protective metallurgical coatings, advanced ceramic/ceramic composites, powders and nanomaterials (pigments, NTC...) usable for their functional properties or as filler in composites. Most of these applications are related to energy and transport areas.

Almost all industrial components and pieces of machinery involve moving parts and frictional contact areas: bearings for wind turbine blades, injection parts of diesel engines, pump and drilling components for oil extraction, airplane gearbox elements... These components are typically coated with DLC coating (for Diamond Like Carbon) that reduces friction and abrasive wear, thus enhancing durability, improving the assembly efficiency and saving energy [114]. PECVD DLC coatings are proposed by the main actors of the market such as Ionbond, Oerlikon Balzers, Richter Precision Inc., Schaeffler or Hauzer. The historical application of CVD coatings for transportation concerns the deposition by pack-cementation of diffusion coatings on superalloys used in aircraft gas turbines in order to improve their resistance to high-temperature oxidation and hot corrosion [115, 116]. Diffusion coatings are dominated by aluminide and aluminide modified coatings, containing additional noble metals like Pt or reactive elements such as Y or Hf [117, 118] that promotes the growth of a protective stable thermally grown oxide layer. New developments are regularly reported in the aerospace industry, for example concerning REACH compliant solutions and particularly alternatives to processes using hexavalent chromium. Hardide company provides for example a tungsten carbide based coating deposited at low temperature CVD, around 350 °C, from a mixture of WF_6, hydrogen and one hydrocarbon [119]. It is used on the cockpit canopy locking mechanism of the Eurofighter Typhoon against corrosion and wear, and was recently approved by other aerospace manufacturers (https://hardide.com/applications/surface-coatings-for-aerospace/). A recent patent [120] mentions the protection of helicopter rotor blades, propeller blades or fan blades against erosion caused by water droplets. For automotive applications, Teijin and Tsukishima Kikai Co. promote a lighter and more performing new window material, based on a top hard $SiO_xC_yH_z$ layer deposited by high-frequency PECVD from an organosiloxane, organosilane or organosilazane on a polymer substrate [121, 122]. This hard layer exhibits a resistance to abrasion and heat close to glass-based products (https://www.teijin.com/news/2017/ebd170308_02.html).

In addition to protective coatings, CVD processes are perfectly suitable for the deposition of functional films. The deposition of low emissivity or thermochromic layers for example helps to increase the thermal energy efficiency of buildings (smart window concept) [123, 124]. The coating step is either integrated off-line or on-line. It is then performed directly and continuously on the glass float and benefits from the float residual thermal energy [125]. Besides the glass coating industry, the semiconductor industry, the III–V high-frequency device industry, the laser diode industry and finally photovoltaics industry are the largest industries using CVD processes in terms of mass production [126]. Two main types of solar cells are industrially produced thanks to CVD: Si based and III–V element multi-junction based [126]. Research focuses on their lower cost, efficiency and durability enhancement, along with the use of less toxic and critical resources [127]. For energy-related applications, such as energy conversion or storage, CVD-related processes show several advantages. It is possible on one hand to grow structures with very high specific area (arrays of nanotubes, nanowires...) [128] and on the other hand to coat in a perfectly conformal way structured substrates (arrays of trenches, pillars...) [129]. These two assets enable the elaboration of two promising solutions, 3D electrodes with higher performances than 2D structures and all solid-state batteries.

3.5 Atomic Layer Deposition

Thomas PROSLIER

Institut de recherches sur les lois fondamentales de l'univers, Commissariat de l'énergie atomique et aux énergies renouvelables, CEA, Université Paris-Saclay, 91191 Gif-sur-Yvette, France

Atomic layer deposition (ALD) is a recent synthesis technique discovered in the 70's in Finland. It is a vapor phase technique capable of producing thin films of a variety of materials on almost any type of substrate. ALD is based on sequential, self-limiting reactions, and offers exceptional conformality on high aspect ratio structures, thickness control at the Angstrom level, and tunable film composition. With these advantages, ALD has emerged as a powerful tool for many industrial and research applications from microelectronics, photovoltaic, catalysis, cosmetics, particle accelerators. For a more comprehensive summary of ALD and its many applications, please refer to existing reviews on the topic [130–134].

The advantages of ALD are precise thickness and composition control at the ångstrom or monolayer level. The self-limiting aspect of ALD leads to excellent step coverage and conformal deposition on high aspect ratio structures. As device requirements push toward smaller and more spatially demanding structures, ALD has demonstrated potential advantages over alternative deposition methods, such as chemical vapor deposition (CVD) and various physical vapor deposition (PVD) techniques, due to its conformality and control over materials thickness and composition. These desirable characteristics originate from the cyclic, self-saturating nature of ALD processes. ALD was popularly introduced as atomic layer epitaxy (ALE) in 1977. As further ALE processes were developed to incorporate metals and metal oxides, many materials were deposited non-epitaxially and the more general name of ALD was adopted to reflect this reference [1]. It should be noted, too, that many ALD procedures were developed from a variety of CVD processes. In contrast to their CVD analogs, the ALD procedures feature alternating exposure of chemical precursors to react to form the desired material, often at significantly lower temperatures.

A schematic ALD process is illustrated in figure 3.5.1. It consists of sequential pulses of gaseous chemical precursors that react with the surface of the substrate. These individual surface reactions are called 'half-reactions' (A and B) and are only part of the final materials synthesis. During each half-reaction, the precursor is pulsed into a chamber under vacuum ($<$1 Torr) for a controlled amount of time to allow the precursor to fully react with the substrate surface through a self-limiting process that leaves no more than one monolayer at the surface. Subsequently, the chamber is purged with an inert carrier gas (typically N_2 or Ar) to remove any unreacted precursor or reaction by-products. This is then followed by the counter-reactant precursor pulse and purge, creating up to one layer of the desired material. This process is then cycled until the appropriate film thickness is achieved.

Some surface areas will react before other surface areas because of different precursor gas fluxes. However, the precursors will adsorb and subsequently desorb

DOI: 10.1051/978-2-7598-2446-5.c053

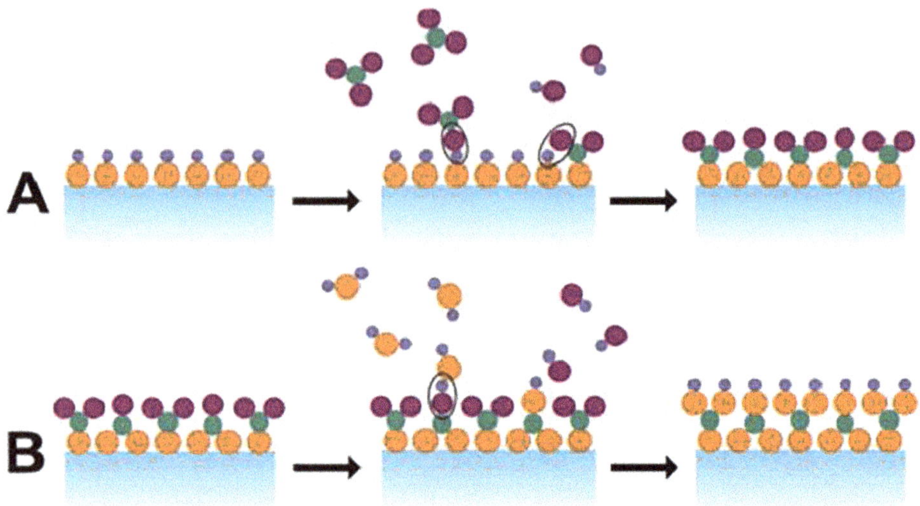

FIG. 3.5.1 – Schematic of ALD process. The substrate surface has natural functionalization or is treated to functionalize the surface. The Precursor A is pulsed and reacts with surface. Excess precursor and reaction by-products are purged with inert carrier gas. Precursor B is pulsed and reacts with surface and the excess precursor and reaction by-products are purged with inert carrier gas. The steps A–B are repeated until the desired material thickness is achieved.

from the surface areas where the reaction has reached completion. The precursors will then proceed to react with other unreacted surface areas and produce a very conformal deposition. The self-limiting nature of the surface reactions also produces a non-statistical deposition because the randomness of the precursor flux is removed as an important factor. As a result, ALD films remain extremely smooth and conformal to the original substrate because the reactions are driven to completion during every reaction cycle. Because no surface sites are left behind during film growth, the films tend to be very continuous and pinhole-free. This factor is extremely important for the deposition of excellent dielectric films.

ALD processing is also extendible to very large substrates and to parallel processing of multiple substrates. The ALD precursors are gas phase molecules, and they fill all space independent of substrate geometry and do not require line of-sight to the substrate. ALD is only limited by the size of the reaction chamber. The ALD process is also dominated by surface reactions. Because the surface reactions are performed sequentially, the two gas phase reactants are not in contact in the gas phase. This separation of the two reactions limits possible gas phase reactions that can form particles that could deposit on the surface to produce granular films. The Atomic layer deposition is capable of depositing a variety of thin film materials from the vapor phase. ALD has shown great promise in emerging semiconductor and energy conversion technologies.

The wide variety of alloys grown by ALD is summarized in figure 3.5.2. The synthesis of new compounds is an active research field and every year the realm of new precursors and chemistries discovered increase.

Typically ALD processes are conducted at modest temperatures (<450 °C). The temperature range where the growth is saturated depends on the specific ALD process and is referred to as the 'ALD temperature window'. Temperatures outside of the window generally result in poor growth rates and non-ALD type deposition due to effects such as slow reaction kinetics or precursor condensation (at low temperature) and thermal decomposition or rapid desorption of the precursor (at high temperature). In order to benefit from the many advantages of ALD, it is desirable to operate within the designated ALD window for each deposition process. The primary advantages of ALD are all derived from the sequential, self-saturating, gas-surface reaction control of the deposition process.

Firstly, the conformality of ALD-deposited films is often the critical factor in choosing ALD over competing deposition techniques such as CVD or sputtering. Conformality of high aspect ratio and three dimensionally structured materials is made possible by its self-limiting characteristic, which restricts the reaction at the surface to no more than one layer of precursor. With sufficient precursor pulse times, the precursor can disperse into deep trenches, allowing for complete reaction with the entire surface. Subsequent cycles allow for uniform growth on high aspect ratio structures, whereas CVD and PVD may suffer from non-uniformity due to faster surface reactions and shadowing effects, respectively. Examples of the superior conformality of ALD films are shown in figure 3.5.3 for NbSi films on trenched Si wafer [137] and MoN films deposited on fused capillary tubes [136] with a diameter

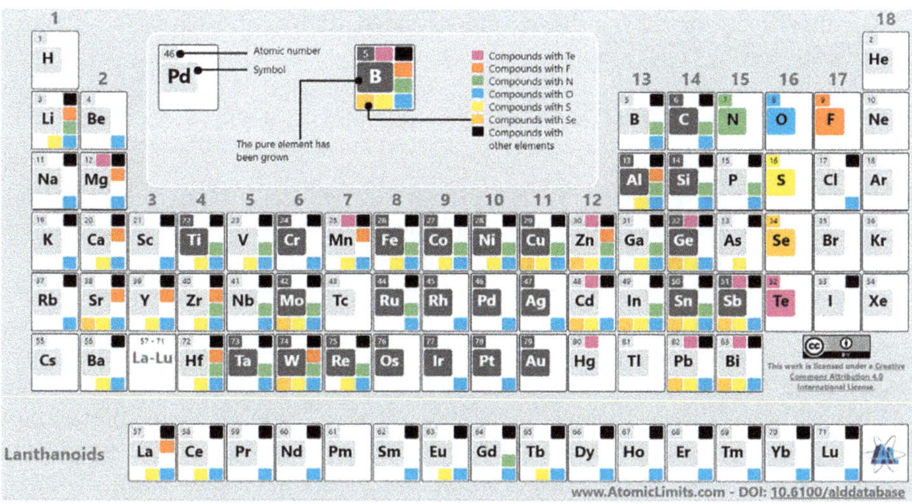

FIG. 3.5.2 – Representation of the various alloys grown by ALD. An exhaustive list of compounds synthesized by ALD and the related chemistries can be found at: https://www.atomiclimits.com/.

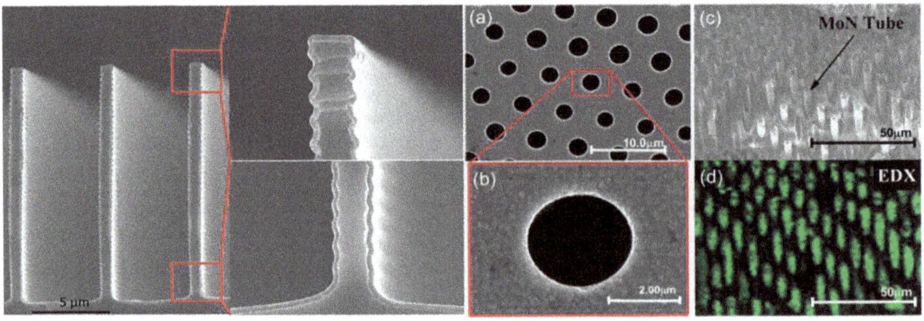

FIG. 3.5.3 – Illustration of the high aspect ratio coating capability of ALD. On the left, NbSi films deposited on a trenched Si wafer with an aspect ratio of ~10, the film thickness is the same at the bottom or at the top on the trench. On the right, 2 microns capillary tubes (a, b) sample with an aspect ratio of 500 coated with 50 nm MoN by ALD. The cross sections in (c, d) reveal an homogeneous MoN thin film deposited all along the capillary tubes.

of 2 μm and a length of 1 mm. From these two example, the process to evenly coat the substrate morphology is evident and many more example of high aspect coating by ALD exits throughout the literature such as aerogel, nanopowders with aspects ratios up to 10 000 [136].

A second advantage of ALD is the thickness control of the deposited thin films. By utilizing layer-by-layer deposition, the thickness of a film can be tailored by the number of ALD cycles. Growth per cycle for many ALD films has been summarized in previous reviews [130, 131] and is typically less than one Å/cycle, depending on the individual process. Another prominent advantage of ALD is composition control. Composition control has been demonstrated with materials such as zinc tin oxide (ZTO) and $SrTiO_3$, among many others.

Conclusion

Deposition processes described in this chapter all belong to more generic families, which have been mostly industrially implemented already for decades. However, several of their inherent limitations were more recently overcome thanks to the development of specific technological variations. Cold spraying, suspension plasma spraying, ionized PVD and specific CVD processes such as ALD are representative examples of emerging deposition processes deriving from more generic families, finally allowing for a better control of coating microstructures and chemistries and subsequent properties. As these couples "coating – process" exhibit better performances compared to conventional systems, industrials found more and more interest into them, as their applications in transportation and energy industries attest. If yet they are usually reserved for high added-value markets, for instance aerospace application in the transportation industry, their continual development helps to reduce implementation costs and manoeuvrability expenses, hence spreading their use. Finally, thanks to the great diversity of deposition processes presented here, it is possible to answer the majority of industrial needs with an adequate "coating – process" couple, not only depending on the required coating specifications (composition, microstructure, thickness…) but also on the substrate nature (material, shape, number…).

References

[1] Cavaliere P. (2018) *Cold-spray coatings, recent trends and future perspectives.* Springer International Publishing.

[2] Alkhimov A., Kosarev V., Papyrin A. (1990) A method of cold gas-dynamic spraying, *Dokl. Akad. Nauk SSSR.* **315**, 1062.

[3] Papyrin A., Kosarev V., Klinkov S., Alkhimov A., Fomin V. (2006) *Cold spray technology.* Elsevier, Oxford.

[4] Maev R., Leshchynsky V. (2008) *Introduction to low pressure gas dynamic spray: physics & technology.* Wiley-VCH, Weinheim.

[5] Assadi H., Gärtner F., Stoltenhoff T., Kreye H. (2003) Bonding mechanism in cold gas spraying, *Acta Mater.* **51**, 4379, https://doi.org/10.1016/S1359-6454(03)00274-X.

[6] Schmidt T., Gärtner F., Assadi H., Kreye H. (2006) Development of a generalized parameter window for cold spray deposition, *Acta Mater.* **54**, 729, https://doi.org/10.1016/j.actamat.2005.10.005.

[7] Grujicic M., Zhao C., DeRosset W., Helfritch D. (2004) Adiabatic shear instability based mechanism for particles/substrate bonding in the cold-gas dynamic-spray process, *Mater. Des.* **25**, 681, https://doi.org/10.1016/j.matdes.2004.03.008.

[8] Hussain T., McCartney D.G., Shipway P.H., Zhang D. (2009) Bonding mechanisms in cold spraying: the contributions of metallurgical and mechanical components, *J. Therm. Spray Technol.* **18**, 364, https://doi.org/10.1007/s11666-009-9298-1.

[9] Rokni M.R., Nutt S.R., Widener C.A., Champagne V.K., Hrabe R.H. (2017) Review of relationship between particle deformation, coating microstructure, and properties in high-pressure cold spray, *J. Therm. Spray Technol.* **26**, 1308, https://doi.org/10.1007/s11666-017-0575-0.

[10] Lugsheider E. (2006) High kinetic process developments in thermal spray technology, *J. Therm. Spray Technol.* **15**, 155, https://doi.org/10.1361/105996306X108246.

[11] Bagherifard S., Monti S., Zuccoli M.V., Riccio M., Kondás J., Guagliano M. (2018) Cold spray deposition for additive manufacturing of freeform structural components compared to selective laser melting, *Mater. Sci. Eng. A.* **721**, 339, https://doi.org/10.1016/j.msea.2018.02.094.

[12] Lupoi R. (2014) Current design and performance of cold spray nozzles: experimental and numerical observations on deposition efficiency and particle velocity, *Surf. Eng.* **30**, 316, https://doi.org/10.1179/1743294413Y.0000000214.

[13] Henderson C.B. (1976) Drag coefficients of spheres in continuum and rarefied flows, *AIAA J.* **14**, 707, https://doi.org/10.2514/3.61409.

[14] Brenner A., Senderoff S. (1949) Calculation of stress in electrodeposits from the curvature of a plated strip, *J. Res. Natl. Bur. Stand.* **42**, 105.

[15] Kuroda S., Clyne T.W. (1991) The quenching stress in thermally sprayed coatings, *Thin Solid Films* **200**, 49, https://doi.org/10.1016/0040-6090(91)90029-W.

[16] Kuroda S., Fukushima T., Kitahara S. (1990) Generation mechanisms of residual stresses in plasma-sprayed coatings, *Sel. Proc. 11th Int. Vac. Congr. IVC-11 7th Int. Conf. Solid Surf. ICSS-7* **41**, 1297, https://doi.org/10.1016/0042-207X(90)93938-F.

[17] Yandouzi M., Sansoucy E., Ajdelsztajn L., Jodoin B. (2007) WC-based cermet coatings produced by cold gas dynamic and pulsed gas dynamic spraying processes, *Surf. Coat. Technol.* **202**, 382, https://doi.org/10.1016/j.surfcoat.2007.05.095.

[18] Marx S., Paul A., Köhler A., Hüttl G. (2006) Cold spraying: innovative layers for new applications, *J. Therm. Spray Technol.* **15**, 177, https://doi.org/10.1361/105996306X107977.

[19] Champagne V.K., Ed. (2007) *The cold spray materials deposition process: fundamentals and applications*. Woodhead Publishing.
[20] Li W., Yang K., Yin S., Yang X., Xu Y., Lupoi R. (2018) Solid-state additive manufacturing and repairing by cold spraying: a review, *J. Mater. Sci. Technol.* **34**, 440, https://doi.org/10.1016/j.jmst.2017.09.015.
[21] Champagne V.K. Jr., Champagne V.K. III, Widener C. (2018) Cold spray applications, *Cold-Spray Coat. Recent Trends Future Perspect* (P. Cavaliere, Ed). Springer International Publishing.
[22] Gitzhofer F., Bouyer E., Boulos M. (1996) Suspension plasma spray deposition, WO9606957.
[23] Fauchais P.L., Heberlein J.V.R., Boulos M. (2014) *Thermal spray fundamentals, from powder to part*. Springer US.
[24] Fauchais P., Vardelle M., Vardelle A., Bianchi L., Léger A.C. (1995) Parameters controlling the generation and properties of plasma sprayed zirconia coatings, *Plasma Chem. Plasma Process.* **16**, S99, https://doi.org/10.1007/BF01512630.
[25] Bianchi L., Denoirjean A., Blein F., Fauchais P. (1997) Microstructural investigation of plasma-sprayed ceramic splats, *Thin Solid Films* **299**, 125, https://doi.org/10.1016/S0040-6090(96)09307-8.
[26] Madejski J. (1976) Solidification of droplets on a cold surface, *Int. J. Heat Mass Transf.* **19**, 1009, https://doi.org/10.1016/0017-9310(76)90183-6.
[27] Goutier S. (2010) Etude expérimentale de l'impact et de la solidification de gouttes céramiques et métalliques de tailles micrométrique et millimétrique sur différents types de substrats: compréhension du mécanisme de formation des dépôts par projection thermique, Ph.D. thesis, University of Limoges.
[28] Wittmann K., Blein F., Coudert J.F., Fauchais P. (2001) Control of the injection of an alumina suspension containing nanograins in a DC plasma, *Therm. Spray 2001 New Surf. New Millenium* (C. Berndt, K.A. Khor, E.F. Lugscheider, Eds). ASM International, pp. 375–382.
[29] Pawlowski L. (2009) Suspension and solution thermal spray coatings, *Surf. Coat. Technol.* **203**, 2807, https://doi.org/10.1016/j.surfcoat.2009.03.005.
[30] Duan Z., Heberlein J., Janisson S., Fauchais P. (1999) Effects of nozzle fluid dynamics on the dynamic characterisitcs of a plasma spray torch, *Therm. Spray 1999*. ASM International, pp. 247–252.
[31] Janisson S., Meillot E., Vardelle A., Coudert J.F., Pateyron B., Fauchais P. (1999) Plasma spraying using Ar-He-H2 gas mixtures, *J. Therm. Spray Technol.* **8**, 545, https://doi.org/10.1361/105996399770350232.
[32] Pilch M., Erdman C.A. (1987) Use of breakup time data and velocity history data to predict the maximum size of stable fragments for acceleration-induced breakup of a liquid drop, *Int. J. Multiph. Flow.* **13**, 741, https://doi.org/10.1016/0301-9322(87)90063-2.
[33] Damiani D. (2012) Elaboration, fitting in shape and characterization of materials for energy by the interaction radiation-matter, HdR, University of Limoges.
[34] Meillot E., Vert R., Caruyer C., Damiani D., Vardelle M. (2011) Manufacturing nanostructured YSZ coatings by suspension plasma spraying (SPS): effect of injection parameters, *J. Phys. Appl. Phys.* **44**, 194008, https://doi.org/10.1088/0022-3727/44/19/194008.
[35] Meillot E., Vincent S., Caruyer C., Damiani D., Caltagirone J.P. (2013) Modelling the interactions between a thermal plasma flow and a continuous liquid jet in a suspension spraying process, *J. Phys. Appl. Phys.* **46**, 224017, https://doi.org/10.1088/0022-3727/46/22/224017.
[36] Caruyer C. (2011) Modélisation de nanomatériaux injectés par voie liquide dans un jet de plasma pour la fabrication de nanostructures, Ph.D. thesis, University of Bordeaux 1.
[37] Girard F., Meillot E., Vincent S., Bianchi L., Caltagirone J.P. (2014) Mechanisms of heat transfer between a droplet and a plasma jet in suspension plasma spraying, *J. Phys. Conf. Ser.* **550**, 12022, https://doi.org/10.1088/1742-6596/550/1/012022.
[38] Basu S., Saha A., Kumar R. (2012) Thermally induced secondary atomization of droplet in an acoustic field, *Appl. Phys. Lett.* **100**, 54101, https://doi.org/10.1063/1.3680257.

[39] VanEvery K., Krane M.J.M., Trice R.W., Wang H., Porter W., Besser M., Sordelet D., Ilavsky J., Almer J. (2011) Column formation in suspension plasma-sprayed coatings and resultant thermal properties, *J. Therm. Spray Technol.* **20**, 817, https://doi.org/10.1007/s11666-011-9632-2.
[40] Bernard B., Quet A., Bianchi L., Joulia A., Malié A., Schick V., Rémy B. (2017) Thermal insulation properties of YSZ coatings: suspension plasma spraying (SPS) versus electron beam physical vapor deposition (EB-PVD) and atmospheric plasma spraying (APS), *7th Rencontres Int. Proj. Therm.* **318**, 122, https://doi.org/10.1016/j.surfcoat.2016.06.010.
[41] Bernard B., Bianchi L., Malié A., Joulia A., Rémy B. (2016) Columnar suspension plasma sprayed coating microstructural control for thermal barrier coating application, *J. Eur. Ceram. Soc.* **36**, 1081, https://doi.org/10.1016/j.jeurceramsoc.2015.11.018.
[42] Bernard B., Quet A., Bianchi L., Schick V., Joulia A., Malié A., Rémy B. (2017) Effect of suspension plasma-sprayed YSZ columnar microstructure and bond coat surface preparation on thermal barrier coating properties, *J. Therm. Spray Technol.* **26**, 1025, https://doi.org/10.1007/s11666-017-0584-z.
[43] Schulz U., Saruhan B., Fritscher K., Leyens C. (2004) Review on advanced EB-PVD ceramic topcoats for TBC applications, *Int. J. Appl. Ceram. Technol.* **1**, 302, https://doi.org/10.1111/j.1744-7402.2004.tb00182.x.
[44] Tan Y., Srinivasan V., Nakamura T., Sampath S., Bertrand P., Bertrand G. (2012) Optimizing compliance and thermal conductivity of plasma sprayed thermal barrier coatings via controlled powders and processing strategies, *J. Therm. Spray Technol.* **21**, 950, https://doi.org/10.1007/s11666-012-9780-z.
[45] Padture N.P., Gell M., Jordan E.H. (2002) Thermal barrier coatings for gas-turbine engine applications, *Science* **296**, 280, https://doi.org/10.1126/science.1068609.
[46] Miller R.A. (1997) Thermal barrier coatings for aircraft engines: history and directions, *J. Therm. Spray Technol.* **6**, 35, https://doi.org/10.1007/BF02646310.
[47] Bauer E. (1958) Phänomenologische theorie der kristallabscheidung an oberflächen. I, *Z. Krist.* **110**, 372, https://doi.org/10.1524/zkri.1958.110.1-6.372.
[48] Thornton J.A. (1977) High rate thick film growth, *Annu. Rev. Mater. Sci.* **7**, 239, https://doi.org/10.1146/annurev.ms.07.080177.001323.
[49] Hoffman D.W., Thornton J.A. (1977) The compressive stress transition in Al, V, Zr, Nb and W metal films sputtered at low working pressures, *Thin Solid Films* **45**, 387, https://doi.org/10.1016/0040-6090(77)90276-0.
[50] Windischmann H. (1992) Intrinsic stress in sputter-deposited thin films, *Crit. Rev. Solid State Mater. Sci.* **17**, 547, https://doi.org/10.1080/10408439208244586.
[51] Billard A., Perry F. (2005) Pulvérisation cathodique magnétron, Tech. Ing. Trait. Surf. Métaux Par Voie Sèche En Milieu Fondu. base documentaire: TIB360DUO, https://www.techniques-ingenieur.fr/base-documentaire/materiaux-th11/traitements-de-surface-des-metaux-par-voie-seche-et-en-milieu-fondu-42360210/pulverisation-cathodique-magnetron-m1654/.
[52] Michiels M., Konstantinidis S., Snyders R. (2013) La pulvérisation cathodique magnétron en régime d'impulsions de haute puissance (HiPIMS), Tech. Ing. Nanomatériaux Synthèse Élabor. base documentaire: TIB195DUO, https://www.techniques-ingenieur.fr/base-documentaire/innovation-th10/nanomateriaux-synthese-et-elaboration-42195210/la-pulverisation-cathodique-magnetron-en-regime-d-impulsions-de-haute-puissance-hipims-in207/.
[53] Anders A. (2011) Discharge physics of high power impulse magnetron sputtering, *PSE 2010 Spec. Issue.* **205**, S1, https://doi.org/10.1016/j.surfcoat.2011.03.081.
[54] Helmersson U., Lattemann M., Bohlmark J., Ehiasarian A.P., Gudmundsson J.T. (2006) Ionized physical vapor deposition (IPVD): a review of technology and applications, *Thin Solid Films* **513**, 1, https://doi.org/10.1016/j.tsf.2006.03.033.
[55] Gudmundsson J.T. (2008) Ionized physical vapor deposition (IPVD): magnetron sputtering discharges, *J. Phys. Conf. Ser.* **100**, 82002, https://doi.org/10.1088/1742-6596/100/8/082002.

[56] Kouznetsov V., Macák K., Schneider J.M., Helmersson U., Petrov I. (1999) A novel pulsed magnetron sputter technique utilizing very high target power densities, *Surf. Coat. Technol.* **122**, 290, https://doi.org/10.1016/S0257-8972(99)00292-3.

[57] Macák K., Kouznetsov V., Schneider J., Helmersson U., Petrov I. (2000) Ionized sputter deposition using an extremely high plasma density pulsed magnetron discharge, *J. Vac. Sci. Technol. A.* **18**, 1533, https://doi.org/10.1116/1.582380.

[58] Sellers J. (1998) Asymmetric bipolar pulsed DC: the enabling technology for reactive PVD, *Pap. Present. Fifth Int. Conf. Plasma Surf. Eng.* **98**, 1245, https://doi.org/10.1016/S0257-8972(97)00403-9.

[59] Kelly P.J., Henderson P.S., Arnell R.D., Roche G.A., Carter D. (2000) Reactive pulsed magnetron sputtering process for alumina films, *J. Vac. Sci. Technol. A.* **18**, 2890, https://doi.org/10.1116/1.1319679.

[60] Helmersson U., Lattemann M., Alami J., Bohlmark J., Ehiasarian A.P., Gudmundsson J.T. (2005) High power impulse magnetron sputtering discharges and thin film growth: a brief review, *Soc. Vac. Coaters 48th Annu. Tech. Conf. Proc.*, 458.

[61] Sarakinos K., Alami J., Konstantinidis S. (2010) High power pulsed magnetron sputtering: a review on scientific and engineering state of the art, *Surf. Coat. Technol.* **204**, 1661, https://doi.org/10.1016/j.surfcoat.2009.11.013.

[62] Bohlmark J., Lattemann M., Gudmundsson J.T., Ehiasarian A.P., Aranda Gonzalvo Y., Brenning N., Helmersson U. (2006) The ion energy distributions and ion flux composition from a high power impulse magnetron sputtering discharge, *Thin Solid Films* **515**, 1522, https://doi.org/10.1016/j.tsf.2006.04.051.

[63] Huo C., Raadu M.A., Lundin D., Gudmundsson J.T., Anders A., Brenning N. (2012) Gas rarefaction and the time evolution of long high-power impulse magnetron sputtering pulses, *Plasma Sources Sci. Technol.* **21**, 45004, https://doi.org/10.1088/0963-0252/21/4/045004.

[64] Lundin D., Sarakinos K. (2012) An introduction to thin film processing using high-power impulse magnetron sputtering, *J. Mater. Res.* **27**, 780, https://doi.org/10.1557/jmr.2012.8.

[65] Alami J., Bolz S., Sarakinos K. (2009) High power pulsed magnetron sputtering: Fundamentals and applications, *14th Int. Symp. Metastable Nano-Mater. ISMANAM-2007* **483**, 530, https://doi.org/10.1016/j.jallcom.2008.08.104.

[66] Sanchette F., Ducros C., Schmitt T., Steyer P., Billard A. (2011) Nanostructured hard coatings deposited by cathodic arc deposition: from concepts to applications, *Surf. Coat. Technol.* **205**, 5444, https://doi.org/10.1016/j.surfcoat.2011.06.015.

[67] Lyubimov G.A., Rakhovskiĭ V.I. (1978) The cathode spot of a vacuum arc, *Sov. Phys. Uspekhi.* **21**, 693, https://doi.org/10.1070/pu1978v021n08abeh005674.

[68] Boxman R.L., Goldsmith S. (1992) Macroparticle contamination in cathodic arc coatings: generation, transport and control, *Surf. Coat. Technol.* **52**, 39, https://doi.org/10.1016/0257-8972(92)90369-L.

[69] Martin P.J., McKenzie D.R., Netterfield R.P., Swift P., Filipczuk S.W., Müller K.H., Pacey C.G., James B. (1987) Characteristics of titanium arc evaporation processes, *Thin Solid Films* **153**, 91, https://doi.org/10.1016/0040-6090(87)90173-8.

[70] Boxman R.L., Goldsmith S., Greenwood A. (1997) Twenty-five years of progress in vacuum arc research and utilization, *IEEE Trans. Plasma Sci.* **25**, 1174, https://doi.org/10.1109/27.650894.

[71] Raizer Y.P. (1991) *Gas discharge physics.* Springer-Verlag Berlin Heidelberg.

[72] Sanders D.M., Anders A. (2000) Review of cathodic arc deposition technology at the start of the new millennium, *Surf. Coat. Technol.* **133–134**, 78, https://doi.org/10.1016/S0257-8972(00)00879-3.

[73] Randhawa H. (1988) Cathodic arc plasma deposition technology, *Thin Solid Films* **167**, 175, https://doi.org/10.1016/0040-6090(88)90494-4.

[74] Ducros C., Cayron C., Sanchette F. (2006) Multilayered and nanolayered hard nitride thin films deposited by cathodic arc evaporation. Part 1: deposition, morphology and microstructure, *Surf. Coat. Technol.* **201**, 136, https://doi.org/10.1016/j.surfcoat.2005.11.102.

[75] Brachet J.-C., Idarraga-Trujillo I., Flem M.L., Saux M.L., Vandenberghe V., Urvoy S., Rouesne E., Guilbert T., Toffolon-Masclet C., Tupin M., Phalippou C., Lomello F., Schuster F., Billard A., Velisa G., Ducros C., Sanchette F. (2019) Early studies on Cr-coated zircaloy-4 as enhanced accident tolerant nuclear fuel claddings for light water reactors, *J. Nucl. Mater.* **517**, 268, https://doi.org/10.1016/j.jnucmat.2019.02.018.
[76] Le Flem M., Ducros C., Sanchette F. (2013) Multilayer material resistant to oxidation in a nuclear environment, WO2013160587.
[77] Schafrik R.E., Ward D.D., Groh J.R. (2001) Application of alloy 718 in GE aircraft engines: past, present and next five years, *Superalloys 718 625 706 Var. Deriv.*, pp. 1–11.
[78] Ezugwu E.O., Wang Z.M., Machado A.R. (1999) The machinability of nickel-based alloys: a review, *J. Mater. Process. Technol.* **86**, 1, https://doi.org/10.1016/S0924-0136(98)00314-8.
[79] Choudhury I., El-Baradie M. (1998) Machinability of nickel-base super alloys: a general review, *J. Mater. Process. Technol.* **77**, 278, https://doi.org/10.1016/S0924-0136(97)00429-9.
[80] Arunachalam R., Mannan M.A. (2000) Machinability of nickel-based high temperature alloys, *Mach. Sci. Technol.* **4**, 127, https://doi.org/10.1080/10940340008945703.
[81] Yashar P.C., Sproul W.D. (1999) Nanometer scale multilayered hard coatings, *Vacuum* **55**, 179, https://doi.org/10.1016/S0042-207X(99)00148-7.
[82] Chu X., Barnett S.A. (1995) Model of superlattice yield stress and hardness enhancements, *J. Appl. Phys.* **77**, 4403, https://doi.org/10.1063/1.359467.
[83] Powell C.F., Campbell I.E., Gonser B.W. (1995) *Vapor-plating: the formation of coatings by vapor-deposition techniques.* John Wiley & Sons, Inc., New York.
[84] Powell C.F., Oxley J.H., Blocher J.M., Klerer J. (1966) *Vapor deposition.* John Wiley & Sons, Inc., New York.
[85] Stringfellow G.B. (1999) *Organometallic vapor-phase epitaxy: theory and practice.* Academic Press, San Diego, https://doi.org/10.1016/B978-0-12-673842-1.X5000-5.
[86] Morosanu C.E. (1990) *Thin films by chemical vapour deposition.* Elsevier, Amsterdam, https://doi.org/10.1016/C2009-0-12728-9.
[87] Hitchman M.L., Jensen K.F. (1993) Chemical vapor deposition: principle and applications. Academic Press, Inc., London.
[88] Kodas T.T., Hampden-Smith M.J. (2007) The chemistry of metal CVD. VCH, Weinheim, https://doi.org/10.1002/9783527615858.fmatter.
[89] Razeghi M. (1989) *The MOCVD challenge.* Adam Hilger Press, Bristol, Vol. I., 0-85274-161-8.
[90] Razeghi M. (1995) *The MOCVD challenge.* IOP Publishing, Bristol, Vol. II., 0-7503-0309-3.
[91] Rees W. Jr., Ed (1996) *CVD of nonmetals.* VCH, Weinheim, https://doi.org/10.1002/9783527614813.
[92] Jones A.C., O'Brien P. (2008) *CVD of compound semiconductors.* John Wiley & Sons, Ltd, https://onlinelibrary.wiley.com/doi/abs/https://doi.org/10.1002/9783527614639.fmatter.
[93] Xu Y., Yan X.-T., Eds (2010) Chemical vapour infiltration, *Chem. Vap. Depos. Integr. Eng. Des. Adv. Mater.*, Springer London, London, pp. 165–213, https://doi.org/10.1007/978-1-84882-894-0_5.
[94] Kääriäinen T., Cameron D., Kääriäinen M.-L., Sherman A. (2013) *Atomic layer deposition: principles, characteristics, and nanotechnology applications.* John Wiley & Sons, Ltd, https://doi.org/10.1002/9781118747407.fmatter.
[95] Matsumura H., Umemoto H., Gleason K.K., Schropp R.E.I. (2019) *Catalytic chemical vapor deposition: technology and applications of cat-CVD.* Wiley-VCH, Weinheim, https://doi.org/10.1002/9783527818655.fmatter.
[96] Bianco R., Rapp R.A. (1996) Pack cementation diffusion coatings, *Metall. Ceram. Prot. Coat.* (K.H. Stern, Ed). Springer Netherlands, Dordrecht, pp. 236–260, https://doi.org/10.1007/978-94-009-1501-5_9.
[97] Pfaff G. (2003) Special Effect pigments, high perform, *Pigments.* John Wiley & Sons, Ltd, pp. 75–101, https://doi.org/10.1002/3527600493.ch7.
[98] Biver C., Maury F., Santucci V., Senocq F., Vinsonneau S. (2018) Method and device for chemical vapor deposition of polymer film onto a substrate, EP2477754B1.

[99] Schropp R.E.I. (2015) Industrialization of hot wire chemical vapor deposition for thin film applications, *8th Int. Conf. Hot-Wire CVD Cat-CVD Process. HWCVD 8* **595**, 272, https://doi.org/10.1016/j.tsf.2015.07.054.
[100] Holleman J. (2002) Selective chemical vapor deposition, *Chem. Phys. Thin Film Depos. Process. Micro- Nano-Technol.* (Y. Pauleau, Ed). Springer Netherlands, Dordrecht, pp. 171–198, https://doi.org/10.1007/978-94-010-0353-7_8.
[101] Scott B.A., Olbricht W.L., Meyerson B.A., Reimer J.A., Wolford D.J. (1984) Homogeneous chemical vapor deposition, *J. Vac. Sci. Technol. A.* **2**, 450, https://doi.org/10.1116/1.572364.
[102] Vahlas C., Caussat B., Serp P., Angelopoulos G.N. (2006) Principles and applications of CVD powder technology, *Mater. Sci. Eng. R Rep.* **53**, 1, https://doi.org/10.1016/j.mser.2006.05.001.
[103] Windhorst T., Blount G. (1997) Carbon-carbon composites: a summary of recent developments and applications, *Mater. Des.* **18**, 11, https://doi.org/10.1016/S0261-3069(97)00024-1.
[104] Su J., Xiao Z., Liu Y., Meng F., Peng Z., Gu L., Li G., Xing R. (2010) Preparation and characterization of carbon/carbon aircraft brake materials with long service life and good frictional properties, *New Carbon Mater.* **25**, 329, https://doi.org/10.1016/S1872-5805(09)60037-8.
[105] Maury F. (1996) Trends in precursor selection for MOCVD, *Chem. Vap. Depos.* **2**, 113, https://doi.org/10.1002/cvde.19960020306.
[106] Auciello O., Krauss A.R. (2002) In situ real-time characterization of thin films, *Meas. Sci. Technol.* **13**, 644, https://doi.org/10.1088/0957-0233/13/4/705.
[107] Maury F., Duminica F.-D. (2007) Diagnostic in TCOs CVD processes by IR pyrometry, *First Int. Symp. Transparent Conduct. Oxides* **515**, 8619, https://doi.org/10.1016/j.tsf.2007.04.007.
[108] Senocq F., Duminica F.-D., Maury F., Delsol T., Vahlas C. (2006) Iron thin films from Fe(CO)[sub 5] and FeCp[sub 2]/H[sub 2]O under atmospheric pressure, *J. Electrochem. Soc.* **153**, G1025, https://doi.org/10.1149/1.2352050.
[109] Maury F., Gue A.M., Viguier N. (2001) Photoemission oscillations observed in chemical beam epitaxy of GaAs using triethylgallium and organoarsines, *Electrochem. Soc. Proc., I*, (M.D. Allendorf, M.T. Swihart, M. Meyyappan, Eds). Washington, DC, pp. 183–190.
[110] Viguier N., Maury F. (1999) Threshold photoemission analysis of the surface reactions of triethylgallium and trimethylgallium on GaAs(100): a promising technique for kinetics studies, *Appl. Phys. Lett.* **74**, 266, https://doi.org/10.1063/1.123276.
[111] Allendorf M.D., Besmann M., Kee R.J., Swihart M.T. (2008), Modeling CVD processes, *Chem. Vap. Depos. Precursors Process. Appl.* (A.C. Jones, M.L. Hitchman, Eds). RSC Publishing, 93, https://doi.org/10.1039/9781847558794.
[112] Etchepare P.-L., Vergnes H., Samélor D., Sadowski D., Caussat B., Vahlas C. (2015) Modeling a MOCVD process to apply alumina films on the inner surface of bottles, *Surf. Coat. Technol.* **275**, 167, https://doi.org/10.1016/j.surfcoat.2015.05.022.
[113] Michau A., Gazal Y., Addou F., Maury F., Duguet T., Boichot R., Pons M., Monsifrot E., Maskrot H., Schuster F. (2019) Scale up of a DLI-MOCVD process for the internal treatment of a batch of 16 nuclear fuel cladding segments with a CrCx protective coating, *Surf. Coat. Technol.* **375**, 894, https://doi.org/10.1016/j.surfcoat.2019.06.101.
[114] Hainsworth S.V., Uhure N.J. (2007) Diamond like carbon coatings for tribology: production techniques, characterisation methods and applications, *Int. Mater. Rev.* **52**, 153, https://doi.org/10.1179/174328007X160272.
[115] Pochet L.F., Howard P., Safaie S. (1997) CVD coatings: from cutting tools to aerospace applications and its future potential, *24th Int. Conf. Metall. Coat. Thin Films* **94–95**, 70, https://doi.org/10.1016/S0257-8972(97)00478-7.
[116] Hetmańczyk M., Swadzba L., Mendala B. (2007) Advanced materials and protective coatings in aero-engines application, *J. Achiev. Mater. Manuf. Eng.* **24**.
[117] Streiff R. (1993) Protection of materials by advanced high temperature coatings, *J. Phys. IV Fr.* **3**, C9, https://doi.org/10.1051/jp4:1993902.

[118] Gialanella S., Malandruccolo A. (2020) Coatings, *Aerosp. Alloys* (S. Gialanella, A. Malandruccolo, Eds). Springer International Publishing, Cham, pp. 387–438, https://doi.org/10.1007/978-3-030-24440-8_7.
[119] Zhuk Y., Aleksandrov S., Lakhotkin Y. (2020) Alloyed tungsten produced by chemical vapour deposition, EP1838896B1.
[120] Zhuk Y. (2019) Water droplet erosion resistant coatings for turbine blades and other components, GB2568063B.
[121] Ekinaka T., Kon T., Kishimoto H., Takeda Y., Suga T., Ogata S., Nakagomi M. (2020) Polymer substrate with hardcoat layer, EP3603954A1.
[122] Ekinaka T., Kon T., Suga T., Kishimoto H., Morita Y., Ogata S., Nakagomi M. (2018) Polymer substrate with hardcoat layer, and manufacturing method for same, EP3354455A1.
[123] Schaefer C., Bräuer G., Szczyrbowski J. (1997) Low emissivity coatings on architectural glass, *Surf. Coat. Technol.* **93**, 37, https://doi.org/10.1016/S0257-8972(97)00034-0.
[124] Kamalisarvestani M., Saidur R., Mekhilef S., Javadi F.S. (2013) Performance, materials and coating technologies of thermochromic thin films on smart windows, *Renew. Sustain. Energy Rev.* **26**, 353, https://doi.org/10.1016/j.rser.2013.05.038.
[125] Gordon R. (1997) Chemical vapor deposition of coatings on glass, *Coat. Glass.* **218**, 81, https://doi.org/10.1016/S0022-3093(97)00198-1.
[126] Leese A.B., Mills A.R. (2008) Commercial aspects of CVD, *Chem. Vap. Depos. Precursors Process. Appl.* (A.C. Jones, M.L. Hitchman, Eds). RSC Publishing, p. 536, https://doi.org/10.1039/9781847558794.
[127] Keshri A.K., Sribalaji M. (2015) Coatings for energy applications, *Thin Film Struct. Energy Appl.* (K. Suresh Babu, Ed). Springer, https://doi.org/10.1007/978-3-319-14774-1.
[128] Wang X., Yushin G. (2015) Chemical vapor deposition and atomic layer deposition for advanced lithium ion batteries and supercapacitors, *Energy Environ. Sci.* **8**, 1889, https://doi.org/10.1039/C5EE01254F.
[129] Chen C., Eichel R.-A., Notten P.H.L. (2017) Metal-organic chemical vapor deposition enabling all-solid-state Li-ion microbatteries: a short review, *J. Electroceramics* **38**, 230, https://doi.org/10.1007/s10832-017-0090-1.
[130] Marichy C., Bechelany M., Pinna N. (2012) Atomic layer deposition of nanostructured materials for energy and environmental applications, *Adv. Mater.* **24**, 1017, https://doi.org/10.1002/adma.201104129.
[131] Principes et applications de la technique ALD (Atomic Layer Deposition), Techniques de l'ingénieur, 2016.
[132] Miikkulainen V., Leskelä M., Ritala M., Puurunen R.L. (2013) Crystallinity of inorganic films grown by atomic layer deposition: overview and general trends, *J. Appl. Phys.* **113**, 21301, https://doi.org/10.1063/1.4757907.
[133] George S.M. (2010) Atomic layer deposition: an overview, *Chem. Rev.* **110**, 111, https://doi.org/10.1021/cr900056b.
[134] Hämäläinen J., Ritala M., Leskelä M. (2014) Atomic layer deposition of noble metals and their oxides, *Chem. Mater.* **26**, 786, https://doi.org/10.1021/cm402221y.
[135] Proslier T., Klug J.A., Elam J.W., Claus H., Becker N.G., Pellin M.J. (2011) Atomic layer deposition and superconducting properties of NbSi Films, *J. Phys. Chem. C.* **115**, 9477, https://doi.org/10.1021/jp201873b.
[136] Groll N.R., Klug J.A., Cao C., Altin S., Claus H., Becker N.G., Zasadzinski J.F., Pellin M.J., Proslier T. (2014) Tunneling spectroscopy of superconducting MoN and NbTiN grown by atomic layer deposition, *Appl. Phys. Lett.* **104**, 92602, https://doi.org/10.1063/1.4867880.
[137] Snyder M.Q., Trebukhova S.A., Ravdel B., Wheeler M.C., DiCarlo J., Tripp C.P., DeSisto W. J. (2007) Synthesis and characterization of atomic layer deposited titanium nitride thin films on lithium titanate spinel powder as a lithium-ion battery anode, *J. Power Sources* **165**, 379, https://doi.org/10.1016/j.jpowsour.2006.12.015.

Chapter 4

New Drivers for Materials Science and Engineering

Abstract

The purpose of this chapter is to illustrate the growing importance of new drivers that are taking an increasingly important place in materials science and engineering. These are generally environmental factors, energy sobriety in the production, and transformation and recycling processes, in addition to societal aspects often linked to risk management for instance nanosafety. It is also about showing how the apparition of digital technologies has a very strong impact on new developments of innovative materials or the optimization of complex production processes. This impact of digital technologies as for example the Artificial Intelligence is reflected in the accelerated discovery of new enhanced solutions, for better process controlling, less waste production, more value addition. Using three examples, this chapter presents these highly transdisciplinary drivers in materials science and engineering where the impact of digital technologies on innovation, the respect for our planet, the preservation of natural resources and human health are at the core of the choices made.

DOI: 10.1051/978-2-7598-2446-5.c004

4.1 Material Resource Efficiency in Low Carbon Energy: Towards a More Circular Economy

E. BOUYER

CEA/DES/DPE 17, rue des Martyrs, 38054 Grenoble, France

Abstract

Development of technologies usually start with the setting up of material composition according to the specifications of the targeted application. Then come the synthesis & shape forming related techniques to make the material, the part, the system. To reduce the impact of technologies one has to pay attention to each step of the development and to their life under operation. This finishes with the product end of life which can induce impact as well if it is carefully managed. This impact could be environmental (*e.g.* waste production of possible toxic substances, use toxic substances during the fabrication steps). Based on examples from new technology for energy sectors, this section aims to show the approaches to minimize impact regarding material resources during the development step of the technologies and to define a path towards a more circular economy in such technological sector. The effort to improve the material resource efficiency in the new technology for energy is driven by the fact that those technologies need critical raw materials (also called "vitamin for the high tech") to bring the specific properties they have to exhibit. Huge efforts have to be taken to limit dependency on those materials by means of reducing their use (*e.g.* thanks to nanotechnology), to replace them (by finding abundant material as substitute) and to tackle with the end of life (by working on the eco-design allowing an easy dismantling & recycling) or a potential second life. The paradigm shifts to be implemented (carbon rich to low carbon energy, linear to circular economy) need to fit with the aspirations of society. Issues related with social acceptance will be discussed.

4.1.1 Introduction & Definition of Strategic, Critical Materials

There is no specific definition of critical Raw Materials but we can try to use such one: from a company point of view, strategic materials correspond to the one necessary for its own development and for the industrial sector. From the state point of view, strategic materials are those that are essential for its economical and/or state sovereign policy. The concept of critical materials is an extension of the previous one: this applies to strategic materials where the supply chain is threatened.

Three items on which the Raw material's national doctrine is based are:

- Energy independence;
- Food self sufficiency;
- Mineral independence.

DOI: 10.1051/978-2-7598-2446-5.c014

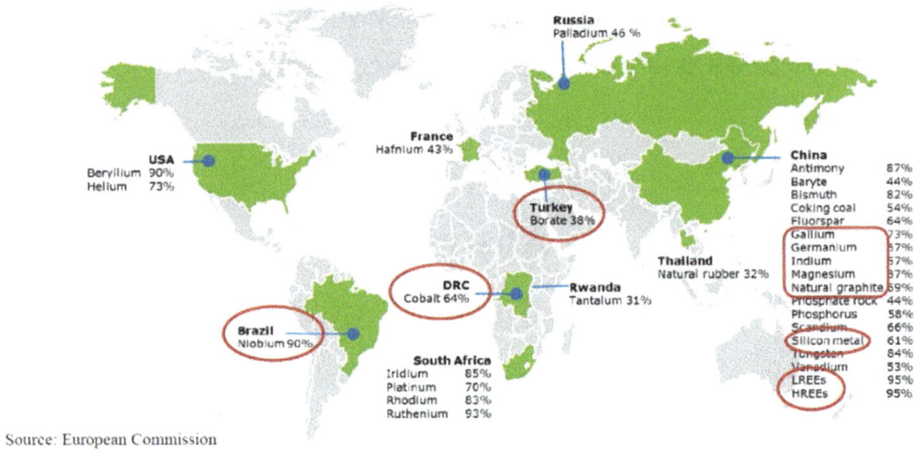

FIG. 4.1.1 – Raw Materials' production monopoly.

The absence of one of these independences prevents any sustainable economic development.

One can use the five following criteria to precisely explain the perimeter of critical materials:

- Quantitative criterion: low quantity production, from few tons to 200 000 tons/a, which distinguish them from industrial metals whose production exceeds millions of tons, even billions of tons for iron ore.
- Technical criterion: often by products of mineral or metallurgical, but there exists rare metals' mine where they are exploited as principal metals.
- Economical criterion: high value products mainly due to high purity required from applications.
- Critical criterion: despite a low turnover, industrial importance could be substantial due to the dependency of several industrial sectors against some of them.
- Monopolistic criterion: several countries (industrial companies) are benefiting from a monopolistic position for a large number of those metals, see figure 4.1.1.

4.1.2 Current Issues in Material Resources

Technologies, like Li-ion batteries, need materials and associated processes to synthesize/prepare, transform and shape them in order to build components, devices in integrated systems. To optimize both upstream & downstream, to overcome the bottleneck shown in figure 4.1.2 and thus reduce technologies footprint, 4 main focus are:

- Use abundant materials (or precursors of materials).
- Minimize or avoid the use of CRM.
- Promote clean and efficient processes (limiting/avoiding waste, effluents).
- Optimize the design to facilitate 2nd life and end of life treatment (recycling).

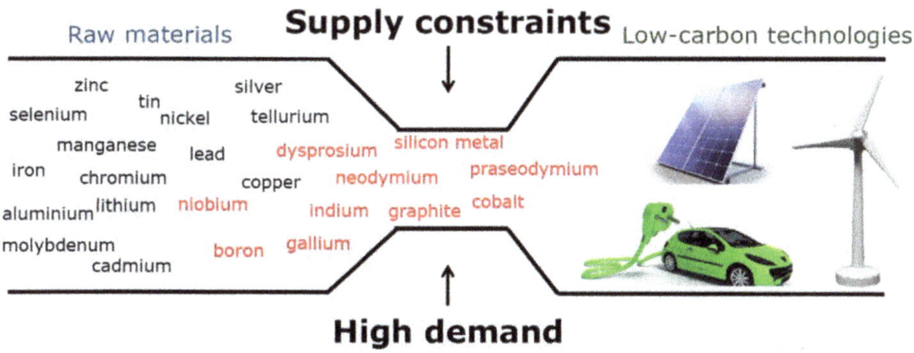

FIG. 4.1.2 – Low carbon technologies RM bottleneck.

4.1.3 From Sobriety to Material Substitution

In practice, 4 main driving forces are initiating works on substitution:

- Potential or existing shortage on given substance leading to a price increase and/or supply difficulties (CRM).
- Banishment of a substance due to regulations (caused by toxicity,...) Cd, Cr VI.
- Novel (abundant & nontoxic) materials/precursors opening new applications.
- Continuous need for materials with enhanced specific properties (performance, durability).

When the two first are handling constraints, the last two fall within opportunities. Substitution can cover different timescale and thus maturity level, from long term vision (starting at very low TRL) to short term approach (high TRL and very close to the deployment at industrial scale).

4.1.4 Case Studies in New Technologies for Energy: Materials and Processes

4.1.4.1 Production

PV is one of the most growing renewable sources for electricity production thanks to the continuous cost reduction in the cells, modules and full panels and to national policies encouraging the deployment of PV solutions. The current worldwide PV technology is mainly based on silicon cells. Despite strong R&D development on disruptive perovskite materials driven by a potential higher efficiency and lower impact (GHG) than silicon based cells during their synthesis and processing, silicon cells have a solid future at least for the next decade.

To illustrate the trends in PV, figure 4.1.3 shows a large portfolio of applications. This covers a lot of way to apply PV, from conventional PV field to PV integrated in buildings (roof and/or façade) or in mobility (roads, cars, airplane, pseudo satellite).

FIG. 4.1.3 – Application range of PV.

The ways to implement "PV everywhere" are multiple. In order to reduce the ground footprint, the building integrated PV (BIPV) approach is promoted. Integration in building or in mobile systems involve a strong weight reduction. Reducing the weight implies a new design where reduction in material quantities is targeted. This virtuous circle drives innovation towards material sobriety to diminish the thickness of the different materials necessary to make a PV panel. Picture shown in figure 4.1.4 illustrates a military application abroad where PV panel weight is reduced up to 4 kg/m^2 *versus* 12 kg/m^2 in classical PV field thanks to an overall design effort.

FIG. 4.1.4 – Light PV panel integrated in building deployed in military field operations.

4.1.4.2 Storage

The booming field of Li-ion battery is strongly driven by the emergence of e-mobility. Issues linked to CRM are of importance for the battery deployment worldwide. Cobalt, nickel, graphite and even lithium are substances where all actors are paying attention to their supply chain. A lot of activities are directed towards a substitution of graphite where most of its production comes from China. Due to its high theoretical specific capacity (3579 mAh/g), silicon appears as an interesting alternative to graphite for high density energy applications in Li-ion batteries. However, due to huge volume changes upon lithiation/delithiation, rapid pulverization occurs with loss of performance. Core shell Si@C materials are synthesized (*e.g.* through by laser pyrolysis in a double stage laser pyrolysis set up allowing the control of the Si core organization (amorphous or crystalline) as well as the shell thickness).

Other relevant examples are currently under progress like the reduction in cobalt content in the cathode driven by large industrial materials for battery suppliers. New generation of Li-ion battery will penetrate gradually the market, the new chemistries for anode, cathode and electrolyte are motivated by a reduction in CRM content, a cost reduction and an increasing demand for safer battery devices (this is the case for solid state batteries). Figure 4.1.5 shows the different technologies that should penetrate the battery market in the next years.

4.1.4.3 Conversion

PEM fuel cell is a device that converts chemical energy into electricity without CO_2 emission, see figure 4.1.6. Both electrochemical reactions occurring at anode and cathode need noble catalysts to allow an operation temperature range between room

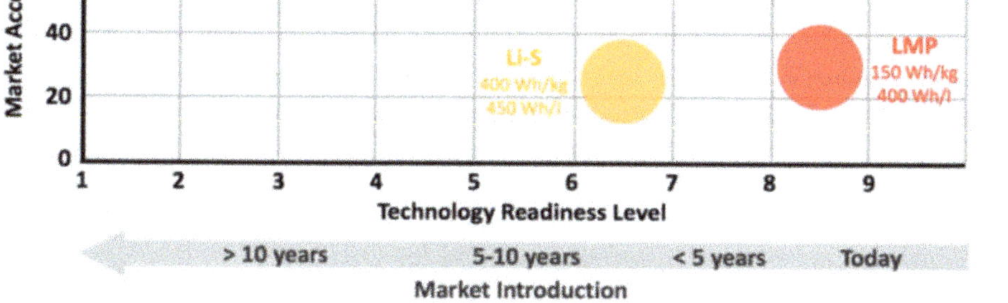

FIG. 4.1.5 – Market introduction of Li-ion battery technologies.

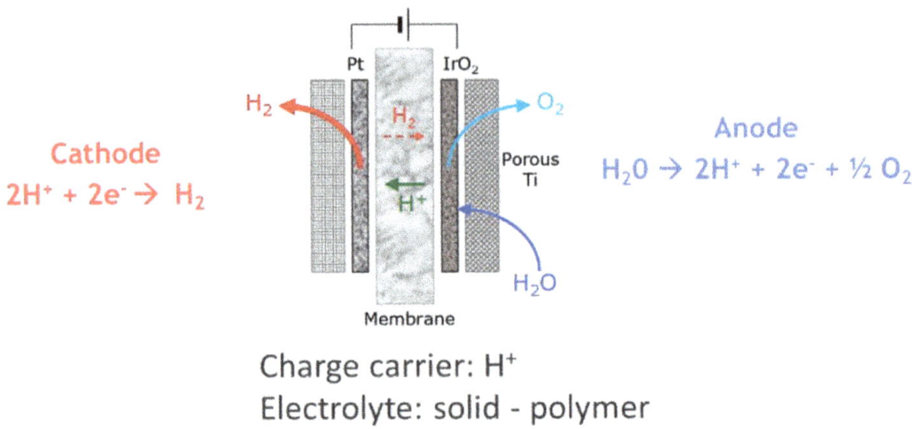

FIG. 4.1.6 – Principle of a PEM fuel cell (PEMFC).

temperature to 90 °C. One of the main issues impeding a larger penetration of such a technology is the content of noble metals. Several strategies are tested to either reduce the PGM (Platinum Group Metal) content or replace them completely.

Taking inspiration from the structure and function of natural enzymes called hydrogenases, novel electrocatalysts for hydrogen production (HER) and oxidation (HOR) have been designed, fully characterized and modeled thanks to theoretical chemistry methods. These molecular catalysts are based on Earth abundant elements such as iron, nickel, molybdenum and cobalt. The combination of this bioinspired molecular approach with nanochemical tools (in particular covalent and non-covalent surface modification techniques) allowed developing the first noble-metal free electrocatalytic materials for reversible H_2 oxidation/production under conditions compatible with PEM technologies. These materials based on a molecular Ni complex attached on carbon nanotubes were indeed tested under half-cell condition and found active with very high stability under operation.

Carbonaceous materials doped with several hetero-elements like nitrogen, or transition metal as iron or cobalt, are able to catalyze oxygen reduction (ORR) in acidic medium with an efficiency close to platinum catalyst. Nano-carbon as graphene or nanotube opens the way of a new era for these catalysts; they offer higher stability, conductivity and surface area compared to carbon black. Moreover these new materials offer a great chemical versatility which allow to use several grafting methods (covalent or non-covalent) or to directly build active catalytic site on the material surface. DRF/INAC develop nano-structurated graphene based catalysts in order to enhance oxygen and reactant diffusion and to increase the density of catalytic active site obtained by doping method. New molecular precursors of high nitrogen content are studied to model highly nitrogen-doped graphene catalysts, and bioinspired approaches are also developed for ORR.

The integration of both noble metal-free ORR and HOR nanomaterials into an operating and fully noble metal-free PEM fuel cell has been achieved. The implementation of Earth-abundant electrocatalysts into PEMFC has to be continued in order to increase cyclability, durability and kinetic *versus* standard PGM based solutions.

4.1.4.4 Transportation

Transportation of electricity involves huge quantities of conductive materials, either for stationary infrastructure (transportation and distribution network) within a country or for mobility applications, especially e-mobility. One of the most used conductive materials is based on copper. Primary resources are there (and limited) but their extraction becomes more difficult (deposit of lower quality) with environmental consequences that could be locally important: in 1910, to get one ton of copper (metal) one needed to extract 55 tons of ores whereas in 2010 125 tons of ores had to be extracted to get one ton of copper.

The larger commercial aircraft (Airbus A380) contains more than 500 km electric cables representing approximately 4 tons of materials, mainly metals like copper. Some novel approaches are trying to substitute copper by lighter and abundant materials while maintaining or even enhancing the electrical properties.

Carbon nanotubes (CNT) are good candidates since they exhibit very good electrical conductivity, especially with no influence on temperature and on frequency, which is not the case for copper. CNT are light material combining in a unique manner both good electrical, thermal and mechanical properties. The question to answer is how to put together at a macroscopic scale these nanomaterials while maintaining their exceptional properties. Based on CVD techniques (PACVD) to grow CNT carpets, then carpet spinning to fabricate yarns up to 15–20 μm diameter, see figure 4.1.7.

Some issues still need to be solved, like the low yarn density (which decreases electrical conductivity) or the limitation to build cable with a higher (mm or cm

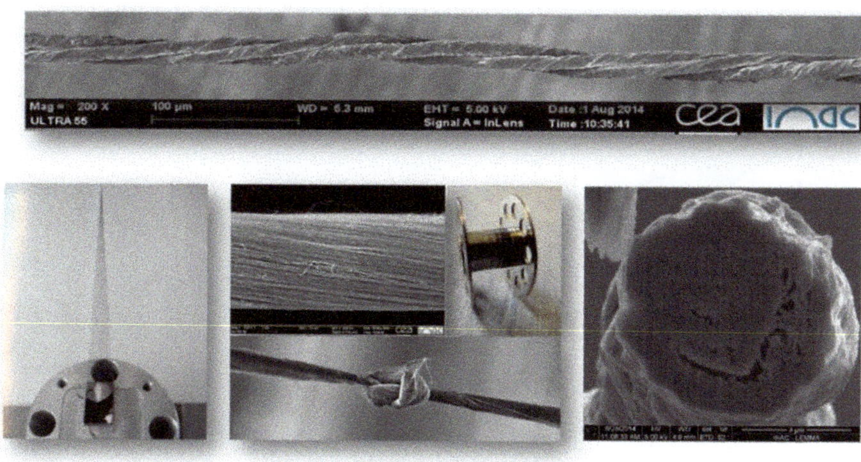

FIG. 4.1.7 – SEM pictures of "Microcables" made of CNT.

scale) diameter. An intermediate option would be to build composite conductive materials by mixing CNT and copper in order to reduce the quantity of the latest.

4.1.5 Circular Economy & Recycling

Introduction of New Paradigms

Circular economy is a great source of change of use that would help to shift from linear to circular economy. As an example, service economy is based on a shift from selling a product to selling a function of a product. This novel approach involves decoupling added value and energy consumption and raw materials. That is the case of Michelin which invoices the distance covered by delivery trucks equipped by its tyres instead of selling them. Similarly, Xerox or Lexmark invoices photocopies per item instead of selling or renting photocopiers.

4.1.5.1 From Linear to Circular Economy, Material & Process

This paradigm shift is promoting a better use of resources (both material and energy) and limit or even avoid the disposal of waste, see figure 4.1.8.

Not only material composition but processes are also of great importance for improving material resource efficiency. Therefore, many efforts are done to move towards more efficient, clean processes to shape materials. Hereafter are listed five key aspects acting as driving force for better processes:

- Limit "substractive" manufacturing and promote "additive" one to make near net shape parts and components (*e.g.* 3D printing, injection molding,...).
- Promote coating techniques (physical, chemical, hybrid) when surface properties need to be exhibited.

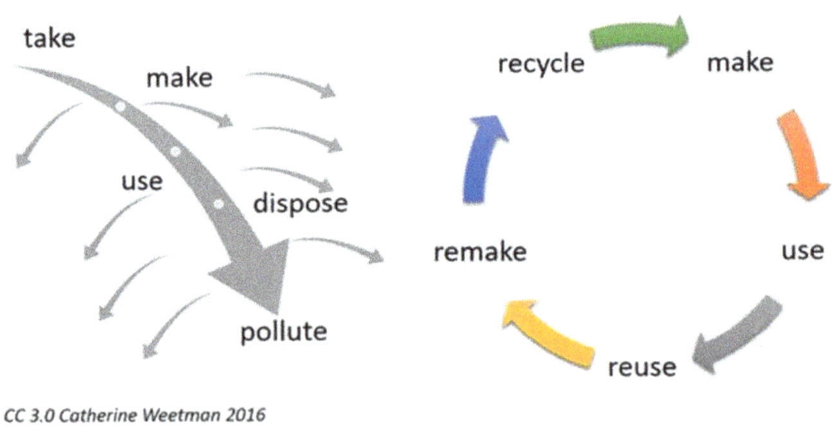

FIG. 4.1.8 – From linear to circular economy.

- Switch processes towards solvent free with low water consumption (*e.g.* SCF).
- Convert current processes or generate novel approaches to operate with novel/alternative feedstock (*e.g.* CO_2, bio-based, recycled materials).
- Continuous need for processes with enhanced performance (yield, throughput, intensification), robust and, with lower energy consumption.

Additive manufacturing (AM) techniques, often referred to as 3D Printing, are different innovative techniques which enhance the freedom in terms of engineered parts design. These techniques employ computer aided design (CAD) data for building layer upon layer the 3D near-net-shape of metal, ceramic, polymer or even composites according to the feedstock employed (powders, wires,...).

The main advantage is the reduction in the quantity of material waste since some CNC machining/milling/welding steps are avoided. In particular, AM reduces the time to market since when the re-design and customization is needed only the CAD model has to be modified.

In addition, these techniques allow the fabrication of complex parts with internal cavities and/or channels in a single-step. In this context, light weight structures (3D-lattices, open-cellular, foams) with tailored internal porosity could be produced. Most of the AM techniques are shown in figure 4.1.9.

To illustrate the potential of AM techniques, figure 4.1.10 shows an Al alloy based valve made by casting (left part). A re-design, thanks to CAD, of the genuine valve has been conducted to allow its shape forming by means of AM/SLM (centre image). While keeping the same Al alloy and maintaining the mechanical properties, a substantial weight gain of approximately 30% has been achieved by redesigning the part for 3D printing (right picture).

4.1.5.2 Recycling

Among most of common metals, a large (and unexploited) potential for recycling is depicted in figure 4.1.11. The penetration of circular economy depends strongly on regulation that will accompany, catalyze the change for both citizens and industries. Upstream change are necessary such as eco-design considering the way to make easier the dismantling and the following recycling steps.

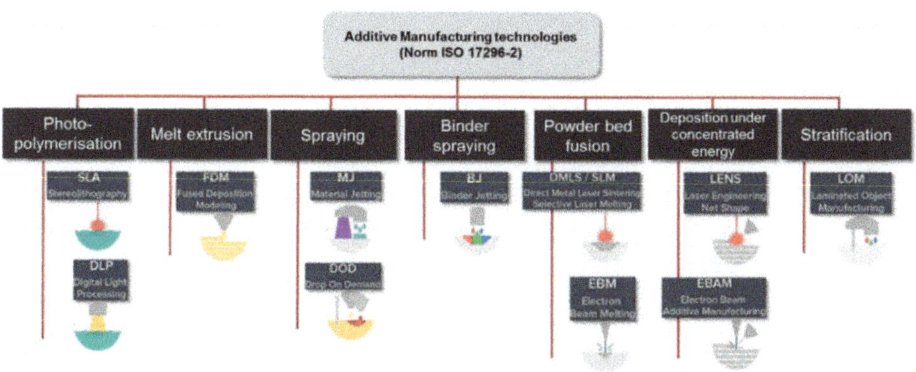

FIG. 4.1.9 – Main families of additive manufacturing techniques.

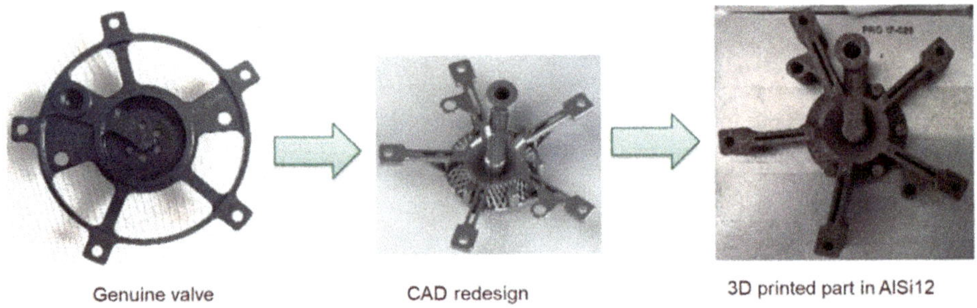

FIG. 4.1.10 – Comparison of parts made by conventional process (left) *versus* AM technique (right).

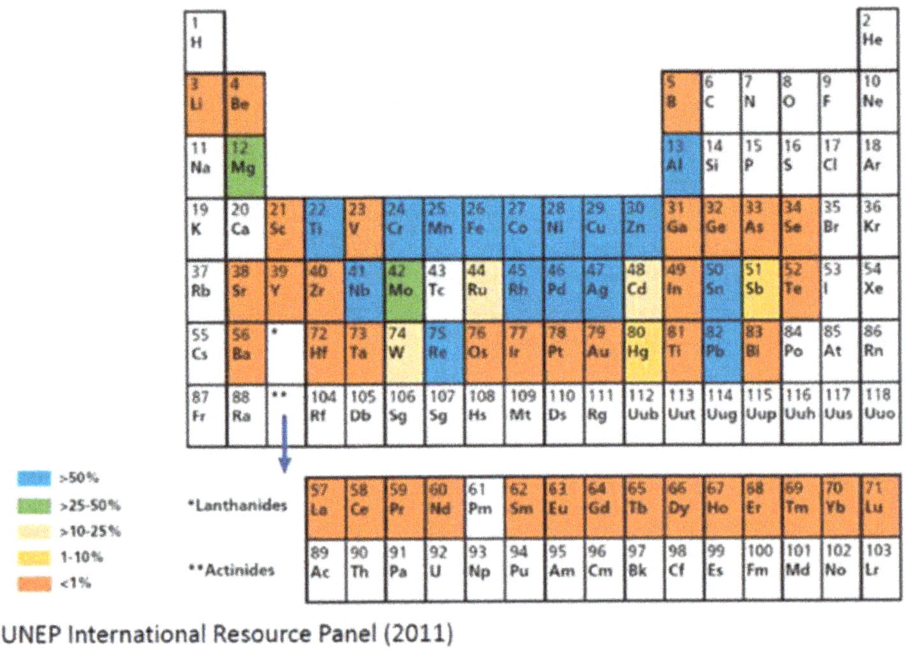

FIG. 4.1.11 – Recycling rate of 60 metals.

4.1.5.3 *Life Cycle Analysis, a Way to Assess the Impact of Technologies*

Life Cycle Assessment (LCA) activity is devoted to understand the environmental characteristics of technologies. To ensure the consistency and integrity of conducted analysis, this task has not been given to researchers directly involved in the development of each technology solution but to qualified individuals mastering the methods of environmental assessment and working transversely and in synergy. The environmental studies conducted aim at understanding the stages of the life cycle that have the greatest impact and the components and processes that significantly weight on the environment for all low carbon technologies. These results help positioning them in regard with competitive technologies and finding new innovation path lowering environmental impacts. New energy technologies' research exists to answer the question of energetic autonomy but also to face the challenge of reducing the contribution of the energy needs to greenhouse gas emissions and climate change.

4.1.6 Social Acceptance, Ethical Issues Versus Education

The following acronyms known as NIMBY (Not In My Back Yard) or BANANA (Build Absolutely Nothing Anywhere Near Anything (or Anyone)) reflect a societal tendency that has to be taken into account while developing new technologies, new approaches. NIMBYISM is a characterisation of opposition by residents to a proposed development in their local area. It carries the connotation that such residents are only opposing the development because it is close to them and that they would tolerate or support it if it were built farther away. The residents are often called Nimbys.

Elements of social acceptance (see figure 4.1.12):

- Awareness of climate change and knowledge of the technology (such as renewable energy).
- Education of citizen.
- Fairness of the decision-making process.
- Overall evaluation of costs, risks and benefits of a technology.

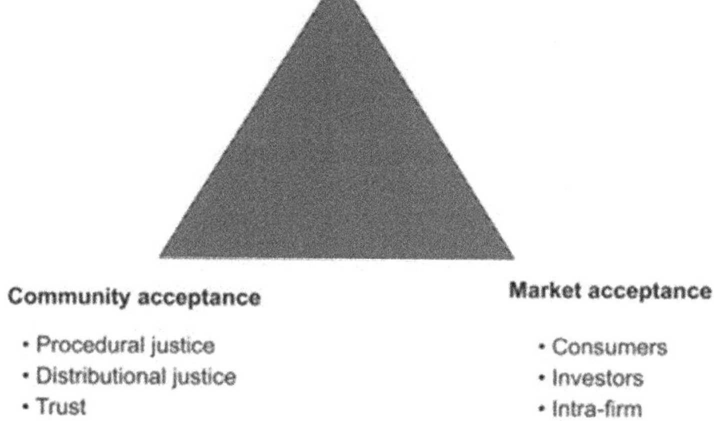

FIG. 4.1.12 – Acceptance aspects.

- Local context.
- Trust in decision-makers and other relevant stakeholders.

Conflict minerals are ores whose exploitation are realized in conditions such as armed conflicts and violations of human rights, affecting further exploitation and trade of the resulting materials. In the USA, the Dodd-Frank Wall Street Reform & Consumer Protection Act (section 1502) signed by President Obama (21/07/2010) "Impose to corporations that operate on American soil to declare the provenance of the minerals they use" (Conflict minerals as define by the US State Department are: tin, gold, tantalum and tungsten). In Europe, a similar law was more recently established but is not a binding legislation as the one in US, the EU legislation will apply from January 21st 2021.

Education of citizens (starting with children) seems to be a part of the response to ethical and social issues. Education would improve the transparency and it will help to provide confidence for new technologies emerging on the market. Therefore, education will reinforce trust which could help strengthen the adherence of the population to changes supported by technologies.

4.1.7 Conclusion

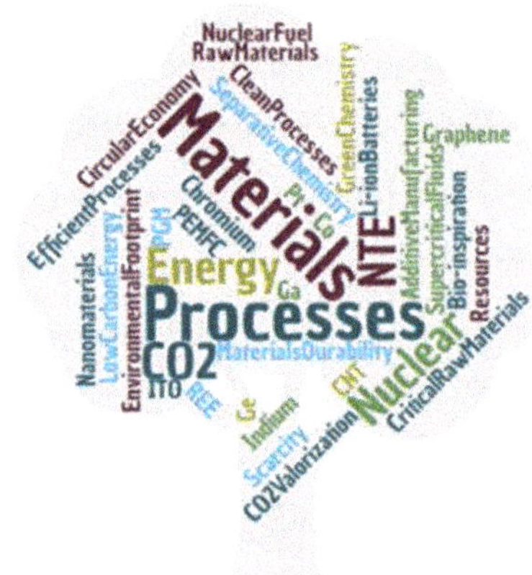

Material efficiency is a necessary approach that one needs to consider while developing energy technology. Material efficiency is needed to accelerate the achievement of the energy transition (leverage effect on material cost). Material efficiency approaches involve short to long term development depending on the level of disruption and on the time to market. Substitution is not only a material issue, what matters is processing and integration of the substitute material in the component/device/complete system. Recyclability of material has to be considered, within a systemic approach of circular economy & material efficiency. Finally Education is a tool that has to be used for increasing the awareness of the people, of the citizens on challenges inside or outside the country (typically the case of CRM) in order to obtain a broader social acceptance shared by a majority of citizens.

4.2 Artificial Intelligence for Materials Science and Engineering

Jean-Philippe POLI, Hiba HAJRI and Laurence BOUDET

Université Paris-Saclay, CEA, List, F-91120, Palaiseau, France

Abstract

Materials science is an active area of research that has been transformed by Computer science several times. Today, Artificial Intelligence is bringing new solutions to discover materials, to improve their performance, and to accelerate the processes. This chapter aims at giving clues to the reader about the current applications of Artificial Intelligence techniques in that field. We will also describe the various methods that are involved in recent researches.

4.2.1 Introduction

Materials science is an interdisciplinary field that focuses on the design and the discovery of new solid materials.

The key factor in promoting advances in materials science relates to Process–Structure–Properties–Performance relationships (P–S–P–P), as overviewed in figure 4.2.1. That is why one of the most important goals of materials scientists is to identify the relationships between the process by which a material is produced, the microstructure or the nanostructure of the material, its properties and its performance. Once uncovered, these P–S–P–P relations can guide discovery and design of new materials, and optimize manufacturing processes.

One important part of these linkages are Processing–Properties–Performance linkages (P–P–P). In fact, each production process has its own parameters and each product has its own properties (*e.g.*, mechanical, physical, etc.). The value of these properties determines the performance of the product. Thus, it is important to detect which parameters have the most important impact on the product properties and then on its performance. For this reason, materials scientists are interested on establishing causal relations between processing parameters and materials' properties.

If Materials science took its roots from Physics and Chemistry, it has been affected several times by Computer science.

Indeed, computational simulations allow researchers to explain complex real world phenomena, based on different techniques as density functional theory [3], Monte Carlo techniques [4] and other methods. Compared with experiments, simulations require less time and researchers have a full control over the configuration settings. However, accurate simulations present also many challenges. For example, a computational simulation requires high performance computing equipment and no explicit use can be made of preceding calculation results when a new system is studied.

DOI: 10.1051/978-2-7598-2446-5.c024

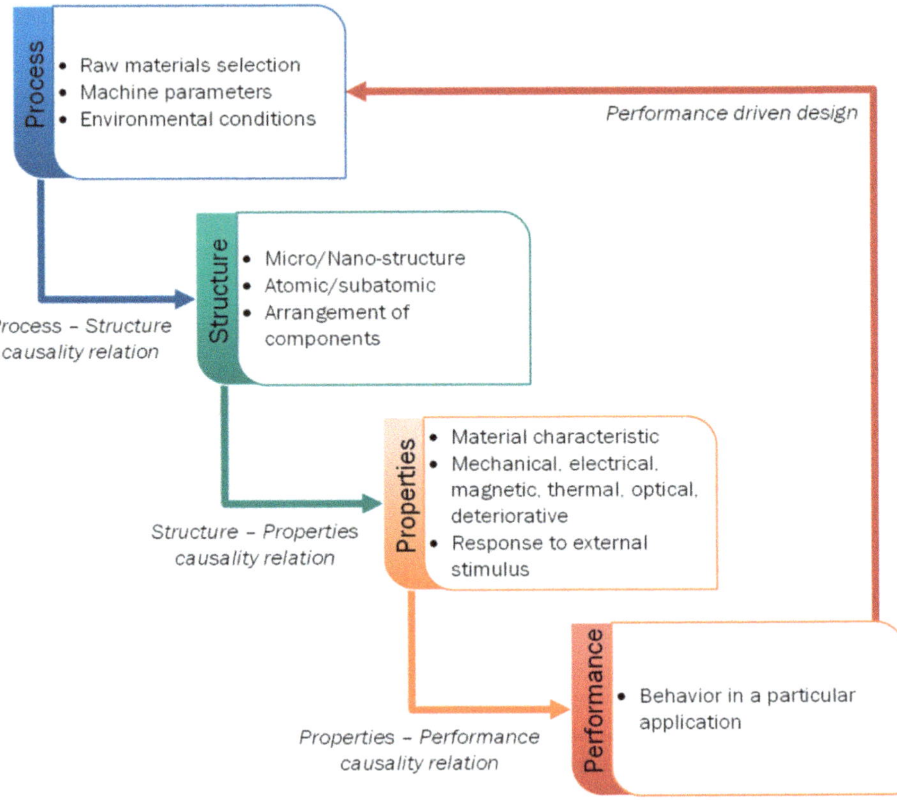

Fig. 4.2.1 – The materials' paradigm [1, 2].

Experiments combined to simulations are widely adopted in the field of materials science to have useful results in less time. However, this combination is often incapable of addressing newly emerging issues. For instance, it is very complicated and inefficient to investigate the transition temperature of glass through experimental measurements because the transition occurs over a wide temperature range [5]. Nevertheless, the glass transition temperature also cannot be exactly simulated using computer programs because it depends on a variety of internal and external conditions, such as pressure, structure, and constitutive and conformational features [6, 7].

The need to study the amount of data generated by this combination has given rise to Material Informatics (MI). In MI, based on available data, P–S–P–P relations are detected and evaluated using statistics. More recently, Artificial Intelligence (AI) approaches have appeared to help in this process.

Artificial Intelligence is a relatively young science that aims at simulating reasoning and learning capacities on machines. Recently, the availability of large volumes of data (and the marketing term "big data") made AI publicized. Indeed, humans cannot process such volumes that push the boundaries of machine learning.

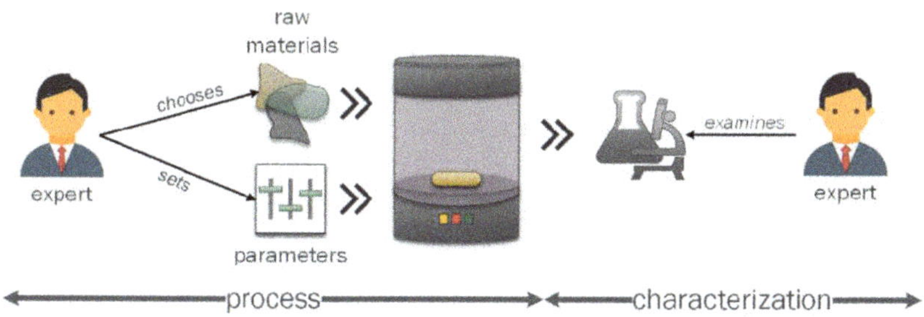

FIG. 4.2.2 – Overview of the design of new materials.

In particular, this huge amount of data meets the incredible progresses of hardware that increase significantly the processing capabilities of computers. The combination of Artificial Intelligence and of big data is sometimes referred to as "the fourth paradigm of science" [8] or "the fourth industrial revolution" [9].

AI approaches to material design do not need a precise mathematical model of the process. The processes are rather modeled as a complex system. Figure 4.2.2 presents a general schema of the modeling that consists in two phases. The first phase is the process and the preparation of this process by an expert. A process can be seen as a black box with parameters and inputs: in the case of material design, the inputs are the different raw materials that will be combined during the process. The result of the process is a new material. It leads to a second phase, which concerns its characterization to determine if it fits the expert's expectations.

AI algorithms can help at several points of the workflow, and more generally can help to develop performance driven strategies to discover new materials. For instance, the prediction of the performance from the process parameters can lead to a new material before confirmation by experimentations, leading to increased productivity.

In this chapter, we propose to give some applications of Artificial Intelligence in the context of materials science. We first present different techniques that have been applied to materials science and whose applications will be presented further. We then focus on the characterization of causal relations on the materials' paradigm in section 4.2.3, presenting the statistics methods that are used in this domain. In section 4.2.4, we present AI applications for materials' properties prediction for process parameters' optimization. We then investigate the selection of raw materials (section 4.2.5). We finally describe applications of AI to characterize the produced materials (section 4.2.6) before drawing some conclusions.

4.2.2 Introduction to Artificial Intelligence

Artificial Intelligence (AI) is a relatively young field of science that was born in 1956, after the workshop at Darmouth College where the founding fathers defined it as the study of the simulation by machines of "every aspect of learning and any other feature of intelligence". Later, computers and robots reveal to be perfect artefacts to demonstrate this artificial intelligence. Artificial Intelligence is the crossing of several sciences as, to cite a few, Mathematics, Philosophy, Economy, Psychology and Computer science.

Artificial Intelligence has had its success in some domains. Very soon, computers can play games and triumph against humans, expert systems can help in diagnostics and neural networks can recognize postal codes. The subfields became numerous: machine learning, knowledge representation, natural language processing, image processing, etc.

Nevertheless, this research field has experienced flip sides, known as AI Winters: it was blamed for the broken promises of a strong artificial intelligence, capable of the same reasoning as humans, on every domain, and capable of learning like humans. Today, it is difficult to ignore Artificial Intelligence, even if the hype can lead sometimes to confusion, and sometimes fears.

In the next subsections, we describe some methods that have been applied in material design. We split them into machine learning and soft computing methods. We invite the curious readers to refer to this reference [10].

4.2.2.1 Machine Learning

Machine Learning (ML) [11] is one of the subfields of AI that consists in creating algorithms that allow a machine to discover how to solve a problem instead of explicitly giving it the instructions. The most common form of learning is called inductive learning because it exploits data. There exist different paradigms and the field of ML usually distinguishes supervised, unsupervised and reinforcement learning, regarding the type of available information.

Supervised learning consists in learning a function from observations of its inputs and outputs. The problem of unsupervised learning amounts to learning patterns when only observations inputs are available. Finally, reinforcement learning rather uses reward: the machine is taught when its decision is good or bad. Halfway between the supervised and unsupervised approaches, some algorithms can process both labelled and unlabeled data: it is called semi-supervised learning [12].

Most of ML techniques address two different problems:

- Classification, that consists in affecting a class (*e.g.*, metal, mineral) to an example described by variables, *i.e.*, the output is a discrete value.
- Regression, that consists in predicting a continuous value from predictive variables.

The data and the output are not the only important ingredients: the representation of the learned information is of paramount importance [10].

The construction of a data driven approach requires a set of generic steps [13, 14], as illustrated in figure 4.2.3. A variety of raw data collected from computational simulations and/or experimental measurements are extracted. The first step consists in preprocessing the extracted data, by understanding the data format and representation, cleaning the data and removing noise and redundant data instances, etc.

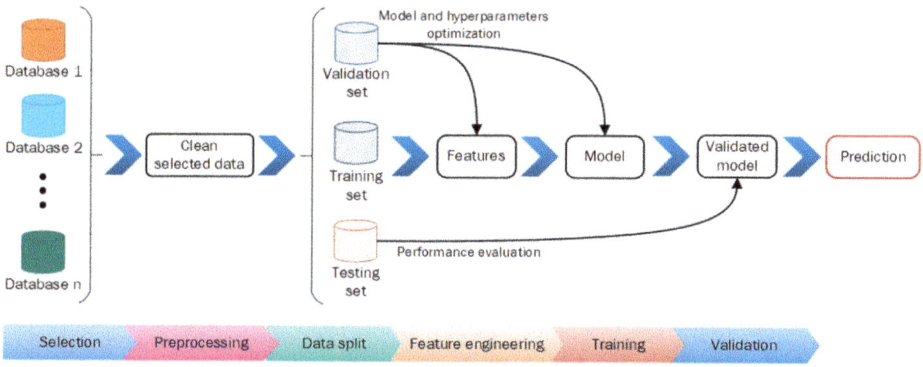

FIG. 4.2.3 – Data driven approach generic workflow.

Then, the next phase is called feature engineering and aims at improving the performances of ML models. A feature is a measurable property of an individual, *i.e.*, of an instance of the dataset. Feature engineering uses domain knowledge to find the properties that constitute an efficient representation of the data. Data mining techniques can be used in that phase. Feature engineering consists also in dimension reduction, with projection such as PCA (Principal Component Analysis). At the end, it is also possible to perform feature selection to keep only relevant features.

Once the data are ready, the clean dataset is usually divided into training, test and validation sets. After the selection of a ML algorithm, the model can be trained on the training set by optimizing its performance, measured by a cost function. A validation set, when needed, is used for the optimization of the hyperparameters. A model can underfit the data when they are not representative enough. At the opposite, it can overfit the data and not be able to predict correctly new data. Before using the model, it has to be validated on the test set. For classification, different metrics are used to evaluate the performance: overall accuracy, True Positive Rate (or sensitivity, recall), False Negative Rate (probability of false alarm) and True Negative Rate (or specificity) amongst others. These measures can be deduced from the confusion matrix. The Receiver Operating Characteristic (ROC) curve and its associated Area Under Curve (AUC) are useful when different algorithms are compared. For regression, performance metrics quantify the errors, such as Root Mean Square Error or Mean Absolute Error functions. Performance on test set is the one that should be used to compare different ML algorithms.

Several data driven approaches are introduced in the literature and several excellent reviews on these works are available in the literature as [3, 7, 8, 14–19].

We now focus on some classical methods of supervised learning ("training" step in figure 4.2.3) that is the main paradigm used in material design.

4.2.2.1.1 k-Nearest Neighbors

The *k*-nearest neighbors (*k*NN) is a non-parametric method used for classification. In order to classify a new instance, this method founds its *k* nearest neighbors in the data set and assigns it the most common class among these neighbors. The metric

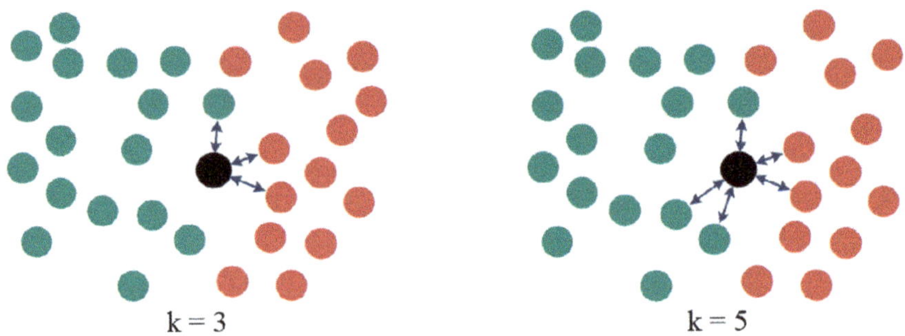

FIG. 4.2.4 – k-Nearest Neighbors principle.

used to find the nearest neighbors and the number of neighbors to consider are critical parameters. Since kNN requires passing through all instances in the data set, its computational cost may be important. Figure 4.2.4 shows the principle of kNN for two different values of k. The instance to be classified is the black point; for $k = 3$, the affected class is red whereas for $k = 5$, it is green.

Despite these drawbacks, kNN can represent very complex decision boundaries and has been applied successfully in many domains. The method is also suitable when few instances of each class are represented in the data set. However, for strongly unbalanced dataset (*i.e.* the numbers of instances of each class have different magnitudes), it can lead to bad results.

4.2.2.1.2 Naive Bayes Classifiers

Naive Bayes Classifiers are a family of probabilistic classifiers that are based on Bayes' theorem. They all assume that all the features are independent from each other. Although they are quite simple, they can perform well on real world applications. They are classically trained with the method of maximum likelihood, and thus do not need a large data set to be trained.

Basically, in a problem of classification with K classes $C_1, \ldots, C_K$ if we consider an input vector $x = (x_1, \ldots, x_n)$, a probabilistic classifier will assess the probability $P(C_k|x)$ for each k from 1 to K. Using Bayes' theorem, this can be decomposed as:

$$P(C_k|x) = \frac{P(C_k)P(x|C_k)}{P(x)}$$

The numerator of that fraction is simply equivalent to the joint probability model $P(C_k, x_1, \ldots, x_n)$ that can be expressed, using the conditional independence assumptions of the model as:

$$P(C_k|x_1, \ldots, x_n) = \frac{1}{Z}P(C_k)\prod_{i=1}^{n} P(x_i|C_k)$$

where $Z = P(x) = \sum_k P(C_k)P(x|C_k)$.

Finally, the decision is just given by $\max_{1 \leq k \leq K} P(C_k) \prod_{i=1}^{n} P(x_i|C_k)$.

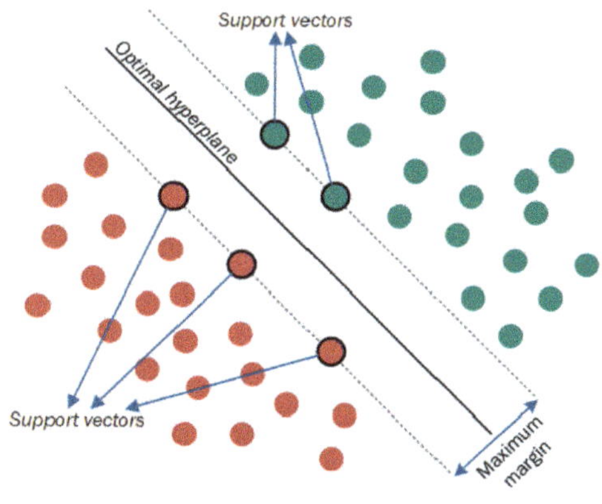

FIG. 4.2.5 – Illustration of SVMs principle.

4.2.2.1.3 Support Vector Machine

Support Vector Machines (SVM) [20] are ML algorithms that were first intended for discrimination (*i.e.* binary classification). They then have been extended to regression and multi-class classification. In the context of discrimination, SVMs find a hyperplane that separates the instances of the two classes. There exist an infinity of hyperplanes that perform this separation, but SVMs choose the one that maximizes the margin distance (*i.e.* the maximum distance between instances of both classes). The idea behind is generalization that will increase the performances on future data.

Support vectors are data points that are closer to the hyperplane. Thus, they have an influence on the location of this hyperplane.

In the case data are not linearly separable, SVMs use the kernel trick. The intuition is that if the data are not linearly separable in the original feature space, they may be in a higher dimensional space (figure 4.2.5).

4.2.2.1.4 Decision Trees

A decision tree is a decision support tool for classification and prediction (figure 4.2.6). It is structured in the form of a flowchart-like tree. Each branch node represents a choice between a number of alternatives and each leaf node represents a decision. The paths from root to leaf represent classification rules.

For predicting a class label from a record using a decision tree, we start from the root of the tree. We compare the values of the root attribute with the record attribute. Then, we follow the branch corresponding to that value and jump to the next node. We continue comparing the record attribute values with other internal nodes of the tree until we reach a leaf node with predicted class value.

Using decision tree offers many advantages. Amongst these advantages, the graphical representation makes easy the understanding of its generated rules

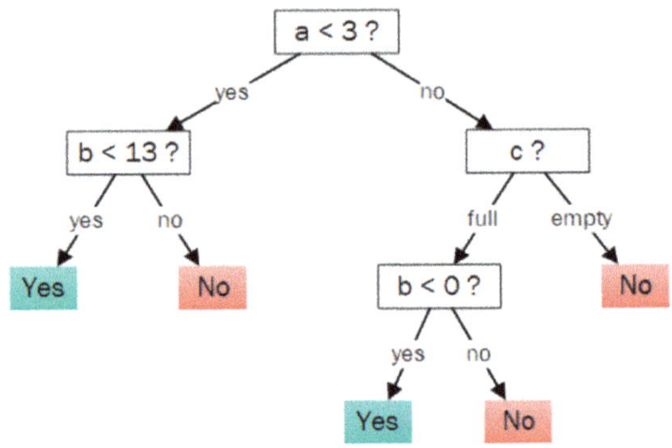

FIG. 4.2.6 – Example of decision tree.

(interpretability). Then, the data type is not a constraint in decision trees since they can handle both numerical and categorical variables. Another strength of decision tree methods is that they are non-parametric methods and have no assumptions about the space distribution and the classifier structure.

Decision trees methods generate axis-parallel decision boundaries that are often perceived as a weakness.

Several algorithms can be used to induce, automatically, decision trees from data: ID3, CART, C4.5 [21–23]. They differ from the type of variables they can handle and their ability to deal with missing values. They all share a top-down strategy to build the decision tree with a greedy approach. The selection of the next node to create is based on the concept of information gain and a measure of separability of the data, usually entropy or Gini criterion. Greedy approaches often lead to overfitting. A pruning phase is often used to avoid this problem.

4.2.2.1.5 Artificial Neural Networks

In 1943, McCulloch and Pitts propose a formalization of the activity of a neuron [24]. Figure 4.2.7 shows a presentation of such a neuron. Several scalar inputs are weighted and summed before being transformed by the activation function.

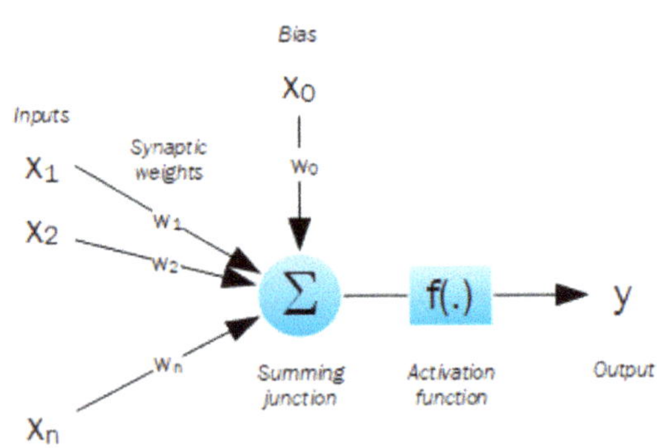

FIG. 4.2.7 – The McCulloch–Pitts neuron.

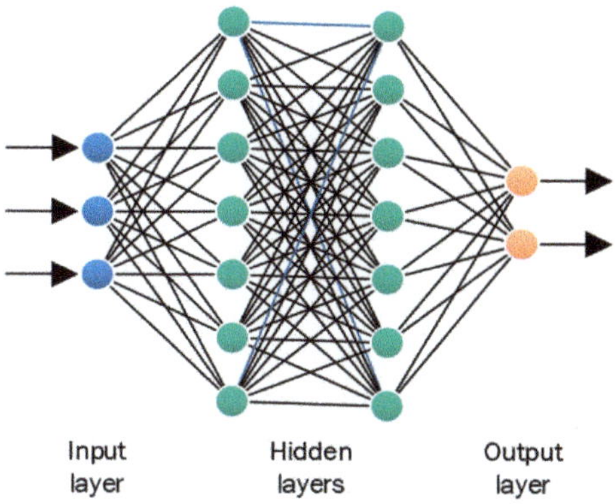

FIG. 4.2.8 – A multilayer neural network.

In the first works, the activation function was simply a threshold function. Intuitively, the neuron activates when a linear combination of its inputs is greater than a threshold [10]. This simple model gave birth to connectionism, *i.e.*, the study of Artificial Neural Networks (ANN). ANN are composed of formal neurons, connected by directed links and arranged into layers.

In [25], Rosenblatt studied perceptrons, *i.e.* ANN with only one layer (the input layer is directly connected to the output). The sigmoid function replaces the threshold function, with the advantage to be differentiable, that helps in training such networks. The learning algorithm scans the learning dataset and adjusts the weights to reduce the error. Each iteration over the dataset is called an epoch.

Multilayer neural networks (figure 4.2.8) have more than two layers of neurons: they have hidden layers. Hidden layers increase the capacity of representation of ANN. Learning algorithms for multilayer neural networks are based on the same principles as for perceptrons, except there can be multiple outputs. Learning algorithms back-propagate the error from a derivation of the overall error gradient.

The interested readers can refer to this book [26].

4.2.2.1.6 Deep Neural Networks

The various methods of ML we have seen suffer from the same problem: they require manual feature engineering, often performed by an expert. Deep Learning, in general, allows going further. In particular, Deep Neural Networks (DNN) do not need this phase anymore and are able to learn their own representation if the data. "Deep" means that there exist several layers between the input and output layers [27]. For instance, one of the famous DNN architectures, AlexNet, counts 650 000 neurons to recognize 1000 different classes [28]. In the last decade, the development of DNN has led to successful applications of data-driven methods and have outperformed traditional ML techniques. In particular, the literature relates applications of Convolutional Neural Networks (CNN) and Recurrent Neural Networks (RNN) to Materials science.

CNN [29] are DNN that can directly takes an image as input. They consider two other types of layers: the convolutional and the pooling layers. The convolution layers apply trainable filters as discrete convolutions. Pooling layers then reduce the dimensionality of the representation by combining neurons into a single output. The other layers are fully connected.

Since most of the network architecture we have described are feedforward network, they can hardly consider related data. RNN [30] are thus suitable to process sequential data. In a RNN, the current output is computed regarding the current inputs and the previous output.

It is important to know that the deeper the networks are, the more data they need to be trained. For instance, AlexNet has been trained on more than 15 million images [28]. In many applications, only simulated data are used to balance the insufficient material data availability. Nevertheless, a model trained on simulated data will not be efficient on real data without adaptation.

4.2.2.2 Expert Systems

In AI, Expert Systems designate systems that emulate the reasoning of human experts to make decisions. Expert Systems are reliable, interpretable and understandable. The logical processes they are based on allow the generation of an explanation of the decision.

Expert Systems have generally three components, as shown in figure 4.2.9: a knowledge base, an inference engine, and a user interface to be consulted.

The knowledge base contains domain-specific information. To guarantee the success of the Expert System, the quality and the accuracy of this knowledge is of paramount importance. The knowledge is formalized by a knowledge engineer into a mathematical formalism, as first order logic, and very often in the form of IF-THEN rules.

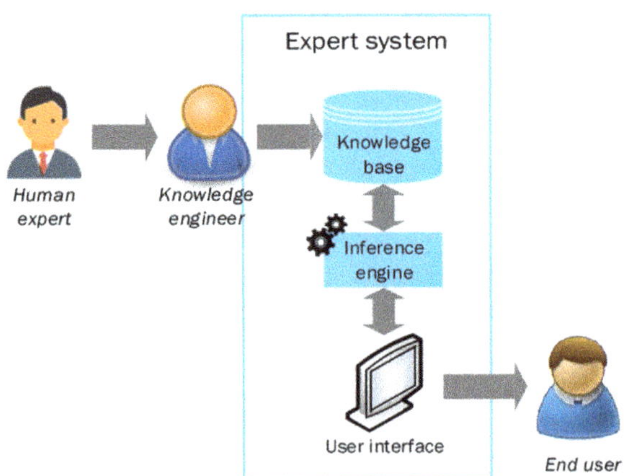

FIG. 4.2.9 – Principle of Expert Systems.

The inference engine collects facts (*e.g.*, observations) and fires rules to produce new facts, until no more rule can be fired. The inference can go forward, as data-driven strategy, to deduce conclusions from the current facts, and it can go backward, as a goal-driven strategy, to find the causes of the current facts.

The user interface plays a major role and can be as sophisticated as natural language understanding and speech recognition. It can also display the explanations, the rules that have been fired, to let the user understand the origins of the decisions.

Expert Systems have been successfully applied in many domains. Nevertheless, the knowledge acquisition has been a huge hindrance to their applications.

4.2.2.3 Soft Computing

Soft computing designates methods that can reason and learn under uncertainty, imprecision and approximation [31]. In this section, we will introduce fuzzy logic and meta-heuristics.

4.2.2.3.1 Fuzzy Logic

In 1965, Zadeh introduced a new logic called fuzzy logic [32] based on fuzzy sets. Fuzzy sets are as usual mathematical sets, except that they define a membership function to define how strongly a point belongs to the set (figure 4.2.10). This membership function maps values of a domain into real values between 0 and 1. 0 means the value does not belong to the set, whereas 1 means it totally belongs to it. From fuzzy sets, fuzzy logic is defined as a multivalued logic in which facts are not true or false, as in Boolean logic, but have degree of truthfulness between 0 and 1.

This fuzzy logic aims at computing with words by representing vague knowledge with fuzzy sets and provides tools that perform inference. Uncertainty and imprecisions are always challenges in real world scenarios, and fuzzy logic provides tools to deal with both of them. Fuzzy expert systems use thus fuzzy logic as formalism to model the knowledge. Fuzzy logic has also lot of applications in control since it replaces mathematical modeling with fuzzy rules.

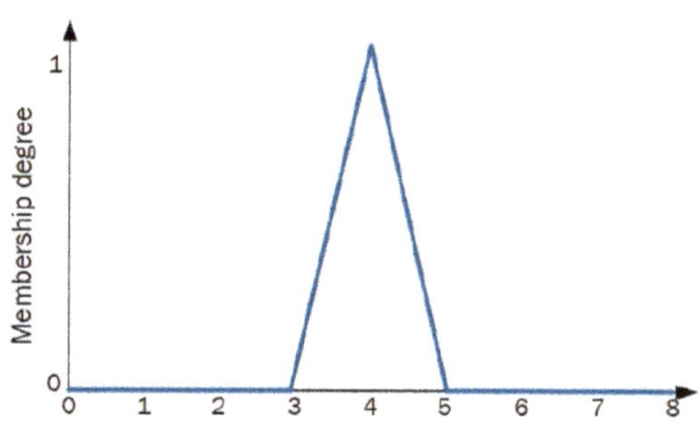

FIG. 4.2.10 – A fuzzy set representing "approximately 4".

Many algorithms have been *fuzzified*, like for instance clustering methods. ANN have been coupled with fuzzy logic in Adaptive Neuro-Fuzzy Inference Systems (ANFIS) [33]. ANFIS are considered as universal estimators.

4.2.2.3.2 Metaheuristics

Metaheuristics are a family of methods of stochastic optimization. Stochastic optimization is performed by algorithms that search for optimal solutions to hard problems, with a certain degree of randomness. Metaheuristics can be used in a wide variety of applications but does not guarantee to find global optimal solutions on some class of problems.

Genetic Algorithms (GA) are metaheuristics [34]. They solve problems that have a huge search space. They are inspired by Darwin's theory of evolution. Each parameter of the problem to solve is considered as a gene, and each individual is a chromosome (a set of genes). The principle of the genetic algorithms is presented in figure 4.2.11.

The first population is constituted randomly. A fitness function has to be chosen adequately to measure the relevance of a chromosome. Some chromosomes can be selected to be muted and the algorithm performs the crossover operation on each pair. Then, some chromosomes are selected and the algorithm applies the mutation operation on them. The new chromosomes are added to the population to constitute a new generation. The algorithm loops until a good solution (possibly, the optimal one) is found.

The crossover operation consists in building two new chromosomes from an existing pair: a crossover point is randomly chosen, and the original chromosomes are cut into two parts, that are then exchanged. The mutation operation selects randomly some genes and changes randomly their values.

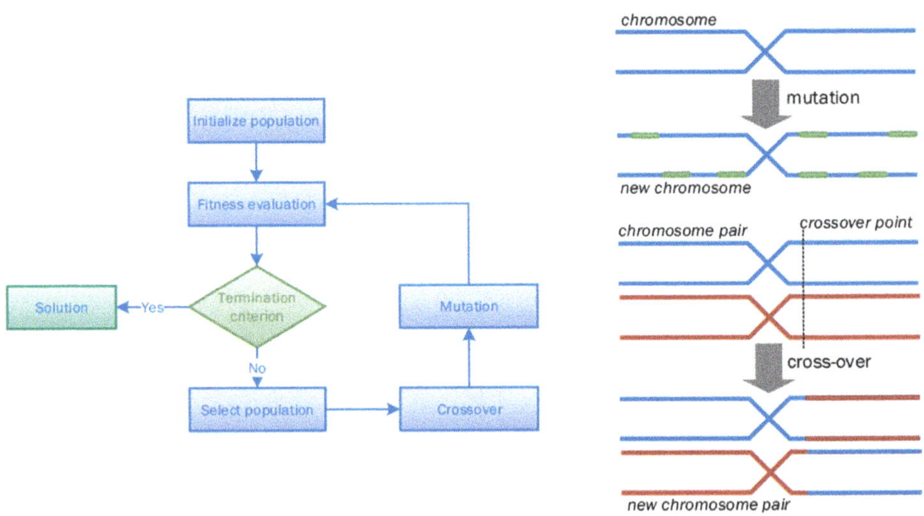

FIG. 4.2.11 – General structure of genetic algorithm.

Particle Swarm Optimization (PSO) [35] is another metaheuristic inspired by social animals like fishes and birds that move together. This algorithm considers candidate solutions evolving in the search space to find the best solution. The movement of the particles depend on:

- Inertia weight: each solution tends to follow its direction;
- Cognitive weight: each solution tends to go back to the best position it has passed by;
- Social weight: each solution tends to go to the best position found by its neighbors.

In case of multi-objective optimization problems, the different objectives could be conflicting. Then, a Pareto front can be used to visualize non-dominated solutions, which are those that cannot be improved for an objective without degrading another.

4.2.3 Qualification of Causality Relations

Traditionally, experiments are used to find and characterize the causality relations shown in figure 4.2.2. However, constructing these relations is a challenging task especially since many parameters may influence the same property and since these relations are not linear in most cases [36]. For these reasons, it is difficult to define those parameters only by experiments. Moreover, experiments need to be conducted over a long period and require efficient equipment, controlled experimental environment and a good expertise of the researcher in order to obtain useful results.

In the field of causality relations qualification, Additive Manufacturing (AM) is a great contributor: in each manufacturing process, many factors may influence the quality of manufactured components. There are process parameters, used materials' properties and environmental properties. Detecting the best parameters and combination of parameters is a challenging task.

Several studies can be found in the literature using statistical analysis for defining the significance of AM's processing parameters. We first introduce the two most popular methods, and we end this section with some application examples.

4.2.3.1 ANOVA

ANOVA (ANalysis Of VAriance) [37, 38] is one of the most used statistical methods and was introduced by statistician and biologist Ronald Fisher in 1918. It is an analysis tool used to assess statistical significance of differences among group means. It is called ANOVA because inferences about group means are drawn by analyzing variance. More precisely, the variability of the observations among groups is compared to the variability within groups. There are several varieties of ANOVA, such as one-factor ANOVA and two-factor ANOVA, which take into account one and two factors, respectively. The factors are the independent variables, which must be categorical variables. The dependent variable should be numerical.

The probability *p_value* is employed to determine whether any differences between the group means are statistically significant. A *p_value* less than a pre-defined significance level (0.05 for instance) leads to the rejection of the null hypothesis which states that the population means are all equal.

There are some assumptions, which need to be met in order to use ANOVA:

- Normality: The values of the variables are normally distributed within each group of samples.
- Homogeneity: All groups have the same variance.
- Independence of case: The observations are sampled randomly and independently of each other.

When these assumptions are not met, one can use non-parametric statistical methods such as Friedmann ANOVA or Kruskal–Wallis ANOVA.

4.2.3.2 Regression Analysis

Regression analysis is a set of statistics tools that are used to estimate the relations between continuous variables. Usually, we consider the relation linear: the linear regression model (figure 4.2.12) considers the relation between two variables, whereas the multiple linear one estimates the relation for several independent variables and one dependent variable.

If we denote Y the dependent variable and X the independent one, the simple linear model is expressed as $Y = a + bX + \varepsilon$, where ε is the residual (error). To be applied, these assumptions must be observed:

- Linearity: X and Y show a linear relationship.
- Autocorrelation: there is no correlation between the residual terms.
- Homoskedasticity: error terms must have constant variance.
- The error terms must be normally distributed.

The multiple linear model is expressed as $Y = a + bX_1 + cX_2 + \cdots + \varepsilon$ and the assumptions are the same as for the simple linear model, and the multicollinearity assumption, that is to say, the independent variables X_i are not correlated.

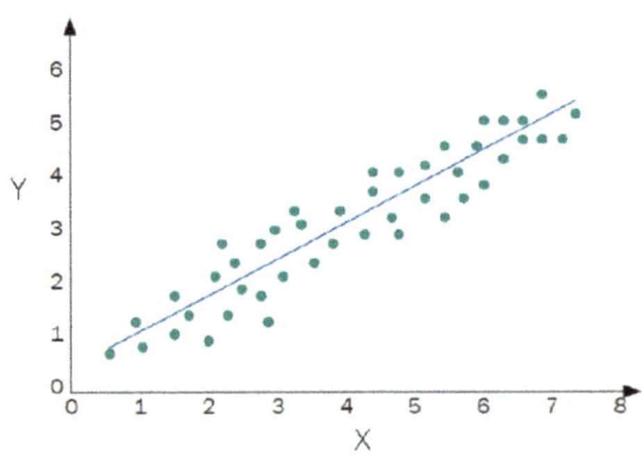

FIG. 4.2.12 – Example of a simple linear regression.

To estimate the parameters of the models, the method of the least squares is generally used. This method simply minimizes the sum of squared residuals.

In more complex datasets, the relation can be non-linear: we talk about non-linear regression.

4.2.3.3 Applications

4.2.3.3.1 Selective Laser Sintering

Selective Laser Sintering (SLS) is an AM method that enables the production of complex non-viable objects. In [39], the authors use fractional factorial Design Of Experiments (DOE) to plan the experiments. Then, response functions were created to study the interaction effects between input parameters and the output (part properties). The processing parameters, selected in this investigation, were the preheating temperature, laser power, scan speed, scan spacing, layer thickness and the part build orientation and consisted in quantitative and qualitative factors. The properties were the tensile strength, tensile modulus, elongation at break, and part density of SLS produced parts. The experiment design allows collecting 480 tensile samples. Regression analysis was employed to determine which effects and interactions have a significant influence on the response values.

In [40], the authors investigate the laser energy density effect on the material properties of the polyamide PA220 like the Flexural modulus, the stress at 10% elongation and the density. They considered two processing parameters (laser power and scan speed) as qualitative factors. A DOE helps to produce 36 samples of polyamide. The ANOVA was used for each property to determine the most significant parameter for it. As results, the most predominant factor for density and stress at 10% was laser power parameter. The scan speed was the most significant parameter for flexural modulus property. Then, a regression analysis was conducted to understand better the influence of each parameter.

The authors of [41] use ANOVA to evaluate which qualitative parameters among the Laser power, the scan spacing and the scan velocity, have significant impact on part density. A regression model was then built using three process parameters laser power, scan spacing and scan velocity. They found as results that Laser power and scan spacing were significant parameters affecting the part density. They also found that the laser power has a significant effect on the part density at the lower settings of the scan velocity.

4.2.3.3.2 Fused Deposition Modelling

Concerning Fused Deposition Modelling (FDM) manufacturing method, [42] investigated the influence of some qualitative manufacturing parameters (build orientation, layer thickness and feed rate) on the mechanical properties of polylactic acid samples. ANOVA, regression models and response surfaces were used to perform a parametric characterization of the mechanical properties. ANOVA was performed to determine the level of statistical significance of the parameters. Then, a regression model based on a multivariable polynomial regression was developed.

The statistical analysis showed that printing time decreases as layer thickness and feed rate increase.

4.2.3.3.3 Friction Stir Welding

The effects of friction stir process parameters (*e.g.*, axial force, tool feed and rotational speed) on mechanical properties of AA6063-T6 aluminum alloy are investigated in [43]. The authors applied the full factorial method to reduce the number of experiments, graphs and comparison with base metal to find the contribution of each parameter for the mechanical properties. As results, they found for example, that speed is the most influent factor to increase the hardness as compared to reinforcement, and that tensile strength decreases as compared to base metal. Other papers relate the use of multivariate non-linear regression analysis, like in [44].

In this section, we studied the two main statistical approaches to identify the causal relations between process parameters and materials' properties. If regression can be used to further predict the properties from the parameters, we will describe application of AI to this kind of prediction in the next section.

4.2.4 Properties and Performance Prediction for Process Optimization

In this part, we make a step further towards goal-driven design of new materials. We describe some methods involving statistical and AI techniques to build predictive models that can tell which properties have to be expected from current process parameters and how to change these parameters to optimize these properties.

The optimization of process parameters is a data-driven task, involving either statistics or AI techniques [45, 46], as shown in figure 4.2.13. They need data, collected from real experiments or simulations. Then, the process has to be modeled, either explicitly or implicitly with data-driven techniques. Then, the optimal parameters are searched and adjusted, either automatically or manually.

4.2.4.1 Statistical Approaches

Most of the works cited in section 4.2.3.3 use a regression model to perform such prediction, and then to choose manually the best parameters.

In [47], a statistical analysis is used to optimize values of laser power, layer thickness, scan speed and hatch spacing, representing qualitative factors, to achieve

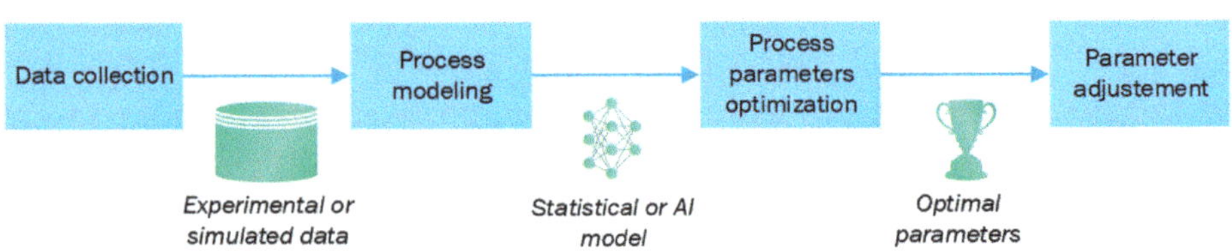

FIG. 4.2.13 – Data-driven parameter optimization workflow [45, 46].

the best compressive strength. ANOVA method was used to evaluate the significance of each parameter and of their combinations. As results, they found that laser power, layer thickness, hatch spacing and interaction between hatch spacing and laser power contribute the most to the value of compressive strength of scaffolds. A regression model was created based on ANOVA results and a trust region algorithm was used based on this model to optimize parameters.

Reference [48] presents a framework for optimization of the powder bed fusion (PBF) process by mapping the quality of a fabricated part into the process parameters and material properties. These mappings are generated from a series of experiments. Authors use sensitivity analysis and ANOVA to measure the influence and sensitiveness of laser power, scan speed, hatch spacing, and beam diameter within the optimal range.

To optimize the press hardening process, [49] employed linear regression combined with least squares optimization to reproduce relations between process parameters and quality parameters and to adjust the process parameters for quality improvement.

4.2.4.2 AI Approaches

4.2.4.2.1 ML Approaches

A recent review specialized on studying ML and optimization approaches for product quality or process improvement in the manufacturing industry from 2008 to 2018 was proposed by [45].

Authors of [48] indicate that their future work consists in developing an ANN for parameter suggestion and build-time estimations, based on detected correlations presented in the section above. To increase their data, they suggest using Generative Adversarial Network, a particular topology of DNN that are used to generate data.

Data from the Japan National Institute of Material Science (NIMS) MatNavi database were used in [50], to make predictive models for fatigue strength of steel. The dataset is composed of 437 instances of steel classes and 25 parameters about materials' compositions and processing. The approach consisted of several steps that included data pre-processing for consistency using domain knowledge, ranking-based feature selection to evaluate the predictive potential of the attributes, predictive modelling using twelve regression models (*e.g.*, SVM, ANN, decision trees, ...), and model evaluation using leave-one-out cross-validation.

In [51], authors model the titanium coating process with an ANN trained with experimental data. The ANN has 6 inputs (*i.e.* ion bombard time, sublayer roughness, material type, sublayer temperature, work and chamber pressure), 2 hidden layers and one output only. The best architecture and parameters for the ANN have been chosen empirically. Other ANNs are then trained to obtain the maximum hardness of the coated thin layer.

4.2.4.2.2 Coupling Neural Networks and Genetic Algorithms

The problem of power system stability (PSS) enhancement was investigated in [52]. On the one hand, a power system stabilizer associated with GA was proposed for damping the small signal oscillations and optimally adjusting the stabilizer parameters. Then, on the other hand, a power system stabilizer based on an ANN was proposed to generate optimized PSS parameters with a learning database. As results, they found that the method using neural networks provides good results compared the method using the GA.

However, in most approaches it is better to combine both ANN and GA (or eventually PSO). In general, these approaches use ANN to predict a property from process parameters, and then use genetic algorithms (GA) to find the optimal parameters. The advantage of ANN is that it is not necessary to know well the underlying process and to model it explicitly. The ANN is then used as the fitness function of the GA.

This workflow has been applied to SLS, which is a nonlinear complex system. Each work aims at addressing one of the SLS drawbacks at a time. Reference [53] focuses on the shrinkage ratio. Their ANN uses layer thickness, hatch spacing, scanning speed, scanning mode, laser power, interval time, and work-surrounding temperature, to assess the shrinkage ratio. The ANN has a single hidden layer of 8 neurons and is trained on 36 samples from experiments. Then a genetic algorithm searches the best values for each parameter to minimize the shrinkage ratio. Another drawback of SLS is warpage. To optimize process parameters for minimized warpage in SLS products, [54] used the finite element method to simulate the warpage of a layer of sintered polymer powder. The ANN is trained on these simulated data to learn relations between process parameters and warpage. Then, a GA is applied to optimize the process parameters for minimum warpage. In [55], authors address the porosity for SLS fabricated parts from self-made powder as a mix of hydroxyapatite and polyamide. Thirty-six samples were manufactured and a genetic programming algorithm is applied to relate a set of qualitative parameters of manufacturing process (layer thickness, laser power and laser scan speed) to open porosity. As results, they found that layer thickness, laser power and laser scan speed have a significant impact on open porosity.

ANNs are not the only way to model a process. Fuzzy expert systems have been successfully applied several times [56–60]. The rules are either given by expert, or deduced from experimental data. In the cited papers, the authors define as simple as possible linguistic variables. Then, they write a complete rule base, covering all the cases. Once the process has been modeled by the fuzzy rule base, it can be then used as fitness function for a metaheuristic. In the field of fused deposition modelling, [61] applied the response surface methodology (RSM) combined with a fuzzy inference system and a GA optimization to improve the FDM process. Then, a GA was used to optimize the responses such as the dimensional error, warp deformation and build time by formulating a single comprehensive response. In [62], the authors perform rule learning with the ANFIS method to constitute a fitness function and then use PSO for minimization.

4.2.4.2.3 Other Nature Inspired Optimization

The work of [63] focused on the optimization problems in the FDM process. They used teaching-learning based optimization (TLBO) [64] algorithm to solve three single-objective optimization problems and non-dominated Sorting TLBO (NSTLBO) algorithm to solve two multi-objective optimization problems. By comparing the results of TLBO algorithm with those provided by GA and quantum behaved particle swarm optimization (QPSO) algorithm, they found that TLBO provided better results. The NSTLBO algorithm showed also better results compared with those provided by the non-dominated sorting genetic algorithm (NSGA-II).

4.2.5 Materials' Recommendation

The manufacturing of new materials often consists in the use of different raw materials. Ashby has proposed materials' selection charts and some methods for compromising between conflicting objectives [65, 66]. Multi-criteria decision analysis techniques have also been proposed. The most used technique in materials' selection is TOPSIS (for Technique for Order of Preference by Similarity to Ideal Solution) [67, 68]. This technique compares different candidate solutions by identifying the weights of each criterion and then calculating the geometric distance with the ideal solution. We can also find classical optimization problem formulations and resolution by solvers [69].

AI can help recommending the material to use in order to obtain certain properties. To our knowledge, the contributions of AI to materials selection are still young.

The first approach relies on ANN. In [70], the authors propose the use of an ANN to optimize both the selection of powder and the process parameters for powder metallurgy. They justify their choice by the large number of inputs due to the number of process parameters, the variation in sintering, and the addition of small amounts of alloying elements. The ANN has two hidden layers and has been trained on experimental data for 200 000 epochs.

However, the most commonly used AI approach for this task is expert systems. In [71], the authors use Certainty Theory to formulate uncertain knowledge about the materials. This consists in considering it with a certainty factor. They proceed to two interviews of experts to collect the data.

The main problem with the methods discussed until now is that they can consider either qualitative or quantitative materials' attributes. Fuzzy logic usually can deal with both, and that is why we can find more papers applying fuzzy expert systems or fuzzy logic to this task.

The fuzzy expert system based approaches compute a performance index for each material and then rank them regarding this index [72]. Fuzzy sets and rules can be defined with expert interviews and an analysis of the data. Authors of [72] underline the high potential of fuzzy logic: the fuzzy sets reduce the sensitivity of the final results, it also reduces the complexity of the mathematics involved compared to the conventional material selection methods, etc.

In [73], the fuzzy version of PROMETHEE, an outranking method introduced in [74], is applied. Their approach is based on trapezoidal fuzzy interval numbers with two different preference functions. They compare their results with different fuzzy approaches that are outperformed.

Finally, [75] proposes a combination of fuzzy logic and TOPSIS for coating selection. A fuzzy expert system using triangular membership functions computes a score for each material that are then ranked by TOPSIS.

4.2.6 Assistance in the Characterization of Materials

Thus far, we have focused on how to set the manufacturing process and to discover quality new materials. Once these materials have been created, it is important to characterize their properties and performance. Material characterization consists in determining the internal structure and composition of the material and its properties.

Depending on materials' nature, different techniques are used to characterize their structure and properties [76, 77]. Different devices can be used to analyze the materials and produce data that can then been analyzed by AI.

Scanning Electron Microscopy is a type of electron microscope that can make images of solid samples by scanning the surface with a focused beam of electrons. X-Ray Photoelectron Spectroscopy has also been applied to a wide range of materials. It is an analytical technique where X-rays are used to irradiate materials and the energies of emitted electrons are simultaneously analyzed. The focused ion beam technique uses focused beam of ion sources, especially gallium, to scan and cut the substrate surface inside a vacuum chamber. X-ray diffraction provides structural information about materials. Information as interatomic distances, bond angles, and crystallinity can be provided. Nanoindentation is an analytical technique for testing mechanical properties of materials at the nanoscale. Properties that can be measured using nanoindentation include Young's modulus, hardness, and fracture toughness. Time-of-flight secondary ion mass spectrometry uses ion beam to bombard the surface of the sample. Then organic and inorganic chemical information is provided from the solid surface. Scanning Probe Microscopy (SPM) forms images of solid samples using a physical probe that scans the specimen and can be used to determine mechanical, electrical, and magnetic properties of said surfaces. Atomic Force Microscopy is a type of SPM with a very high resolution. Transmission Electron Microscopy (TEM) provides techniques as Selected-area electron diffraction and High-resolution TEM to understand information of materials at very high spatial resolution.

Experimental measurements and theoretical simulations used to play an important role to characterize materials' properties and structure. The characterization is thus a tedious task in the process of materials' discovery. Moreover, characterization often needs an interpretation that can differ from one expert to the other. Recently, advanced techniques are used for the characterization based on AI techniques, in particular ML algorithms. The goal is to accelerate the characterization and to prevent the results from human interpretation.

ANNs have been widely applied to the characterization of properties. An inverse method based on ANN is used in [78] to determine the elastic properties of films from laser-generated surface waves. As results, they found that the proposed ANN technique was useful to determine the critical parameters that affect the properties of layered materials. ANNs have also been used in [79] to estimate the tip-sample distance and dielectric constant of a system composed of a tip scanning over a metallic nanowire. They demonstrated that the proposed technique has the ability to estimate quantitatively the dielectric constant for an even more undetermined system where the tip shape is unknown. Authors of [80] developed a neural network approach to test and validate the computational envelope in parametric investigations of materials using molecular simulations. They conclude from the study that ANN can be used in combination with simulations to achieve faster computation times and enables to gather a wide set of data. Finally, [81] deployed ANN to examine the surface morphology and curvature of carbon nanotubes (CNTs). They found that their proposed ANN model can explain the variation of surface morphology of CNTs with the Raman spectra.

In [82], the elastic characteristics of CNTs are modelled using two AI methods: genetic algorithms and SVM. These methods use geometry and temperature as inputs. As results, they found that genetic algorithms have shown better performance than SVM to model the compressive strength of Single Walled CNTs.

Reference [83] presented an online Fatigue Stress Concentration Factor (FSCF) predicting method using Extreme Learning Machine (ELM). Seven parameters (*i.e.* tensile strength, yield strength, fatigue strength, theoretical stress concentration factor, notch root radius, samples size and notch fatigue limit) were adopted as inputs, and the corresponding FSCF value is used as the output. With the randomly generated hidden neuron parameters, the ELM-based predictor can be fast trained.

The prediction of mechanical properties of ultra-high performance concrete was investigated in [84]. A case-based reasoning (CBR) approach was proposed to find similar cases, conduct some adaptation and finally predict desirable mechanical properties. As results, they found that the considered CBR approach is capable of accurately predicting mechanical properties and recommending new testing conditions and parameter.

To predict the compressive strength of fabricated geopolymer concrete, [85] introduced hybrid AI approaches:

- PSOANFIS, which is a combination of the ANFIS and PSO methods, and
- GAANFIS, which is a combination of ANFIS and the genetic algorithms.

They validated models provided by these approaches using root-mean-square error (RMSE), mean absolute error (MAE), and correlation coefficient (R) methods. They found that the proposed models performed well for the prediction of the compressive strength of GPC and the PSOANFIS model outperformed the GAANFIS model.

AI techniques have also been applied to crystal- and micro-structure characterization. The microstructure of a material presents its very small structure whereas the crystal structure is the ordered arrangements of ions, atoms or molecules in a crystalline material. Both structures play a crucial role in determining materials' properties. That is why it is important to characterize the micro and the crystal structure of materials.

Amongst the works of microstructure characterization, [86] applies computer vision and machine learning methods to develop a system characterizing powder feedstock materials for metal additive manufacturing. Feature detection and description algorithms are applied to create a microstructural scale image representation. The system classifies powder images into the correct material systems with an accuracy greater than 95%.

To investigate the crystal structures of materials, [87] used ML combined with quantum mechanical calculations to predict the crystal structures of binary alloys. A ML model was used to predict several candidate crystal structures, and DFT calculations were used to determine their thermodynamic stability.

By also combining machine learning and quantum mechanical methods, [88] constructed an informatics-based structure suggestion model for structure prediction, called Data Mining Structure Predictor. It mines correlations embodied within experimental data and uses them to direct efficiently quantum mechanical techniques toward stable crystal structures.

4.2.7 Conclusion

Recently, Artificial Intelligence has played an important role in Materials science. Progresses in Artificial Intelligence, specifically in Materials science, are promoted every days. It was not possible to be exhaustive in this chapter, but the goal was to initiate the readers to this field and give recent literature as a starting point.

The data generated by the machines and the simulations are now accessible and allow data driven approaches to be developed. As in many other fields, Machine Learning shows impressive results. It allows predicting the properties and performance of a new material from the process parameters, optimizing these parameters for performance driven design, and even helping in the choice of raw materials to mix and in the characterization of new materials. All these tasks allow researchers to find new materials faster than ever, and obviously to decrease the costs.

However, real world data are expensive to produce and cannot feed many models that need large datasets to be performant, such as deep learning. The gap between simulated and real data is still important, as in many other application domains. Moreover, materials scientists also want to understand the decision of Artificial Intelligence algorithms. As for other sciences, we can bet that interpretable models will be preferred to black box models (such as neural networks).

Acknowledgements

We want to thank the CEA Cross-Cutting Program on Materials and Processes Skills, for giving us the occasion to be acquainted with Materials science and to study the application of Artificial Intelligence in this domain.

References

[1] Fowler D., Arróyave R., Ross J., Malak R., Banerjee S. (2017) Looking outwards from the "central science": an interdisciplinary perspective on graduate education in materials chemistry, in *ACS symposium series*, pp. 65–89.

[2] Suhonen T., Laukkanen A., Andersson T., Pinomaa T., Holmberg K. (2016) VTT ProperTune: computational multiscale materials modeling concept, *Therm. Spray Bull.* **68**, 102.

[3] Saal J.E., Kirklin S., Aykol M., Meredig B., Wolverton C. (2013) Materials design and discovery with high-throughput density functional theory: the open quantum materials database (OQMD), *JOM* **65**, 1501.

[4] Binder K., Heermann D., Roelofs L., Mallinckrodt A.J., McKay S. (1993) Monte Carlo simulation in statistical physics, *Comput. Phys.* **7**, 156.

[5] Camacho-Zuniga C., Ruiz-Trevino F. (2003) A new group contribution scheme to estimate the glass transition temperature for polymers and diluents, *Ind. Eng. Chem. Res.* **42**, 1530.

[6] Hautier G., Jain A., Ong S.P. (2012) From the computer to the laboratory: materials discovery and design using first-principles calculations, *J. Mater. Sci.* **47**, 7317.

[7] Liu Y., Zhao T., Ju W., Shi I. (2017) Materials discovery and design using machine learning, *J. Materiomics* **3**, 159.

[8] Agrawal A., Choudhary A.N. (2016) Perspective: materials informatics and big data: realization of the "fourth paradigm" of science in materials science, *APL Mater.* **4**, 53208.

[9] Schwab K. (2016) The Fourth Industrial Revolution.

[10] Russell S., Norvig P. (2009) *Artificial intelligence: a modern approach*, 3rd edn.

[11] Bishop C.M. (2006) *Pattern recognition and machine learning* (Information Science and Statistics). Springer-Verlag.

[12] Chapelle O., Schlkopf B., Zien A. (2010) *Semi-supervised learning.* The MIT Press.

[13] Schmidt J., Marques M.R., Botti S., Marques M.A.L. (2019) Recent advances and applications of machine learning in solid-state materials science, *Nat. Comput. Mater.* **5**.

[14] Wei J., Chu X., Sun X.-Y., Xu K., Deng H.-X., Chen J., Wei Z., Wei M. (2019) Machine learning in materials science, *InfoMat* **1**, 338.

[15] Smith J., Xiong W., Yan W., Lin S., Cheng P., Kafka O.L., Wagner G.J., Cao J., Liu W.K. (2016) Linking process, structure, property, and performance for metal-based additive manufacturing: computational approaches with experimental support, *Comput. Mech.* **57**, 583.

[16] Kalidindi S.R., Medford A.J., Mcdowell D.L. (2016) Vision for data and informatics in the future materials innovation ecosystem, *JOM* **68**, 2126.

[17] Hill J., Mulholland G., Persson K., Seshadri R., Wolverton C., Meredig B. (2016) Materials science with large-scale data and informatics: unlocking new opportunities, *MRS Bull.* **41**, 399.

[18] Datta S., Chattopadhyay P.P. (2013) Soft computing techniques in advancement of structural metals, *Int. Mater. Rev.* **58**, 475.

[19] Bock F.E., Aydin R.C., Cyron C.J., Huber N., Kalidindi S.R., Klusemann B. (2019) A review of the application of machine learning and data mining approaches in continuum materials mechanics, *Front. Mater.* **6**, 110.

[20] Cortes C., Vapnik V. (1995) Support-vector networks, *Mach. Learn.* **20**, 273.

[21] Breiman L. (1983) Classification and Regression Trees.

[22] Quinlan J.R. (1986) Induction of decision trees, *Mach. Learn.* **1**, 81.

[23] Quinlan J.R. (1993) C4.5: Programs for Machine Learning.

[24] Fitch F.B. (1944) McCulloch Warren S. and Pitts Walter. A logical calculus of the ideas immanent in nervous activity. Bulletin of mathematical biophysics, vol. 5 (1943), 115–133, *J. Symbolic Logic* **9**, 49.

[25] Rosenblatt F. (1958) The perceptron: a probabilistic model for information storage and organization in the brain, *Psychol. Rev.* **65**, 386.

[26] Bishop C. (1995) *Neural networks for pattern recognition.* Oxford University Press.

[27] Goodfellow I., Bengio Y., Courville A. (2016) *Deep learning.* MIT Press.

[28] Krizhevsky A., Sutskever I., Hinton G.E. (2017) ImageNet classification with deep convolutional neural networks, *Commun. ACM* **60**, 84.

[29] Rawat W., Wang Z. (2017) Deep convolutional neural networks for image classification: a comprehensive review, *Neural Comput.* **29**, 2352.

[30] Pascanu R., Gulcehre C., Cho K., Bengio Y. (2014) How to construct deep recurrent neural networks, in *Proceedings of the Second International Conference on Learning Representations (ICLR 2014)*.
[31] Kumari U. (2017) Soft computing applications: a perspective view, in *2nd International Conference on Communication and Electronics Systems (ICCES)*.
[32] Zadeh L.A. (1965) Fuzzy sets, *Inf. Control* **8**, 338.
[33] Jang J.-S.R. (1993) ANFIS: adaptive-network-based fuzzy inference system, *IEEE Trans. Syst. Man Cyber.* **23**, 665.
[34] Goldberg D.E. (1989) *Genetic algorithms in search, optimization, and machine learning.* Addison-Wesley Professional.
[35] Kennedy J., Eberhart R. (1995) Particle swarm optimization, in *Proceedings of ICNN'95 – International Conference on Neural Networks.*
[36] Qi X., Chen G., Li Y., Cheng X., Li C. (2019) Applying neural-network-based machine learning to additive manufacturing: current applications, challenges, and future perspectives, *Engineering.*
[37] Stahle L., Wold S. (1989) Analysis of variance (ANOVA), *Chemom. Intell. Lab. Syst.* **6**, 259.
[38] Gelman A. (2005) Analysis of variance – why it is more important than ever, *Ann. Stat.* **33**, 1.
[39] Hofland E.C., Baran I., Wismeijer D.A. (2017) Correlation of process parameters with mechanical properties of laser sintered PA12 parts, *Adv. Mater. Sci. Eng.* **2017**.
[40] Beal V., Paggi R., Salmoria G., Lago A. (2009) Statistical evaluation of laser energy density effect on mechanical properties of polyamide parts manufactured by selective laser sintering, *J. Appl. Polym. Sci.* **113**, 2910.
[41] Singh A.K., Prakash R.S. (2009) Simulation based two level factorial designed experiments for optimisation of selective laser sintering process, *Int. J. Comput. Aided Eng. Technol.* **1**, 322.
[42] Chacón J., Caminero M.A., García-Plaza E., Núñez P.J. (2017) Additive manufacturing of PLA structures using fused deposition modelling: effect of process parameters on mechanical properties and their optimal selection, *Mater. Des.* **124**, 143.
[43] Rani C. (2015) Relationship between process parameters and mechanical properties of friction stir processed AA6063-T6 based composite.
[44] Karthikeyan L., Senthil Kumar V.S. (2011) Relationship between process parameters and mechanical properties of friction stir processed AA6063-T6 aluminum alloy, *Mater. Des.* **32**, 3085.
[45] Weichert D., Link P., Stoll A., Rüping S., Ihlenfeldt S., Wrobel S. (2019) A review of machine learning for the optimization of production processes, *Int. J. Adv. Manuf. Technol.* **104**, 1889.
[46] Köksal G., Batmaz I., Testik M.C. (2011) A review of data mining applications for quality improvement in manufacturing industry, *Expert Syst. App.* **38**, 13448.
[47] Singh J.P., Pandey P.M., Verma A.K. (2016) Fabrication of three dimensional open porous regular structure of PA-2200 for enhanced strength of scaffold using selective laser sintering, *Rapid Prototyping J.*
[48] Marrey M., Malekipour E., El-Mounayri H., Faierson E.J. (2019) A framework for optimizing process Parameters in powder bed fusion (PBF) process using artificial neural network (ANN), *Proc. Manuf.* **34**, 505.
[49] Stoll A., Pierschel N., Wenzel K., Langer T. (2019) Process control in a press hardening production line with numerous process variables and quality criteria, in *Machine learning for cyber physical systems.* Springer, pp. 77–86.
[50] Agrawal A., Deshpande P.D., Cecen A., Basavarsu G.P., Choudhary A.N., Kalidindi S.R. (2014) Exploration of data science techniques to predict fatigue strength of steel from composition and processing parameters, *Integrating Mater. Manuf. Innov.* **3**, 8.
[51] Yazdi M.R.S., Khorasani A.M., Faraji M. (2011) Optimization of coating variables for hardness of industrial tools by using artificial neural networks, *Expert Syst. App.* **38**, 12116.
[52] Jebali M., Kahouli O., Abdallah H.H. (2017) Optimizing PSS parameters for a multi-machine power system using genetic algorithm and neural network techniques, *Int. J. Adv. Manuf. Technol.* **90**, 2669.

[53] Rong-Ji W., Xin-Hua L., Qing-Ding W., Lingling W. (2009) Optimizing process parameters for selective laser sintering based on neural network and genetic algorithm, *Int. J. Adv. Manuf. Technol.* **42**, 1035.

[54] Dastjerdi A.A., Movahhedy M.R., Akbari J. (2017) Optimization of process parameters for reducing warpage in selected laser sintering of polymer parts, *Addit. Manuf.* **18**, 285.

[55] Garg A., Lam J.S.L., Savalani M. (2015) A new computational intelligence approach in formulation of functional relationship of open porosity of the additive manufacturing process, *Int. J. Adv. Manuf. Technol.* **80**, 555.

[56] Tanyildizi H. (2009) Fuzzy logic model for prediction of mechanical properties of lightweight concrete exposed to high temperature, *Mater. Des.* **30**, 2205.

[57] Gencel O., Ozel C., Koksal F., Martinez-Barrera G., Brostow W., Polat H. (2013) Fuzzy logic model for prediction of properties of fiber reinforced self-compacting concrete, *Mater. Sci.* **19**, 203.

[58] Zalnezhad E., Sarhan A.A.D.M., Hamdi M. (2013) Surface hardness prediction of CrN thin film coating on AL7075-T6 alloy, *Int. J. Precis. Eng. Manuf.* **14**, 467.

[59] Jaya A.S.M., Hashim S.Z.M., Rahman M.N.A. (2010) Fuzzy logic-based for predicting roughness performance of TiAlN coating, in *10th International Conference on Intelligent Systems Design and Applications.*

[60] Kuram E., Ozcelik B. (2016) Micro-milling performance of AISI 304 stainless steel using Taguchi method and fuzzy logic modelling, *J. Intell. Manuf.* **27**.

[61] Peng A., Xiao X., Yue R. (2014) Process parameter optimization for fused deposition modeling using response surface methodology combined with fuzzy inference system, *Int. J. Adv. Manuf. Technol.* **73**, 87.

[62] Tofigh A.A., Rahimipour M.R., Shabani M.O., Davami P. (2015) Application of the combined neuro-computing, fuzzy logic and swarm intelligence for optimization of compocast nanocomposites, *J. Compos. Mater.* **49**, 1653.

[63] Rao R.V., Rai D.P. (2016) Optimization of fused deposition modeling process using teaching-learning-based optimization algorithm, *Eng. Sci. Technol. Int. J.* **19**, 587.

[64] Rao R.V., Savsani V.J., Vakharia D.P. (2011) Teaching–learning-based optimization: a novel method for constrained mechanical design optimization problems, *Comput. Aided Des.* **43**, 303.

[65] Ashby M.F. (1989) Materials selection in conceptual design, *Mater. Sci. Technol.* **5**, 517.

[66] Ashby M.F. (2016) *Materials selection in mechanical design*, 5th edn. Butterworth-Heinemann.

[67] Hwang C.L., Yoon K. (1981) *Multiple attribute decision making: methods and applications.* Springer-Verlag.

[68] Jee D.-H., Kang K.-J. (2000) A method for optimal material selection aided with decision making theory, *Mater. Des.* **21**, 199.

[69] Chen R.W., Navin-Chandra D., Kurfess T., Prinz F. (1994) A systematic methodology of material selection with environmental considerations, in *Proceedings of 1994 IEEE International Symposium on Electronics and the Environment.*

[70] Cherian R.P., Smith L.N., Midha P.S. (2000) A neural network approach for selection of powder metallurgy materials and process parameters, *Artif. Intell. Eng.* **14**, 19.

[71] Trethewey K.R., Wood R.J.K., Puget Y., Roberge P.R. (1998) Development of a knowledge-based system for materials management, *Mater. Des.* **19**, 39.

[72] Khabbaz R.S., Manshadi B.D., Abedian A., Mahmudi R. (2009) A simplified fuzzy logic approach for materials selection in mechanical engineering design, *Mater. Des.* **30**, 687.

[73] Gul M., Celik E., Gumus A.T., Guneri A.F. (2018) A fuzzy logic based PROMETHEE method for material selection problems, *Beni-Suef Univ. J.. Basic Appl. Sci.* **7**, 68.

[74] Brans J.-P., Vincle P. (1985) A preference ranking organization method, *Manage. Sci.* **31**, 647.

[75] Athanasopoulos G., Riba C.R., Athanasopoulou C. (2009) A decision support system for coating selection based on fuzzy logic and multi-criteria decision making, *Expert Syst. App.* **36**, 10848.

[76] Sibilia J.P. (1988) *A guide to materials characterization and chemical analysis.* John Wiley & Sons.

[77] Kaufmann E.N. (2003) *Characterization of materials, 2 volume set* (E.N. Kaufmann, Ed.). Wiley-VCH, p. 1392. ISBN: 0-471-26882-8.
[78] Xu B., Shen Z., Ni X., Wang J., Guan J., Lu J. (2004) Determination of elastic properties of a film-substrate system by using the neural networks, *Appl. Phys. Lett.* **85**, 6161.
[79] Sacha G., Rodriguez F., Varona P. (2009) An inverse problem solution for undetermined electrostatic force microscopy setups using neural networks, *Nanotechnology* **20**.
[80] Asproulis N., Drikakis D. (2009) Nanoscale materials modelling using neural networks, *J. Comput. Theor. Nanosci.* **6**, 514.
[81] Al-Khedher M., Pezeshki C., McHale J., Knorr F. (2007) Quality classification via Raman identification and SEM analysis of carbon nanotube bundles using artificial neural networks, *Nanotechnology* **18**.
[82] Vijayaraghavan V., Garg A., Wong C., Tai K. (2014) Estimation of mechanical properties of nanomaterials using artificial intelligence methods, *Appl. Phys. A* **116**, 1099.
[83] Wang B., Zhao W., Du Y., Zhang G., Yang Y. (2016) Prediction of fatigue stress concentration factor using extreme learning machine, *Comput. Mater. Sci.* **125**, 136.
[84] Khosravani M.R., Nasiri S., Anders D., Weinberg K. (2019) Prediction of dynamic properties of ultra-high performance concrete by an artificial intelligence approach, *Adv. Eng. Softw.* **127**, 51.
[85] Dao D.V., Trinh S.H., Ly H.-B., Pham B.T. (2019) Prediction of compressive strength of geopolymer concrete using entirely steel slag aggregates: novel hybrid artificial intelligence approaches, *Appl. Sci.* **9**, 1113.
[86] DeCost B.L., Jain H., Rollett A.D., Holm E.A. (2017) Computer vision and machine learning for autonomous characterization of am powder feedstocks, *JOM* **69**, 456.
[87] Curtarolo S., Morgan D., Dane K., Rodgers J., Ceder G. (2003) Predicting crystal structures with data mining of quantum calculations, *Phys. Rev. Lett.* **91**, 135503.
[88] Fischer C.C., Tibbetts K.J., Morgan D., Ceder G. (2006) Predicting crystal structure by merging data mining with quantum mechanics, *Nat. Mater.* **5**, 641.

4.3 Integrative Approach for Safe Manufacturing

Claude EMOND[1,2] and Frédéric SCHUSTER[3]

[1] *University of Montreal, School of Public Health, Montreal, Quebec, Canada*
[2] *BioSimulation Consulting Inc., Newark, Delaware, USA 19713*
[3] *Université Paris Saclay, CEA, Cross-cutting Program on Materials Science and Engineering, 91191 Gif-sur-Yvette, France*

Abstract

Nanotechnology is an emerging science, characterised by the manipulation of atoms and molecules less than 100 nm, and it offers new ways to explore new materials using new physical and chemical properties. However, the hazards of nanotechnology are often unknown and are therefore unpredictable because few assessments are available for determining the most important function. Since 2000, the interest in nanotechnology has quickly risen among those in science-related disciplines, industry and academia because of two important phenomena: (1) the public's awareness of the dangers that could result from the use of nanotechnology and (2) the involvement of toxicology in nanotechnology. This momentum of the second phenomenon has substantially decreased the promotion of industrial nanotechnology. The number of toxicology publications has grown over the years, which spurred the industry to decrease its own publicity regarding nanotechnology. Toxicology was quickly implemented to address this new challenge, first by employing the analytical techniques used for fine particles. However, this approach was uncertain, particularly because of the specific nature of these nanoparticles or because of their characteristics to react differently during various experimental conditions. Consequently, many published toxicology studies, in the decay of the 1990s, are now unreliable because of poor chemical characterisations of nanoparticles, resulting in poor replicability of some studies. Over time, toxicology has yielded a better methodology to reproduce the experiments by controlling the exposure variables and providing better characterisation analysis. Although nanotechnology offers attractive opportunities for industries, the new properties from these manipulations at the atomic or molecular scale can also present new potential risks, as well as unknown health risks. Currently, there are only a few, if there is, any health risk assessments that governments can use to regulate nanotechnology efficiently and accurately. *The aim of the section is to present an analytical framework for identifying important parameters to be considered for an evaluation reference assessment of nanotechnology.*

4.3.1 Introduction

In this section, we present an approach to improve knowledge of the hazards associated with the nanoparticle synthesis at the time of manufacture to offer potential uses or at least identify usage limitations. The interest focused on the nanoparticle itself during an analysis of its reactivity approach. Thus, the approach developed in this section is intended to complement the one used for occupational health. For

DOI: 10.1051/978-2-7598-2446-5.c034

example, an industry created a new nanoparticle and wants to use it in different manufacturing processes (*e.g.* new uses for additives, products or other items). The approach discussed here is to assess the reactivity of the nanoparticle to deliver a security dimension of its own, while protecting workers at their workplaces. Specifically, the approach consists of the following three components: (1) conducting a literature review of health issues related to nanotechnology assessment, (2) developing an analytical framework that identifies the important parameters to be considered in an evaluation reference of nanotechnology and (3) optimising this analysis grid through the collaboration of companies in Canada and Europe, both well-established and emerging ones.

4.3.1.1 Methodology and Literature Review

The research involving the physicochemical parameters for characterising nanomaterials is one of the key stages of this work. Because knowledge is limited regarding occupational exposure to nanomaterials, our approach includes a literature review, which identified the reports, documents and articles that discussed the risk assessment of exposure or described the factors that influence exposure to nanomaterials. Unlike conventional particles, manufactured nanoparticles have more reactive properties, which make them difficult to characterise. However, the literature found during our review listed several determinants (*e.g.* size, chemical composition, surface reactivity, solubility, formation of aggregates or agglomerates) that influence exposure and are essential for a better characterisation of nanomaterials [1–3]. These parameters listed in figure 4.3.1 are not an exhaustive list; however, they will influence the risk of exposure and the potential impacts from exposure.

Apart from the physical and chemical parameters, the exposure pathways are also significant factors that influence the toxicity regarding nanomaterials.

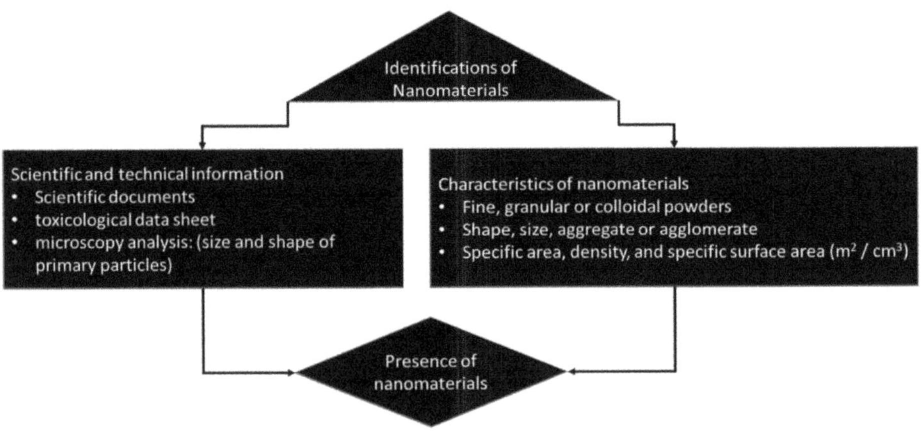

FIG. 4.3.1 – The first phase of the preliminary assessment of exposure is to identify the nanomaterials.

The populations that are potentially exposed to nanomaterials are primarily employees who work in the nanotechnology industry, followed by those who manufacture consumer products that contain nanoparticles. Because of the nanoparticles' size, in particulate form, the respiratory tract is the main entrance into the body. The limited data available show that when a person inhales the nanoparticles, they are distributed to different areas of the respiratory tract depending on their size [4]. Nanomaterials can also be found in the gastrointestinal tract during swallowing after inhalation. Although the penetration of nanomaterials through the skin is still under investigation, according to the stages of handling and maintenance procedures, workers can be exposed to nanomaterials [5]. However, the route of exposure is influenced by many factors, including the nature of the nanomaterials (*e.g.* powder suspended in a liquid), the types of synthetic methods used, the degree of containment, the quantities used, the means for protection that were put in place and the ability of nanomaterials to mix in the air or on work surfaces [6, 7].

Many studies are published regularly regarding the use of toxicological tests to determine the effects of nanomaterials on health. Several *in vivo* and *in vitro* tests have been selected *a priori* to be introduced into the decision gate; however, even if a test selection is made based on how frequently the topic occurs in publications and/or on the usefulness of the nanomaterials, more work should be conducted to verify their relevance before becoming a reference. The detection and determination of nanomaterials require several specific tools, some of which have not yet been harmonised or developed. This data collection, as well as the preparation of samples for lead testing, closely depends on the collaboration between industry and those involved with conducting research [5, 8, 9].

4.3.1.2 Characteristics of Nanomaterials

The literature discusses several methods for assessing nanomaterial exposure [10–13]. These methods are either based on the standard model, developed by the National Research Council (NRC) in 1983, for assessing health risks of chemicals [14] or on a graded risk management method known as control banding (CB). It is important to understand the different methods described in the literature to provide a new methodology for meeting the criteria in accordance with the study objectives and with the current state of knowledge.

The Lux Research method is briefly described as an approach for qualitatively evaluating the nano-specific risks of a finished product throughout its life cycle [15]. However, several important evaluation criteria are not included, in particular the absence of risk rating criteria and toxicity evaluation criteria to differentiate risks to humans from those to the environment or to consider products containing several types nanomaterials. Moreover, as suggested in a report from the French Agency for Environmental and Work Health Safety (*Agence Française de Sécurité Sanitaire de l'Environnement et du Travail* [AFSSET]), the various shortcomings of the evaluation approach prevent updating the dimension settings to the rhythm of advancing scientific knowledge [10, 15].

The "DuPont environment" method is used to evaluate nano-specific risks to humans and the environment, as well as for products containing manufactured nanomaterials during their entire life cycle [11]. People use this method because they want to make their decisions based on the findings from risk assessments, so they include management support elements (*e.g.* evaluating the effectiveness of risk management, methodological framework for actions to be taken based on an analysis of the situation) [10, 15]. This method is interesting, but it is applicable only in the perspective of a life cycle, which far exceeds the type of decisional grid to be developed in the framework of this project.

We also analysed and provided comments about another method based on an evaluation grid proposed in Switzerland by the Federal Office for the Environment (FOEN) and the Federal Office of Public Health (FOPH) [12, 15]. This method is a precautionary matrix for synthetic nanomaterials to distinguish and estimate the nano-specific risks to workers (occupational hazards), consumers and the environment from products that contain manufactured nanomaterials. The principle proposed by this evaluation grid is to combine the scores for each criterion considered using an equation that generates an overall risk result. The various risks (*e.g.* consumer, worker, environmental) are then characterised in terms of the results. Thus, if a nano-specific risk is considered to be low, then the situation does not require additional risk studies. However, if the result of nano-specific risk exceeds the action level, then additional steps are required (*e.g.* more risk assessments, risk reduction actions). However, this method is limited in its ability to estimate risk situations when nanomaterials are present [15]. A criticism discussed in the AFSSET report is that the scoring method and the final scores that were used to derive the sum of these potential effects account for only two criteria during the entire life cycle: (1) the level of redox activity and/or catalytic nanomaterial and (2) the stability in the relevant environments (physiological and environmental conditions). Indeed, this tool considers that the toxicity level remains constant throughout the entire product life cycle and that the mechanism of toxicity is restricted only to the formation of reactive oxygen species (ROS). Another limitation of this evaluation grid is that it ignores the inherent characteristics of exposure. Moreover, as noted in the description of the previous method proposed by FOEN and FOPH, it is not our aim to consider the entire life cycle of the nanoparticle, but only its responsiveness in order to define the limits of use and exposure.

The risk assessment model and the controlled exposure to nanoparticles or "control banding" proposed by [13] are based on the use of a limited number of factors to assess the risks to reduce the complexity of the assessment process and increase applicability. This model uses three levels of control bands, also referred to as bands of engineering controls, that are based on the fundamentals of occupational hygiene, plus a fourth control band for the most at risk. However, it is essential that additional measures occur and that new knowledge is regularly updated because these two (assessment model and control measure) help to ensure that risks are determined with the greatest possible precision for a given situation and ensure continuous and effective control of exposure [6].

After conducting a critical review of the main approaches published, we noticed that none of them is satisfactory because we are looking for tools such as objective evaluations and characterisations that provide accurate results, which can then be used to provide recommendations. However, this review was essential because it identified existing tools, which subsequently inspired us to propose an approach to meet our objectives. Before presenting our evaluation matrix, it is important to note that this study scrutinises the identified literature to ensure that the data are accurate and promote a better understanding of nanoparticles, during various stages of evaluation as discussed in the next section development of the evaluation grid. Thus, it should be clear that the interest is not to directly protect workers or the population, but to assess the nanoparticles as an entity in order to determine whether there are limitations regarding the applicability and use of this technology as industry is planning to do. This distinction is important because our approach complements the one used by the French National Institute for Industrial Environment and Risks (*Institut National de l'Environnement Industriel et des Risques* [Ineris]), which focuses on the workers. However, if we limited the risk to just the nanoparticle, then this would have an obvious impact on workers. Moreover, it should be noted that the evaluation matrix developed in this part of the section is not focused on using the results from quantitative assessments of toxic risks of nanomaterials to assign characteristic scores based on the level of risk. In particular, the evaluation matrix focuses on using the results from qualitative assessments of manufactured nanomaterials to determine whether they are safe for use.

4.3.1.3 Development of the Evaluation Matrix Grid

The findings from the analysis of various health risk assessments of nanotechnology tools allowed us to reflect on an evaluation matrix grid pattern that has been developed, taking into account that it is currently laborious – if not impossible – to develop a tool to evaluate the quantitative and qualitative risks of exposure to nanomaterials. As previously mentioned, current data regarding the toxicity of nanomaterials are limited, they are often not reproduced and they do not allow a quantitative assessment of risk exposures. Therefore, we decided to develop our intelligence grid based on a health risk assessment widely used for environmental issues. This approach is based on the one developed by the U.S.-based NRC [16].

The purpose of the matrix grid is to help improve our understanding of nanomaterials in a finished product in order to anticipate their potential effects. This evaluation grid is intended for all stakeholders who are responsible for ensuring the health and safety of workers and consumers and for all workers and consumers of nanotechnologies.

The matrix grid's design is based on recommendations regarding characterising the potential exposure to nanomaterials that are available in the scientific literature focusing on nanotechnologies. The grid's design is based on a stepwise approach and considers proposals on techniques and methods to ensure better characterisations of nanomaterials. Moreover, this tool is designed to promote the safety of nanomaterials in a safe marketing perspective that most certainly leads to better responsiveness and acceptability of these materials by the company. This grid was developed in the following four stages: (1) preliminary assessment; (2) characterisation of the exposure, including the dose–response relationship and exposure assessment; (3) risk characterisation of health; and (4) report findings.

4.3.1.3.1 Preliminary Assessment of Exposure

An important step in evaluating the health risks of nanomaterials is to conduct a preliminary assessment of exposure. First, this assessment involves an evaluation of the technical characteristics of the materials to determine whether they are nanomaterials, and, if so, an analysis of the situation is conducted to determine whether there is a potential risk from a potential exposure. It is important to identify nanomaterials because it confirms the presence or absence of nanoparticles generated during the industrial synthesis process, identifies the types of nanomaterials and helps alert us that we are near nanoparticles, as identified and defined by recognised experts in the field. Manufacturers of nanoparticles may view this information as absurd, but it makes sense to those in the industry who purchase and use nanoparticles to make finished products. Frequently, members of the industry will not reveal whether they use nanoparticles during their manufacturing processes. In the interest of transparency, it is advisable as a first step to make the preliminary assessment of exposure, regardless of the type of industry using the nanotechnology.

4.3.1.3.1.1 Hazard Identification. It is important to identify nanomaterials based on criteria specific to this type of particle (figure 4.3.1). Materials that do not meet the definition of a nanoparticle will therefore not be affected by the decision grid. This first phase of hazard identification includes a search for scientific and technical information about the product in question and the collection of information about the characteristics of nanomaterials.

4.3.1.3.1.2 Scientific and Technical Information. These data refer to any scientific and technical documentation about the nanomaterials. These data include the following:

- Name of the nanomaterials in the sample.
- Unique registration number, if available, from the Chemical Abstracts Service database [17].
- Structural composition and molecular formula of the chemical.
- Composition of nanomaterials used, including purity and additives.
- Catalytic activity, if known.
- Basic morphology of the nanoparticle.
- Description of the surface chemistry (*e.g.* coating, modification).
- Main commercial uses (if the nanomaterial was ever used in the past).

4.3.1.3.1.3 Characteristic of Nanomaterial. The previous first phase was intended to express a more qualitative analysis of nanomaterial. The present second phase is a quantitative characterisation of the sample content. Because this is a preliminary assessment exposure for evaluating a new sample of nanoparticles, the emphasis here is mainly on the data provided when purchasing the nanopowder or suspension and on the data provided by the industry. This characterisation of nanomaterials is an almost entirely theoretical exercise and is *a priori* with the information that is available. The data collected relate to the following parameters:

- Raw materials and intermediate products and finished products.
- Characteristics of materials (*e.g.* nanopowders, granular or colloidal).
- Chemical composition of metals, oxides or polymers.
- Shape (*e.g.* sphere, cube, aggregate or agglomerate).
- Size (*e.g.* dimension, the number of nanoparticles per unit area).
- Physicochemical properties (*e.g.* surface area, surface reactivity, solubility).
- Type of nanomaterials (*e.g.* carbon nanotubes, quantum dots, dendrimers).
- Dispersion medium (*e.g.* sprays, gels, colloids).

4.3.1.3.1.4 Interpretation of Hazard Identification. This different information is not intended to be an exhaustive list, but a combination of the essential parameters for identifying nanomaterials. In addition, the parameters also help to provide insight regarding the possible risk of exposure, which is the subject of a subsequent step to develop a decisional grid. If an evaluation of the received information confirms that nanomaterials are present, then the second phase of the preliminary assessment of exposure begins. However, if an evaluation of the received information determines that the material is not at the nanoscale level, then the results are compiled and given to the industrials.

4.3.1.3.1.5 Potential Exposure to Nanomaterials. To proceed to the second phase of the preliminary evaluation, an evaluation must have confirmed that nanomaterials are analysed in the sample. Then, research is conducted to determine the potential exposure of workers and consumers to nanomaterials. This second phase has three subdivisions: the analysis of the different manufacturing processes, the analysis of the conditions of workers at their workstations and the analysis of the finished products (figure 4.3.2). This second phase consists of visiting the workplace to document the processes used to manufacture the nanoparticles, cataloguing the

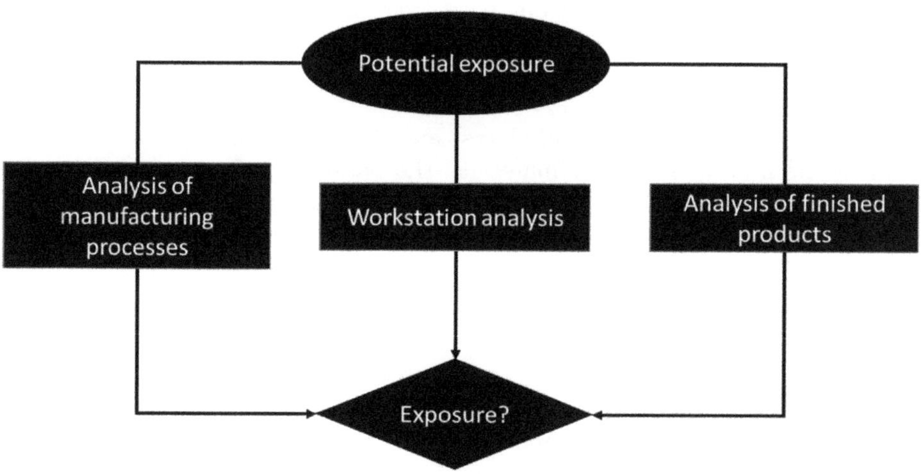

FIG. 4.3.2 – The second phase of the preliminary assessment of exposure: Exposure research for potential nanomaterials.

methods used for conducting the work, identifying the workstation and developing a summary analysis of the product finishes that could contain nanoparticles.

4.3.1.3.1.6 Analyse Different Manufacturing Processes. Several techniques that are used for manufacturing nanomaterials have been designed to control the characteristic parameters such as size, shape, composition and the degree of particle agglomeration for dispersed systems and grains as the bulk materials. These manufacturing processes will influence the routes of exposure, the potential risk of exposure and the reactivity of nanoparticles. An analysis of manufacturing processes is therefore necessary for the preliminary assessment of exposure. The many creative techniques able to structure nanostructures can be classified into two main routes: (1) the ascending (or bottom up) approach is used after steam condensation or crystallisation and (2) the downlink (or top down) approach employs mechanical grinding. Both approaches tend to yield nano-sized structures (tables 4.3.1 and 4.3.2).

Using the ascending (or bottom up) approach to manufacture nanomaterials consists of building structures atom by atom or molecule by molecule. This approach is used for metals, ceramics, semiconductors, polymers, fullerenes and carbon nanotubes. These structures are then used directly (*e.g.* by catalysis) or they serve as building blocks for constructing nano-structured architectures of varying complexities. Metal building blocks, ceramics, hybrid systems and self-organising structures are obtained by chemical transformations called soft chemistry or sol–gel processes. Metal nanopowders, ceramics, nanotubes and fullerenes are also obtained by physical processes of gas condensation.

The downlink (or top down) approach is used to produce nanomaterials from successive divisions of "classic" microstructured materials. This approach essentially applies mechanical stress, called "severe" (or strong) shocks, which produces significant deformations (using mechanical grinding techniques) and alloying effects, followed by mechanical stress from heat treatments, consolidation or chemical reactions. Industry also uses the metasynthesis intermediate method, which allows workers to handle larger amounts of atoms by modifying the nanoparticles so they rely on their self-organisation.

Visits of production sites (workplaces) are used to evaluate all work processes, including the movement of raw materials, the packaging of nanomaterials, which nanoparticles are used or produced, how finished goods are stored and methods for waste disposal. The analysis must also consider the following aspects:

- Raw materials and intermediate and finished products.
- Amounts of handled goods and their nanomaterials' ability to remain suspended in the air.
- Potential sources of emissions of nanoparticles (*e.g.* type, number, size).
- Exposure routes (*e.g.* respiratory, skin, gastrointestinal).
- Extent and duration of exposure.
- Technical work activities and proximity to the target source (*e.g.* time, usually).

Visits to workplaces are also used to assess safety measures by observing workplace configurations (*e.g.* space, size), evaluating the general ventilation system

TAB. 4.3.1 – Different nanomaterials' manufacturing methods (Part A)[a].

Processing method	Principle of operation	Type of nanomaterials	Size of nanomaterials
Physical methods			
Evaporation or condensation under inert or reactive pressure	Heating is used for evaporation of a metal; metal vapour condensation is used to collect dispersed nanoparticles.	Al, Co, Cu, Fe, Ni, Pd, Pt and oxides	20 up to 100 nm
Laser pyrolysis	A cross-jet interaction between a carbon dioxide (CO_2) beam and a stream of reagents is used.	Si, SiC, SiCN, SiC(NAlY), SiCO, Si_3N4, TiC and TiO_2 fullerenes, soot, carbonaceous, etc	15 up to 20 nm
Supercritical fluid (without chemical reaction)	Supercritical fluid could be employed as a solvent, anti-solvent, plasticiser, nebulizer, emulsion phase, etc.	Materials for vectorisation of active substances	From nm up to micrometres (μm)
Microwaves	Nanoparticle processing routes coupling, thereby allowing faster and more homogeneous medium heating.	Ni, Ag	20 up to 200 nm
Ionic or electronic irradiation	Several methods are used, such as high-energy heavy ions deposition, which allows the production of nanoporous polymers; and ionic mobility irradiation inside vitreous matrices for the metal aggregates production.	Nanoporous structures inside the material: macroscopic and nanostructured	No data available
Low-temperature annealing of amorphous "bulk" phases	Nanostructures are obtained from amorphous phases by recrystallisation.	Metallic and inter-metallic alloys: Al, Fe and Zr	Full nanometric range
Thermal spray	The vaporised or melted particles are quenched and condense in the form of ultra-fined grains.	Ceramic nanopowders (TiC, TaC, SiC), silicides ($MoSi_2$) and doped-oxides (TiO_2)	No data available
Physical vapour deposition	Physical vapour processing routes from a metallic source.	TiN, CrN and (Ti,Al)N coatings	No data available

[a]From reference [18].

TAB. 4.3.2 – Different nanomaterials' manufacturing methods (Part B)[a].

Processing method	Principle of operation	Type of nanomaterials	Size of nanomaterials
Chemical methods			
Vapour phase reactions	The chemical vapour deposition (CVD) process allows thin layers to be produced by a chemical reaction between a volatile compound and the substrate's surface to be coated.	Materials from different nature (*e.g.* carbides, nitrides, oxides, alloys, metallic)	From nm up to millimetre (mm)
Liquid reactions	The reagent solution leads to the nanoparticles' formation by physicochemical conditions (*e.g.* pH, temperature, concentration).	Most metals and oxides	From nm up to few hundred nm
Solid reactions	This process covers numerous synthetic pathways that control the size of the nanoparticles (*e.g.* by the size of the initial products, by diffusion, by specific reaction sites or by volume reaction).	Most metals and oxides	From nm up to micrometres (μm)
Sol–gel techniques	This process involves nanomaterials produced from alkoxides or colloidal solutions as crystallised nanopigments or thin films. These techniques are based on inorganic polymerisation techniques.	Most oxides	From nm up to few hundred μm
Supercritical fluid with chemical reactions	This process involves continuous adjustment of the physicochemical properties of the reaction medium by adjusting the pressure and the temperature parameters on small scales, with the aim of controlling the chemical reactions and production of the material.	Most metals and oxides and some nitrides	From nm up to millimetre (mm)
Mechanical methods			
Mechanical milling and mechanical activation (powder metallurgy)	The mechanical milling consists of grinding micrometric powders between (1 and 50 μm) of several alloys.	All types of materials (*e.g.* ceramics, metals, polymers, semiconductors)	From nm up to nanostructured micronic particles
Consolidation and densification	Techniques used to produce solid parts from powders (*e.g.* metallic, ceramic, semiconducting and organic powders).	No data available	No data available
Strong deformation	The structure of a crystallised material (metal, ceramic) should be refined until the grain size is approximately a few nanometres. The employed techniques include torsion and extrusion, among others.	No data available	No data available
Specific methods for carbon nanotubes			
Several methods (*e.g.* laser vaporisation, CVD plasma, thermal CVD)	Different methods are used to grow carbon nanotubes.	Carbonaceous	From nm up to few hundred mm

[a]From reference [18].

(*e.g.* efficiency, failure), determining whether personal and collective protection equipment are properly used by workers and determining whether safety procedures and other preventive measures are followed by all employees when conducting work.

4.3.1.3.1.7 Analyse the Quality of Finished Products. Because nanomaterials have many properties, including surface reactivity, they can be used for different applications and innovations in areas such as health, information technology and industry (table 4.3.3). We must therefore consider the scenario in which an unpredictable release of nanomaterials in finished products could cause a health risk to those who manufacture, handle or use the product. A thorough evaluation of the product before it is marketed to companies and the public, would help to reduce this risk.

If, as a result of these investigations, it appears that there has been a potential risk of exposure, then step 2 of exposure characterisation should be performed. If there has not been a potential risk of exposure, then the results will be compiled and returned to the clients requesting this assessment. Although interest in finished products is well beyond the objectives of this report, it became essential to note. As we evaluate the reactivity of nanoparticles generated for industrial use, it is important to provide feedback and recommendations to workplaces about the precautions they can take to ensure safe application and use, while protecting the health and safety of their employees. It is also important to question the stability of nanomaterials in the finished products to which workers and consumers may be exposed.

4.3.1.3.2 Exposure Assessment

Because nanoparticles are chemicals, workplaces must ensure that precautions and care are consistently taken to protect the health and safety of their workers, and even if the toxicity of the nanoparticles has not yet been established. In order to characterise the exhibition of the nanoparticles, the concentration of the nanoaerosols must be measured by conducting a physicochemical analysis of samples, and then by performing a toxicological evaluation of the particles (figure 4.3.3). Thus, having a better understanding of the physical and chemical characteristics of nanomaterials and the dosage of exposure for biological indicators will help increase surveillance. A standardised method or an international consensus to characterise the exposure to nanoparticles does not currently exist. This is also true for characterising the toxicity when no reference method exists until now.

A recent article [21] highlighted some questions that must be asked and issues that must be considered when an evaluation of nanoparticles is conducted to determine the toxicity. The first question involves the issue of the size of the nanoparticles. Specifically, are nanoscale particles more toxic than micrometric particles with the same chemical nature? This issue is currently unresolved because it is difficult to obtain single nanoparticles or disperse without changing the nature of the environments in which they are located. Another issue focuses on the ability of nanoparticles to penetrate into cells. To address this issue, the surface chemistry of nanoparticles must be known. In general, the nanoparticles do not penetrate the

TAB. 4.3.3 – Application of nanomaterials by industry[a].

Sectors	Actual and potential applications
Aerospace and automotive	Light-weight structures, self-lubricating coatings, lubricant on mechanical components, more durable and recyclable tires, catalytic converters, self-cleaning glass, paints, anti-scratch coatings, anti-corrosion and anti-soiling
Food industry	Airtight coatings on plastic (better preservation), membranes for treating food liquids and traceability markers
Chemistry and materials	Pigments, printer inks, ceramic powders, corrosion inhibitors, multi-functional catalysts, anti-fouling and self-cleaning windows, antibacterial nanoproducts, highly resistant textiles and coatings, thermal insulation, membranes for water treatment and wood treatment
Cosmetics	Anti-ultraviolet sunblock creams, more abrasive toothpaste, makeup (lipstick) and anti-aging, shine and colouring products
Defence	Lightweight textiles, detectors and correctors for chemical and biological agents and miniaturised monitoring systems
Electronics & communication	Insulating materials for transistors, solar cells, handheld electronic libraries, computers and high-speed electronic games, wireless technologies, flat screens and hard disk polishing
Energy	Rechargeable battery electrode (<1 nm), photovoltaic cells, structural light-weight wind-power stations and low consumption bulbs
Environment and ecology	Pollution and effluent treatments, production of ultra-pure water, pesticides and fertilisers (more effective and less damaging) and chemical analysers
Health, pharmacy, and biomedical	Drugs and active agents, oral vaccines, tissue, bone and organ regeneration, self-diagnostic kits, pregnancy tests, biocompatible implants, gene therapy and DNA analysis and membranes for dialysis
Textile, clothing	Thin films (self-cleaning and hydrophobic coatings), antibacterial nanoproducts, fire retardants, ultraviolet and infrared filters, special effects (color, antistain) and non-deformable and anti-fouling fabrics
Various applications	Biomimetic material (improved performance), material by self-assembly and secure paper (anti-fraud)

[a]Adapted from reference [18–20].

cells passively; however, active transport is possible. There is also a question about toxicity that would be dependent on the cell. Specifically, because the cells do not have all the same properties or the same functions, could the impact on them also vary? Another issue focuses on the influence of aggregation. The phenomenon of attraction between the nanoparticles induces an accumulation, or cluster, of nanoparticles of several tens of nanometres, resulting in changes in the kinetics of nanoparticles compared to their unit form. To gain a better understanding of the nanoparticles, it is important to ask another question, which is the same that was

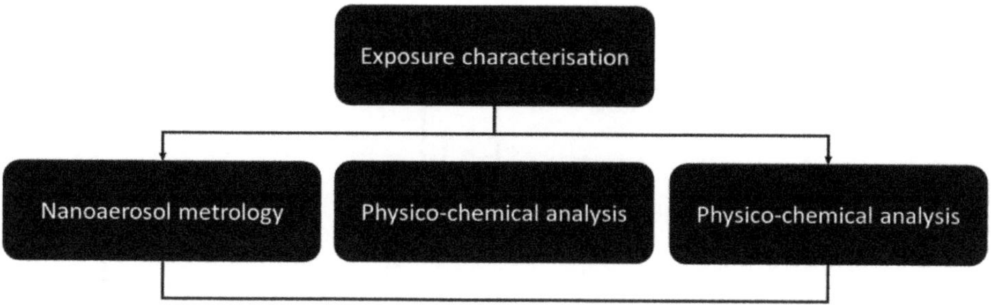

FIG. 4.3.3 – Steps to characterise the exposure characterisation.

raised in a study of the mechanisms of action: is it the dominant property that mostly influences their toxicity? The last question is as follows: How much of an influence do technical interference and positive controls play on the exposure to nanoparticles? Because these basic questions and issues are still unresolved [21], we experienced challenges when selecting the appropriate chemical and biochemical tests.

It is important to consider several criteria when evaluating occupational exposure. The evaluation should focus on characterising the nanoparticles dispersed into the air (*i.e.* in aerosol form). Similarly, the submicron range (diameters between 1 and 1000 nm) of the fraction of the aerosol size should be measured to integrate the measurement of the agglomerates and aggregates of nanoparticles [22]. Different technologies offering several opportunities are currently available for characterising the various properties of nanoparticles in the sample. It is not the aim of this section to provide or promote a specific technique or device over another. However, the findings from the literature review identified many parameters that must be considered when characterising nanoparticles. It is important to note that selection of the characterisation technique or device will be based on criteria that are acceptable to the manufacturer. It is, after all, the manufacturer who must bear the costs for the professional specialising in performing health risk assessments and who will endorse any recommended compromises based on findings that could jeopardise the health and safety of workers.

4.3.1.3.2.1 Measurement of Exposure. Samples must be collected at several measurements points using different available sampling techniques, because doing so will help identify and characterise the target nanoaerosol throughout among potential emission sources. The nanoaerosol target can be measured at the source, in the respiratory zone or in the near field or far field in relation to the workstation. The respiratory zone usually corresponds to a hemisphere with a radius of 30 cm starting from the face of a person [23]. The near field is typically a sphere of a 1-m radius centred on the head of the person, whereas the far field extends to the rest of the working space [24] (figure 4.3.4).

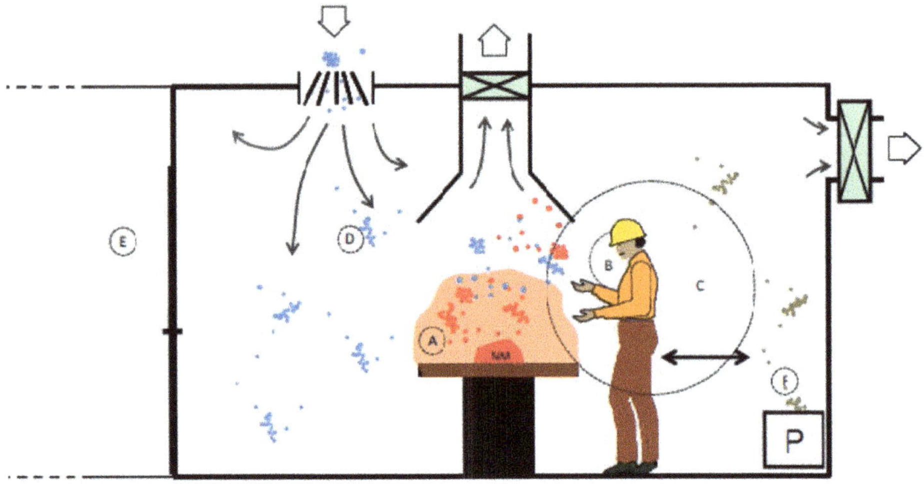

FIG. 4.3.4 – The positioning of the measuring points in a nanomaterial processing room. Blue particles represent the background aerosol from the outside. Green particles represent the background aerosol from a point source (P) of particles. Red particles represent the target aerosol resulting from using the nanomaterial during manufacturing. The measurement points are as follows: at the source (A), in the respiratory zone (B), in the near field (close to the operator) (C), in far field of the operator at airway height inside the room (D) or outside the room (E) and in the vicinity of the point source (P) of particles (F) (Adapted from reference [23–25]).

4.3.1.3.2.2 Physicochemical Assessment. After identifying the various probable sources of emissions of nanoparticles, many types of additional measures should be considered to determine the number concentration (number per cubic centimetre [cm^3]) or surface (square micrometres [μm^2] or square nanometres per cubic metre [nm^2/m^3]), the particle size distribution, chemical composition, surface area and surface reactivity of the nanoparticles; their shape, their size, crystal structure, solubility or their ability to form aggregates or agglomerates (table 4.3.4). A wide range of existing instruments are discussed in the literature, but others are being developed. The choice of instrument should be based on the type of study to be undertaken.

The physicochemical analysis is one of the crucial steps in evaluating the risks from nanoaerosols because it helps detect the particles on the samples, which can consist of groups such as ions, metals, organic compounds and others that have a relevant chemical reactivity. The composition of the particles is another parameter that affects the ability to react with biological materials. For example, the production of free radicals, which are unstable groups with short half-lives but are highly reactive, could influence different biological processes such as oxidative stress or an inflammatory process [26–28]. Moreover, the presence of metals such as copper, chromium and vanadium is important in the synthesis of some nanoparticles. However, these metals can catalyse the production of free radical species [26].

TAB. 4.3.4 – Main physicochemical parameters to be considered in the characterisation of exposure[a].

Principal parameters	[29]	[27, 28]	[30]	[31]	[2]
Size and distribution (granulometry)	+	++	++	++	++
Aggregates/agglomerates	+	++	+	++	+
Crystal structure	+	+	+	++	
Chemical composition	+	+	+		++
Specific surface area and porosity	+	++	+	++	++
Surface reactivity		++	+	++	+
Zeta potential		+	+	++	+
Form and morphology		++	++		+
Solubility/dissolution		+	+		++
Physical and chemical properties (*e.g.* purity)		++		++	
Processing methods				++	

Note: + = important; ++ = priority.
[a]Adapted from reference [3].

All of these features are important to consider when assessing the health risks from exposure to nanoparticles.

4.3.1.3.3 Dose–Response Assessment

Like any chemical substance, the potential toxicity of nanoparticles when they are found in the human body will depend on their physicochemical properties (*e.g.* size, surface reactivity, solubility in biological fluids). Few data are currently available regarding the toxicology of engineered nanomaterials; however, knowledge of the different body exposure routes is an essential prerequisite for toxicological evaluations [32]. During the different phases represented by pharmacokinetic absorption, distribution, metabolism and excretion, the nanoparticles, because of their specific characteristics, seem to adopt different behaviours in response to those chemicals.

The ability of nanoparticles to cross membrane barriers and end up in the general circulation of the human body has already been demonstrated [16, 17]. Although overall nanoparticle toxicity mechanisms are similar to those of larger sized particles, the surface they occupy per unit weight increases their reactivity. For example, gold, which is an inert metal on the micrometre scale, is considered a catalyst in the nanoscale. Thus, many modes of action are similar, but exaggerated at the nanoscale. This property is very important because the nanoparticles can produce ROS [26], resulting in oxidative stress phenomena and inflammatory responses, which can induce DNA damage and lead to genotoxicity [33]. Thus, the size of the nanoparticles promotes their distribution in organelles inaccessible for the micrometric particles and the induction of oxidative stress [34, 35].

During a toxicological assessment of nanoparticles, special attention should be given to the most sensitive target organs (*e.g.* liver, kidneys, lungs, central nervous system, spleen) to avoid a precocious effect. Thus, the choice of cells to be evaluated in the assay protocol must be decisive. The toxicity of nanomaterials cannot be solely attributed to their physicochemical properties; therefore, the toxicity analysis

TAB. 4.3.5 – The main *in vitro* or *in vivo* toxicological tests adapted according to the type of nanoparticles and the pursued objectives[a].

Toxicological tests	Available suggested tests
Cell viability or proliferation tests	Resazurin test, neutral red test, trypan blue counting test: cell counts
Membrane integrity tests	Lactate dehydrogenase (LDH) activity assay
Mitochondrial activity tests	MTT or WST1 tests
Metabolic activity test	SrB test
Oxidative stress and inflammation test	Activity SOD, GPx, CAT, GST, NF-kb, interleukins (IL-1, IL-6, IL-8)
Genotoxicity test	Comet assay and micronucleus test
Apoptosis test	Caspase activity tests, membrane asymmetry assay
Location testing	Transmission electron microscopy or confocal microscopy
In vivo tests	Biokinetic, carcinogenesis Acute or sub-acute toxicity tests Toxicokinetics: absorption, distribution, metabolism and excretion

[a]From reference [40, 41].

must also consider their ability to interact with the components associated with a matrix based on a set of sensitive cell lines. Thus, cell lines of human origin, derived directly from organs exposed to nanomaterials *via* inhalation (lung: A549, CaLu3), dermal application (skin HaCaT) or ingestion (colon: CaCo-2) can be used for tests *in vitro* [36–39]. Similarly, the cell lines of animal origin, such as lung epithelial cells (RLE-6TN), fibroblasts (NIH-3T3) and macrophages (RAW264.7), as kidney epithelial cells (MDCK, MDCK II, NRK52E), can be adapted to detect biomarkers of exposure or effects in these target organs [37]. These *in vitro* tests allow for a more complete view of the biological impacts of nanomaterials [37]. Toxicology tests can also be oriented to examine the reproductive toxicant and immunotoxic potential of nanoparticles. The details of tests that could be performed are presented in table 4.3.5. Overall, the completion of extensive toxicological testing is essential for characterising the hazards of exposure to nanoparticles. The preference here regarding *in vitro* testing is justified by the need to obtain an answer about biological reactivity response, while ensuring that the assessment is an affordable cost for the manufacturer.

4.3.1.3.4 Risk Characterisation

After a risk characterisation of health has been conducted, an analysis of the results can produce a toxicity profile based on various tests performed. A risk characterisation of health considers the general approach of the health risk assessment because the conditions and methods of preparation and the toxicity are analysed and then the results are compared with historical data. This is the synthesis of the data acquired throughout the previous steps to get to wear an expert judgment. If the nanoparticles are documented after the preliminary evaluation stages of exposure (*e.g.* scientific and technical information, characteristics of nanomaterials) and characterisation of exposure, then the more expert judgement will be fair (figure 4.3.5).

In this evaluation, we are primarily interested in characterising the nanoparticle as an entity that has a reactive potential in relation to a biological material. This approach differs greatly from that of other European, Canadian or American agencies that will evaluate and characterise the risk to workers or the public. Thus, a nanoparticle can be (1) declared as safe, regardless of the type of use; (2) declared as restricted, for uses other than food; or (3) used only under conditions that require

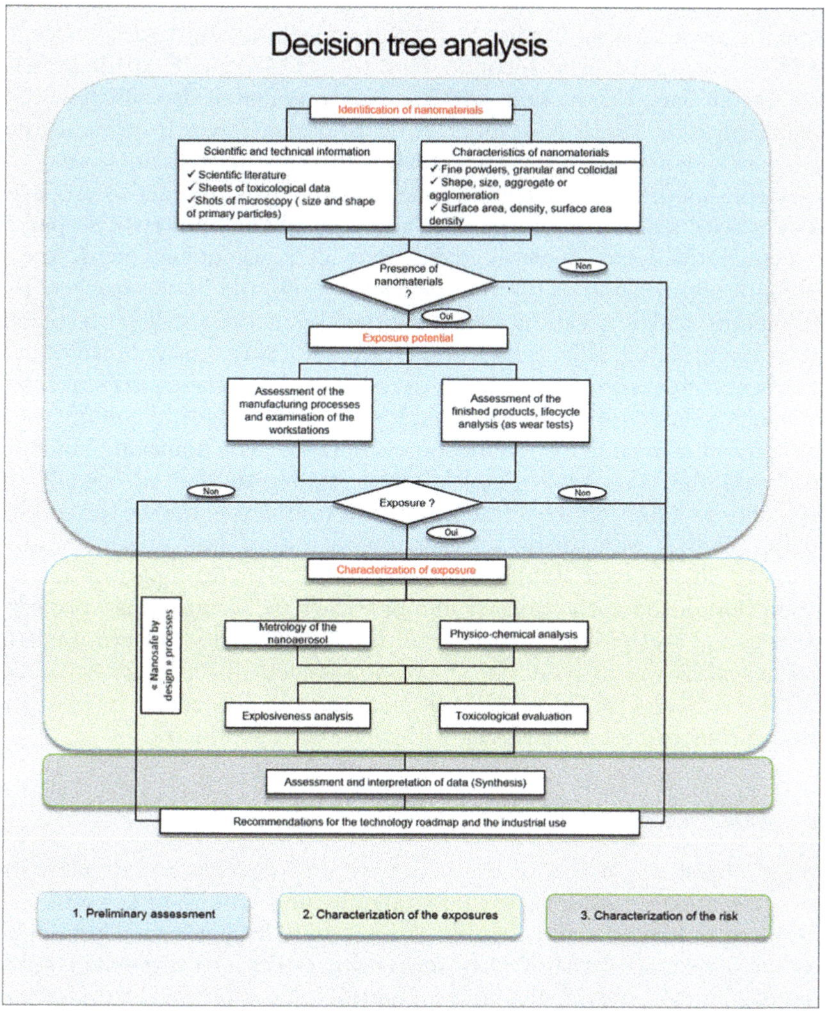

FIG. 4.3.5 – Decision tree for assessing the health risks of nanomaterials.

special care. The comprehensive approach to quantitative risk assessment of exposures to nanoparticles is summarised in tables 4.3.6 and 4.3.7.

4.3.1.4 Assessment of this Methodology

The evaluation of this approach began, first, by contacting two companies with international scopes: (1) Arkema, which is based in Colombes, France; and (2) Raymor Nanotech Industries, which is based in Quebec, Canada. Arkema was the first chemical industry in France, and Raymor Nanotech is considered to be the pride of nanotechnology in Quebec. Representatives from both companies have made presentations at two international conferences. The first event was the 2012 International Conference on Responsible Development of Nanotechnology: Challenges and Perspectives, presented by Nanotechnology for Ethical, Environmental, Economic, Legal and Social Impact (Ne3LS) and held in Montreal, Quebec, Canada on 1–2 November 2012. The second event was NANOSAFE 2012: International Conferences on Safe Production and Use of Nanomaterials, which was held in Grenoble, France on 13–15 November 2012. Representatives from nanotechnology companies attended these conferences. All conference participants recognised the need to establish a protocol such as the one discussed here in order to properly assess the nanoparticles generated at facilities and sold in the market. The participants also welcomed the idea of offering this type of methodology, with the aim of establishing a certification of the nanoparticles to be sold in the market. However, the participants had a question about another point: who will be responsible for issuing or maintaining this certification? In addition, representatives from the industrial sector answered participants' questions about the sources of funding for this assessment. Based on the discussions, it seems that a neutral committee, funded by industry and governments, would be encouraged as a solution. Manufacturers have also said that this type of initiative is similar to that of organic or green products, and its adoption would help improve confidence among the public.

Finally, changes were made to the approach that was presented about the description of the process (tables 4.3.6 and 4.3.7). For example, we changed the description that outlined the toxicity of a nanoparticle because this word represents a negative image and promotes distrust of the industry. Instead, we used the term "nanosafety" when we talked about the responsiveness of a product with biological material. For toxicologists, this variation in terminology does not have any impact, but it does promote better dialogues with those in the industry.

4.3.2 Discussion

The overall objectives of this methodology were to develop a matrix for assessing exposure to nanomaterials for their safe handling and to initiate the first steps that will lead to establishing a certification of nanoparticle-synthesised products in the industry, in partnership with health impact specialists. An important criterion to help convince the industry of the need for certification was linked to the adoption of a simple and gradual approach that will stop the expertise provided at different

TAB. 4.3.6 – Evaluation grid for the analysis of manufactured nanomaterials (Step 1).

Step	Objective	Approach	Materials and means employed
Preliminary assessment of exposure	Nanomaterials identification	*Technical and scientific information* • Scientific documentation • Toxicological data sheets • Microscopy micrographs (particles size and shape) *Nanomaterial characteristics* • Fine, granular or colloidal powders • Shape, size, aggregate or agglomerate • Specific area, density, volume and specific surface area (m^2/cm^3)	– Condensation nucleus counter (CNC) – Portable optical particle counter – Bibliographic reviews
	Potential exposure	*Processing methods analysis* • Manufacturing processes and containment degree of each processing step • Raw materials and intermediate or final products • Total quantities of nanomaterials handled • Nanomaterial application • Materials' ability to be suspended in air *Examination of workstations* • Materials' ability to contaminate the working atmospheres • Potential sources of aerosol emissions (*e.g.* type, number, significance) • Technical work activities (*e.g.* proximity to the target source, time, habits) • Exposure routes (*e.g.* inhalation, skin, ingestion) • Significance and exposure time • Workplace configuration (*e.g.* space) • Generalised and localised ventilation (*e.g.* type, system, efficiency) • Prevention, safety measures and procedures *Final products analysis*	

TAB. 4.3.7 – Evaluation grid for the analysis of manufactured nanomaterials (Steps 2–4).

Step	Objective	Approach	Materials and means employed
Exposure characterisation	Quantification of exposure	*Nanoaerosol metrology* • Aerosol measurement (*e.g.* background, noise) • *In situ* metrology • Number, mass, and geometric area concentration	– Portable optical particle counter – Condensation kernel counter – Optical spectrometer – Transmission electron microscopy and scanning electron microscopy – Cascade impactor – Fixed sampling
		Physicochemical analysis • Physicochemical properties (*e.g.* size, morphology, specific surface area, surface reactivity) • Morphological properties (*e.g.* size distribution, shape) • Particle size and distribution • Chemical composition (inorganic, organic and ionic compounds) • Materials solubility in liquids	
		Toxicological evaluation • Acute toxicity, cytotoxicity • Membrane alteration • Changes in blood parameters • Oxidative stress, inflammatory reaction • Genetic modification	– Enzyme-Linked ImmunoSorbent Assay reader, high performance liquid chromatography Haematological system – Western blot detector – Flow cryometry

TAB. 4.3.7 – (continued).

Step	Objective	Approach	Materials and means employed
Risk characterisation	Data analysis and interpretation	*Health risks assessment* • Method and preparation measurements • Observations and analysis (*e.g.* mode, conditions) • Dose–response relationship • General assessment procedure (exposure [*e.g.* background noise, aerosol]). • Comparison of the results with data from the literature	– Technical expertise – Bibliographic – Scientific expertise
		Report of the results	

stages, thereby decreasing the costs. However, to convince evaluators of the potential health risks, the approach involved providing expert judgement based on measures that are as accurate as possible. This is the reason why we developed the decisional grid presented in tables 4.3.6 and 4.3.7.

We conducted the literature review to identify the methodology for assessing nanomaterials, so we could build on that. Readers must understand that characterising nanoparticles is an essential step. The authors of two publications also reported the use of many toxicological tests and chemical characterisation tests [3, 17]; however, it was important to identify a minimum number of tests that can generate the maximum amount of information, according to the criteria previously discussed. The toxicology tests listed in this section and in the report of the Organisation for Economic Co-operation and Development (OECD) [40, 41] (see table 4.3.6) are most commonly used to study nanotechnology.

The review of nanomaterial assessment methodologies also revealed that quantifying the risk based on to a safety assessment of nanomaterials is still very complex because there is a lack of data regarding potential toxicity, especially in humans. Although in the literature, several methods of risk quantification are proposed: a precautionary matrix for nanomaterials [11, 12], the application of a pilot control banding tool for risk-level assessments and control of nanoparticle exposures [13]. However, most of these approaches do not consider all parameters of the nanomaterials at each stage of their life cycle. Indeed, the parameters considered for quantifying the risks (*e.g.* weighting of scores in the calculation) are not necessarily the same as those used in the proposed methodology, but it is precisely during the life cycle when the levels of hazard and exposure that define the risk are assessed. AFSSET's report proposed the establishment of a repository that would consider the essential parameters for risk assessment [15]. The repository aims to define the nature of expected responses, through these parameters in relation to the various standard tests, and then the results of which would contribute to the rating scale. To our knowledge, this standard has not yet been finalised.

It appears that none of the methods journals have included studies that quantified the risk of nanoparticles investigated, with a view toward certification of nanotechnology products. However, an analysis of the literature was used to help design a methodology so a manufacturer will know the product and how to properly apply it during the production process. The proposed approach is based on the health risk assessment approach proposed by the NRC in 1983 [16]. The approach is to assess the potential source, exposure and dose–response relationship and to characterise the health risk based on factual findings.

The proposed decision grid analysis includes: (1) a preliminary risk assessment, (2) characterisation of exposure, (3) a risk characterisation and (4) results and report. The first step of the analysis is critical because a recent study conducted in part by our team showed that several industrialists, especially those who use nanopowders are unaware whether they use or are exposed to nanoparticles [42, 43]. The preliminary assessment combines all of the available information about the rated nanoparticle. This type of theoretical evaluation, which is inexpensive for those in the industry, can be used to determine whether the information is sufficient to take the next step: exposure characterisation. The preliminary assessment is still fairly comprehensive because it includes a literature review and the characteristics of the nanoparticle studied, a site visit and a review of the manufacturing processes. Once this step is completed, the manufacturer can use the findings to make an empowered decision about the type and nature of this nanoparticle and whether it should be used.

The next step is the characterisation of exposure. This step requires the use of specialised equipment to characterise the nanoparticle and its potential toxicity (tables 4.3.6 and 4.3.7). The aim of characterising the exposure is to reassure workers, management and owners about possible contamination of the workplace by the produced nanoparticles. The toxicological evaluation will establish a linkage between the nanoparticle studied and biological material. Most tests suggested in the grid Biological assessment are *in vitro* tests. These tests are limited because they do not provide the information needed to answer a question other than on the mechanism of action, but they can quantify a level of responsiveness when compared with the control sample. The primary purpose of these tests is not to extrapolate the results *in vitro* to *in vivo*, but rather to identify a degree of reactivity and, if possible, to quantify the level of hazard. We are suggesting the use of *in vitro* testing because it offers a reproducible response, helps limit the cost to a manufacturer and provides reliable information about a manufacturer's product. However, this stage of the decision grid must be discussed to define an approach for establishing relevant quantitative or qualitative thresholds for the studied parameters, thresholds that can then be linked with representative scores of the level of risk. This analysis is much more complex and requires knowledge of metrology of aerosols and of exposure assessment in the workplace.

After the hazard and exposure have been characterised, the next step is risk characterisation (see table 4.3.7). This step aims to analyse and interpret the data received, with a view to quantify and qualify the level of risk attached to the nanoparticle of interest. To qualify or quantify that risk, we are proposing an approach for synthesising previous characterisations (exposure and hazard) to make a judgement about the level of risk. Tools have recently been developed to establish a risk scale. Thus, the French Agency for Food, Environmental and Occupational Health and Safety (*Agence Nationale de Sécurité Sanitaire de l'Alimentation, de l'Environnement et du Travail* [ANSES]) issued a document that proposes to characterise the level of risk of nanotechnologies in situations where the uncertainties are *a priori* higher [44]. In that document, ANSES proposes using a graduated graphic risk management, more commonly known as "Control Banding." A Swiss collaboration between FOPH and FOEN discussed a method of precautionary matrix with the same objectives [12]. This precautionary matrix can estimate a non-specific but potential risk from using nanotechnologies. In addition, this method is based on criteria of danger associated with chemical and biological reactivities of the nanoparticle that represents the nanoparticle on its wider environment. This matrix enables a warning level, as is the case for the method proposed by ANSES.

New strategic approaches to health risk assessment use sophisticated characterisation equipment and also an activity in silico that is comparable to the *in vitro* approach by the Quantitative Structure–Activity Relationship (QSAR) [45, 46]. As described in one publication [47], the use of in silico methods, such as nano-QSAR and Pseudo

Structure–Activity Relationship (PSAR) is highly desirable to accelerate and streamline the steps in toxicological research. However, these methods are not an alternative, but rather an introduction to experimental work [47]. The *in silico* approaches allow for the correlation and exploit the entire potential of the *in vitro* studies.

Since of lingering doubts regarding the issue of genetically modified organisms (GMO), the consumers and the general population are more sceptical to new technologies. In addition, nanotechnology, including the expected financial impact of $100 billion (U.S. dollars) per year, all sectors [18] must develop very detailed plans on how it will be used and managed and, above all, whether the processes are conducted in the same manner as the remainder of the industry. Even if the money invested by governments and business is important, the lack of information and the amount of misinformation about nanotechnology are sufficient enough to raise the fears of the population regarding the present and future development of this industry. As science discovers more and more deleterious effects from nanoparticles, companies must mitigate the information conveyed and build a relationship of trust with the population to promote responsible development.

Establishing the nanoparticle certification will allow manufacturers to manage the marketing of their products. It is conceivable to claim that the nanoparticles certification scheme based on health risk assessment criteria would have beneficial effects on the safety of nanoparticles used, which would help reassure the population. In addition, establishing a certification protocol would help manufacturers identify the level of toxicity of nanoparticles so that they can establish a working methodology adapted to the risks that could occur when handling nanoparticles in order to promote nanosafety and controlled workstations. As a result, the workers would be better protected and, by extension, less exposed to the potential risks of nanoparticles.

The originality of this work comes from the desire of authors to initiate the first steps leading to the certification of products from the synthesis of nanomaterials, in partnership with industry and health impact specialists. Thus, a voluntary assessment approach used by independent experts would establish the constraints of use and would probably help to increase the public's confidence. Because of the transparency of this methodological approach, the industry could include its certification in its development and marketing plan, thereby demonstrating that it is a responsible corporate citizen.

4.3.3 Conclusion

This evaluation work is a huge step towards a nanoparticle certification programme, despite its limited current range, it retains an original character. Discussions at the Nanosafe (2012) conference and in Ne3LS in Montreal (2012) conference where this work was presented did showcase the agencies in Europe, but also large industries, because similar initiatives are being prepared. Many governmental organisations such as the OECD, the ISO Technical Committee TC 229, NanoQuébec, the National Nanotechnology Initiative and the National Institute for Occupational Safety and Health are interested in nanotechnologies. However, no protocol has been proposed to regulate the synthesis of nanoparticles, with a view to establishing a voluntary restriction framework for the industry. This desire for regulation demonstrates the impact of the transparency so industry can show a spirit of entrepreneurship while ensuring nanosafety. This work is only the first step towards a structure that satisfies two groups: (1) the assessor of the health risk because it is based on NRC-developed principles and (2) the industry that is generating the nanoparticles. A second step of this work will be to select tests (chemical and biological characterisation) with a view towards standardising the approach.

Ultimately, this work led to the development of a methodology to prioritise the reactivity of nanoparticles, understand the interactions of nanoparticles with biological materials and establish a usage limit for the type of nanoparticle evaluated. By establishing this methodology, we wanted to be transparent in order to increase the confidence of workers and the public regarding industrial development and use of nanoparticles in the sectors.

Although we have described a methodology for evaluating health risks in order to establish an integrated approach for nanosafety, we decided not to discuss some outstanding issues because we wanted to focus on proposing a simple and practical method for risk assessment, which was the objective of this preliminary work. For instance, there is an outstanding issue regarding the selection of biological and chemical tests to be used in this methodology. Other issues involve the parameters that refer to the criteria of health risk assessment and the ability of industry to absorb the costs of certification. In addition, a tool has not yet been developed that will quantify or qualify the risk characterisation based on biological and chemical tests and also based on potential exposure.

A second phase of this work will be conducted to develop the functional certification approach. Thus, as a subsequent step towards developing the standardised approach, we are refining characterisations of health risks by adapting a matrix similar to those of ANSES and FOPH and FOEN. These agencies have merit in calling them "alert levels" or "security-related exposures" to identify potentially dangerous nanoparticles. The aim here is to develop a matrix or an algorithm that will assign a point-level risk to nanoparticles based on their potential to impact health and then scale the risk. This matrix will include the required characteristics to accurately establish a level of safety based on a restricted use or an overall potential use of nanoparticles.

References

[1] Helland A., Scheringer M., Siegrist M., Kastenholz H.G., Wiek A., Scholz R.W. (2008) Risk assessment of engineered nanomaterials: a survey of industrial approaches, *Environ. Sci. Technol.* **42**, 640.

[2] Klaine S.J., Alvarez P.J., Batley G.E., Fernandes T.F., Handy R.D., Lyon D.Y., Mahendra S., McLaughlin M.J., Lead J.R. (2008) Nanomaterials in the environment: behavior, fate, bioavailability, and effects, *Environ. Toxicol. Chem.* **27**, 1825.

[3] Stone V., Nowack B., Baun A., van den Brink N., Kammer F., Dusinska M., Handy R., Hankin S., Hassellov M., Joner E., Fernandes T.F. (2010) Nanomaterials for environmental studies: classification, reference material issues, and strategies for physico-chemical characterisation, *Sci. Total Environ.* **408**, 1745.

[4] Oberdorster G., Stone V., Donaldson K. (2007) Toxicology of nanoparticles: a historical perspective, *Nanotoxicology* **1**, 2.

[5] INRS (2008) *Les nanotubes de carbone: Quels risques, Quelle prévention?* INRS ND 2286.

[6] Ostiguy C., Roberge B., Ménard L., Endo C.A. (2008) *Guide de bonnes pratiques favorisant la gestion des risques reliés aux nanoparticules de synthèse.* IRSST.

[7] Ostiguy C. (2004) Echantillonnage et analyse de surface, in *Manuel d'hygiène du travail-Du diagnostic à la maîtrise des facteurs de risque.* Modulo-Griffon, Mont-Royal, QC.

[8] Brown D.M., Donaldson K., Borm P.J., Schins R.P., Dehnhardt M., Gilmour P., Jimenez L. A., Stone V. (2004) Calcium and ROS-mediated activation of transcription factors and TNF-alpha cytokine gene expression in macrophages exposed to ultrafine particles, *AJP – Lung Cell. Mol. Physiol.* **286**, L344.

[9] Chevillard S. (2010) *Risque toxicologique des nanoparticules "Effets sur la santé"?*

[10] Nordan M.M., Holman M.W. (2005) A prudent approach to nanotechnology environmental, health, and safety risks, *Ind. Biotechnol.* **1**, 146.

[11] Dupont C. (2007) *Environmental defense – Dupont nano partnership.* DuPont.

[12] F. FOPH (2008) *Guideline on the precautionary matrix for synthetic nanomaterials (version 1.0).* Federal Office for Public Health (FOPH)/Federal Office for the Environment (FOEN).

[13] Paik S.Y., Zalk D.M., Swuste P. (2008) Application of a pilot control banding tool for risk level assessment and control of nanoparticle exposures, *Ann. Occup. Hyg.* **52**, 419.

[14] NRC (1983) *Risk assessment in the Federal Government: managing the process.* 0-309-59880-X.

[15] AFSSET (2010) *Évaluation des risques liés aux nanomatériaux pour la population générale et dans l'environnement.*

[16] NRC (1983) *Risk assessment in the Federal Government: managing the process.* National Academiy Press, Washington.

[17] Oberdorster G., Maynard A., Donaldson K., Castranova V., Fitzpatrick J., Ausman K., Carter J., Karn B., Kreyling W., Lai D., Olin S., Monteiro-Riviere N., Warheit D., Yang H. (2005) Principles for characterizing the potential human health effects from exposure to nanomaterials: elements of a screening strategy, *Part. Fibre Toxicol.* **2**, 8.

[18] AFSSET (2006) *Les matériaux: Effets sur la santé et sur l'environnement.*

[19] Ricaud M., Witschger O. (2012) *Les nanomatériaux: définitions, risques toxicologiques, caractérisations de l'expositionprofessionnelle et mesures de prévention.* INRS. 978-2-7389-2010-2.

[20] Patlolla A., Knighten B., Tchounwou P. (2010) Multi-walled carbon nanotubes induce cytotoxicity, genotoxicity and apoptosis in normal human dermal fibroblast cells, *Ethn. Dis.* **20**, S1.

[21] Schrurs F., Lison D. (2012) Focusing the research efforts, *Nat. Nanotechnol.* **7**, 546.

[22] Witschger O., Le Bihan O., Reynier M., Durand C., Marchetto A., Zimmermann E., Charpentier D. (2012) *Préconisations en matière de caractérisation des potentiels d'émission et d'exposition professionnelle aux aérosols lors d'opérations mettant en oeuvre des nanomatériaux.* INRS, Hygiène et sécurité au travail.

[23] Ineris (2011) *Recommandations en matières de caractérisation des potentiels d'émission et d'exposition professionnelle aux aérosols lors d'opérations mettant en oeuvre des nanomatériaqux: Guide méthodologique.*

[24] Tielemans E., Schneider T., Goede H., Tischer M., Warren N., Kromhout H., Van T.M., Van H.J., Cherrie J.W. (2008) Conceptual model for assessment of inhalation exposure: defining modifying factors, *Ann. Occup. Hyg.* **52**, 577.

[25] Emond C., Kouassi S., Schuster F. (2015) *Approche intégrée pour une conception et une manipulation sécuritaires des nanomatériaux - Un programme basé sur une concertation entre l'industrie et des évaluateurs des risques sanitaires.*

[26] Brown D.M., Stone V., Findlay P., MacNee W., Donaldson K. (2002) Increased inflammation and intracellular calcium caused by ultrafine carbon black is independent of transition metals or other soluble components, *Occup. Environ. Med.* **57**, 685.

[27] Powers K.W., Brown S.C., Krishna V.B., Wasdo S.C., Moudgil B.M., Roberts S.M. (2006) Research strategies for safety evaluation of nanomaterials. Part VI. Characterization of nanoscale particles for toxicological evaluation, *Toxicol. Sci.* **90**, 296.

[28] Powers K.W., Palazuelos M., Moudgil B.M., Roberts S.M. (2007) Characterization of the size, shape and state of dipersion of nanoparticles for toxicological studies, *Nanotoxicology* **1**, 42.

[29] Oberdorster G., Oberdorster E., Oberdorster J. (2005) Nanotoxicology: an emerging discipline evolving from studies of ultrafine particles, *Environ. Health Perspect.* **113**, 823.

[30] Thomas T., Thomas K., Sadrieh N., Savage N., Adair P., Bronaugh R. (2006) Research strategies for safety evaluation of nanomaterials, part VII: evaluating consumer exposure to nanoscale materials, *Toxicol. Sci.* **91**, 14.

[31] Warheit D.B., Borm P.J., Hennes C., Lademann J. (2007) Testing strategies to establish the safety of nanomaterials: conclusions of an ECETOC workshop, *Inhal. Toxicol.* **19**, 631.

[32] Yah C.S., Simate G.S., Iyuke S.E. (2012) Nanoparticles toxicity and their routes of exposures, *Pak. J. Pharm. Sci.* **25**, 477.

[33] Singh N., Manshian B., Jenkins G.J., Griffiths S.M., Williams P.M., Maffeis T.G., Wright C. J., Doak S.H. (2009) NanoGenotoxicology: the DNA damaging potential of engineered nanomaterials, *Biomaterials* **30**, 3891.

[34] AshaRani P.V., Low Kah M.G., Hande M.P., Valiyaveettil S. (2009) Cytotoxicity and genotoxicity of silver nanoparticles in human cells, *ACS Nano* **3**, 279.

[35] Sharma V., Anderson D., Dhawan A. (2012) Zinc oxide nanoparticles induce oxidative DNA damage and ROS-triggered mitochondria mediated apoptosis in human liver cells (HepG2), *Apoptosis* **17**, 852.

[36] Gavini E., Spada G., Rassu G., Cerri G., Brundu A., Cossu M., Sorrenti M., Giunchedi P. (2011) Development of solid nanoparticles based on hydroxypropyl-beta-cyclodextrin aimed for the colonic transmucosal delivery of diclofenac sodium, *J. Pharm. Pharmacol.* **63**, 472.

[37] Kroll A., Dierker C., Rommel C., Hahn D., Wohlleben W., Schulze-Isfort C., Gobbert C., Voetz M., Hardinghaus F., Schnekenburger J. (2011) Cytotoxicity screening of 23 engineered nanomaterials using a test matrix of ten cell lines and three different assays, *Part. Fibre Toxicol.* **8**, 9.

[38] Banga A., Witzmann F.A., Petrache H.I., Blazer-Yost B.L. (2012) Functional effects of nanoparticle exposure on Calu-3 airway epithelial cells, *Cell Physiol. Biochem.* **29**, 197.

[39] Jaeger A., Weiss D.G., Jonas L., Kriehuber R. (2012) Oxidative stress-induced cytotoxic and genotoxic effects of nano-sized titanium dioxide particles in human HaCaT keratinocytes, *Toxicology* **296**, 27.

[40] Aitken R.A., Bassan A., Friedrichs S., Hankin S.M., Hansen S.F., Holmqvist J., Peters S.A.K., Poland C.A., Tran C.L. (2011) *Specific Advice on Exposure Assessment and Hazard/Risk Characterisation for Nanomaterials under REACH (RIP-oN 3) Final Project Report.*

[41] OECD (2009) *Guidance manual for the testing of manufactured nanomaterials.* OECD.

[42] Ostiguy C., Emond C., Dossa I., Malki Y., Boily C., Roughley D., Plaski A., Endo C.A. (2013) Mapping the use of engineered NM in Quebec's industries and research laboratories, *J. Phys.: Conf. Ser.* **429**, 1.

[43] Endo C.A., Ostiguy C., Dossa I.N., Emond C. (2015) *Portrait de la nanotechnologie au Québec dans les milieux industriels et de la recherche universitaire et publique.* IRSST.

[44] Lacour S. (2011) Des tenants et aboutissants de l'étiquetage des nano-produits, *Bulletin veille scientifique, ANSES* **17**, 3.

[45] Pulskamp K., Diabate S., Krug H.F. (2007) Carbon nanotubes show no sign of acute toxicity but induce intracellular reactive oxygen species in dependence on contaminants, *Toxicol. Lett.* **168**, 58.

[46] Cohen Y., Rallo R., Liu R., Liu H.H. (2012) In silico analysis of nanomaterials hazard and risk, *Acc. Chem. Res.*

[47] Cattaneo A.G., Gornati R., Sabbioni E., Chiriva-Internati M., Cobos E., Jenkins M.R., Bernardini G. (2010) Nanotechnology and human health: risks and benefits, *J. Appl. Toxicol.* **30**, 730.

Conclusion

From Innovation to Eco-innovation: A New Challenge for R&D and the Industry in the Circular Economy

Francois TARDIF

CEA-Grenoble, 17 Avenue des Martyrs, 38054 Grenoble Cedex 09, France

As one can see in the previous pages of this book, the innovation process for advanced manufacturing dedicated to energy and transportation, like in almost all the vast applied research area, is sometime iterative and incremental, by association of existing technologies and even sometimes disruptive. It is most of the time an approach driven mainly by purely technological considerations to which we associate a technical and economic analysis to verify the relevance in relation to the markets.

This way of innovation has shown for centuries a formidable effectiveness leading humanity towards the Progress considered until the past decades as intimately linked with the scientific and technological development.

The impact of our so-called "modern" civilizations on the available resources, the environment, the biodiversity and the climate, induces serious threats for future generations. This introduces a sudden change of paradigm in terms of priority for humanity that must henceforth associate the pursuit of the improvement of well-being to sustainability. This evolution leads to deep consequences in almost all the society organization, until our innovation objectives and methods!

The awareness of the risks associated with the overconsumption of resources formally appeared at the Club of Rome in 1972 in these terms: "how to avoid a shortage of oil and minerals to maintain economic growth, despite the exponential expansion of the world population"?

The prestigious MIT was commissioned for a study that modeled the sustainable population level according to the available resources, worries about pollution and social inequalities.

One of the proposed solutions consisted in making our linear model of society (extraction of raw materials, product manufacturing, use and landfill) more circular

DOI: 10.1051/978-2-7598-2446-5.c902

which consists in reusing the products, then the components and finally the residual materials while always maintaining them at their maximum value level. The principle of Circular Economy that will be then improve thereafter was born.

This same issue also led the United Nations to define in 1987 a model of sustainable growth for humanity, the so-called Sustainable Development, which considers that to be sustainable on the long term a society must be stable on its 3 pillars which are: the economic growth, the environmental preservation and societal well-being.

Starting from a simple systemic model of recirculating loop for the manufacturing sector, the Circular Economy has evolved into a global and sustainable model incorporating the same three pillars of the Sustainable Development, of which it becomes the method of economic development. The European Green Deal, which defines our new civilization development strategy for the future, comprises also the Circular Economy through the "New Circular Economy Action Plan" edited in March 2020.

According to this European Circular Action Plan, the success of the transition to the Circular Economy relies in a large extent on digital technologies and ... innovation. Here we are and one can easily understand that innovation for the Circular Economy must of course address the same objectives than the Circular Economy, that is to say to maximize the Global Impact on the same 3 pillars expressed here more precisely in terms of climate neutrality in 2050, health preservation plus environmental regeneration and business competitiveness improvement.

Maximizing the Global Impact during the innovation process defines the Eco-innovation. The Eco-innovation is, therefore, the way to innovate in and for the Circular Economy and is, therefore, our next innovation method.

Of course, computer scientists will always have to bring their scientific excellence in the field of computer science and mechanical engineers in mechanics, but they must now adopt a different approach both in the field of the maturation of new ideas and in projects management.

As depicted in figure 1, this requires reversing the current method of innovation: it is now necessary to start from the finality, that is to say thinking in terms of Global Impact, to deduce the necessary and sufficient technologies, not necessarily high-tech, neither low-tech but right-tech.

Researchers no longer have to think strictly in terms of improving pure technological performances but in terms of interest for humans and the planet.

The pending question is how to consider economy, the environment and the society?

The most mature methodology is concerning the environmental issues thanks to the Ecodesign approach that is governed by the international standard ISO 14006. This is a well-known method for designing products and processes while minimizing the consumed resources and the effects on the environment at each stage of their life cycle. Ecodesign makes it possible to improve, redesign or even imagine completely new concepts thanks to reflections carried out early in the design process by involving all stakeholders. This is also necessary to consider the eventual rebound effects *e.g.* a new way of transportation that is a priori more virtuous and more practical can, for example, encourage more trips.

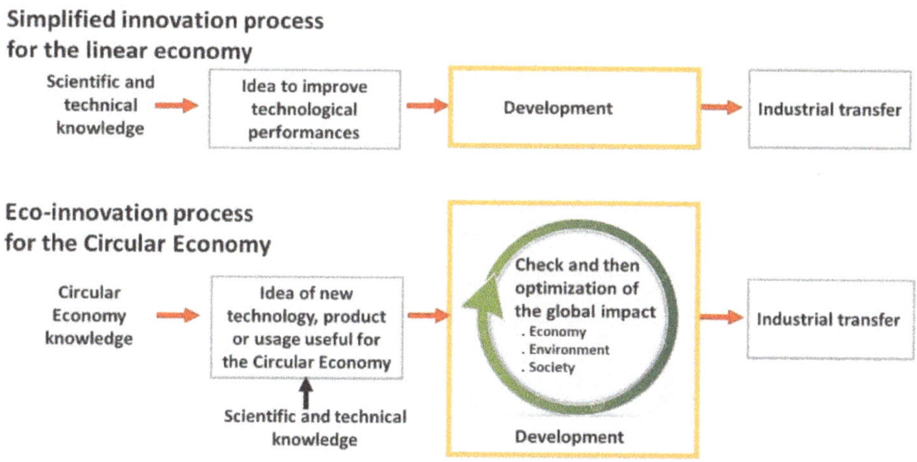

FIG. 1 – Paradigm shift from innovation process to Eco-innovation.

The Ecodesign method can eventually rely on Life Cycle Analysis (or LCA), a standardized method through ISO 14040 which can quantify the environmental impacts of a product at each stage of its life cycle from the extraction of raw material to end-of-life treatments, including manufacture, transport and use. LCA is carried out with software and environmental databases. LCA makes it possible to identify which stage of the life cycle presents the most impact, which environmental aspect is preponderant or which substance is an issue. LCA can, therefore, help the Ecodesign process by identifying possible avenues for improvement for the product without degrading other aspects.

For the economic issues, it is necessary to perform a technical and economic study which is a process well mastered by the specialists in the field.

Social aspects are more difficult to deal with and require sociological experts to assess the impact in terms of common interest, inclusivity and acceptability of citizens.

Finally, the Eco-innovation process can be summarized by the schematic of figure 2.

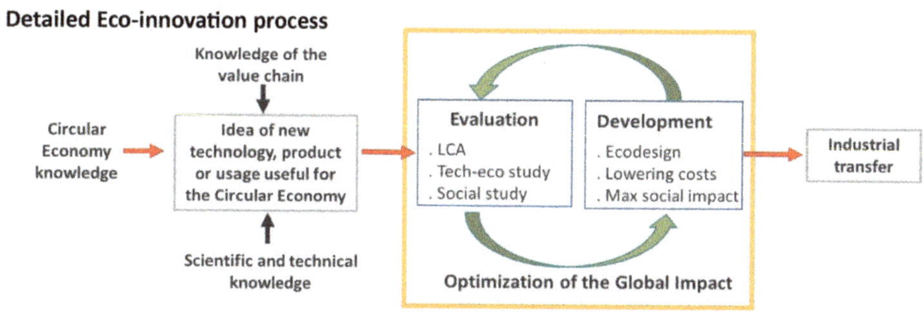

FIG. 2 – Detailed Eco-innovation process.

The New European Circular Action Plan outlines Eco-innovation axes such as the design of "sustainable, competitive and safe" products, the development of new and more "efficient" production processes, the design of technologies allowing "industrial and territorial symbiosis", technologies allowing the circularity of the production generating "less waste and more value", information technologies making it possible to empower consumers and public buyers such as digital product passports.

Eco-innovation opens up a vast new universe for research and technological developments!

www.ingramcontent.com/pod-product-compliance
Ingram Content Group UK Ltd.
Pitfield, Milton Keynes, MK11 3LW, UK
UKHW020712071026
15590UKWH00010B/473

9 782759 824465